MICROOPTICS TECHNOLOGY

OPTICAL ENGINEERING

Series Editor

Brian J. Thompson

Distinguished University Professor
Professor of Optics
Provost Emeritus
University of Rochester
Rochester, New York

Additional Volumes in Preparation

MICROOPTICS TECHNOLOGY

FABRICATION AND APPLICATIONS OF LENS ARRAYS AND DEVICES

NICHOLAS F. BORRELLI
Corning Inc.
Corning, New York

MARCEL DEKKER, INC.

NEW YORK • BASEL

Library of Congress Cataloging-in-Publication Data

Borrelli, Nicholas F.
　　Microoptics technology: fabrication and applications of lens arrays and devices /
Nicholas F. Borrelli.
　　　　p.　cm. — (Optical engineering; v. 63)
　　Includes bibliographical references and index.
　　ISBN 0-8247-1348-6
　　1. Lenses—Design and construction.　2. Optics.　I. Title.　II. Series: Optical
engineering (Marcel Dekker); v. 63.
　　TS517.3.B67　　1999
　　621.36—dc21　　　　　　　　　　　　　　　　　　　　　　　　　　98-56646
　　　　　　　　　　　　　　　　　　　　　　　　　　　　　　　　　　　CIP

This book is printed on acid-free paper.

Headquarters
Marcel Dekker, Inc.
270 Madison Avenue, New York, NY 10016
tel: 212-696-9000; fax: 212-685-4540

Eastern Hemisphere Distribution
Marcel Dekker AG
Hutgasse 4, Postfach 812, CH-4001 Basel, Switzerland
tel: 41-61-261-8482; fax: 41-61-261-8896

World Wide Web
http://www.dekker.com

The publisher offers discounts on this book when ordered in bulk quantities. For more information, write to Special Sales/Professional Marketing at the headquarters address above.

Current printing (last digit):
10　9　8　7　6　5　4　3　2　1

PRINTED IN THE UNITED STATES OF AMERICA

To
Kaye Caroline

From the Series Editor

For many years there has been a significant initiative to make things smaller. The great success of the microelectronics industry has fostered this point of view in many other fields, including mechanics, acoustics, and, most importantly for this series, optics. The initiative can and has resulted in savings in space, weight, and money, but also can and does improve stability, reliability, and performance. In optics, as in many other fields, success in microminiaturization can allow for new applications and new devices and systems. Invasive medical technology is one example that readily comes to mind.

It is not surprising that microoptics has become an important subfield of optics that has now reached a certain maturity. The current volume in our Optical Engineering Series, *Microoptics Technology: Fabrication and Applications of Lens Arrays and Devices,* brings this subfield into focus in an integrated way.

It is certainly worthwhile to find ways to make classical optical components smaller, whether they be lenses, prisms, mirrors, or gratings. Miniaturization, however, brings opportunities to introduce new technologies, new fabrication techniques, and new materials; new concepts for manipulating light in either reflection, refraction, or diffraction, gradient index optics, and diffractive optics are also excellent examples. There are many valuable microoptical elements other than lenses, mirrors, and prisms, such as gratings, polarizers, waveplates, isolators, beam splitters, and separators.

The motivation to develop microoptics is to solve application problems in such areas as switching of optical beams, beam steering (particularly of laser beams), optical image processing, signal processing, optical sensing, three-dimensional imaging, etc.

If the reader is not convinced of the importance of this subfield of optical science and engineering, a check of the indexes of the leading journals in our field will provide convincing evidence of activity, by the sheer number of entries under such headings as microoptics, integrated optics, and diffractive optics. A detailed study of these entries will reveal the breadth of applications of microoptics.

Finally, it should be noted that the development of microoptics allows for hybrid systems such as integrated microoptical systems (IMOS), optical micro-electro-mechanical systems (optical MEMS), and micro-opto-electro-mechanical systems (MOEMS).

Remember, you do not have to be Lilliputian to participate in this growing subfield: being Brobdingnagian is similarly no detriment. Read on.

Brian J. Thompson

Preface

This book arose out of almost 20 years of my experience in the various topics that appear in this book under the term *microoptics*. This experience includes not only basic work in the major areas of refractive, diffractive, and gradient index optics, but also research and application engineering and manufacturing. It was through the engineering and manufacturing experiences that I came to realize that there are really two different technical worlds operating in the microoptic field. One has to do with how and why optical elements behave the way they do, essentially the world of geometric and physical optics, and how they may be applied in the real world, but it does not involve understanding of the way elements are made. The other has to do with the diverse technologies that are brought to bear on the fabrication, and ultimate commercial manufacture, of any type of optical element, such as a diffractive lens. I have found that often persons making an element by using their technical competence are not at all versed in the optical principles on which the element is based, and have even less knowledge of its applications. It was from this realization that I decided to write a book to bridge this gap.

This book was written for the optical scientist who is not familiar with the methods for producing microoptic elements and for the material scientist who would like to understand more about the underlying physical principles of the optical elements. I realize that there is a certain risk in trying to do two things at once. The book may be considered needlessly rigorous, or not nearly rigorous enough. Nonetheless, because microoptics is such a new and growing technology, this book should be of use as an optics primer for the material scientist and as a handbook for the optical scientist.

I am grateful to the S&T Division of Corning Incorporated for giving me the opportunity and allowing me the time to write this book—in particular, to David L. Morse, Director of Glass Research, for his active encouragement, help, and support throughout the effort. I would also like to acknowledge the other members of the research staff for their help, especially the people in the operating divisions who made the technology possible and allowed me to learn from another point of view. I would like to thank the following: David Dawson-Elli for his help and suggestions, L. Button for his effort in the work presented in the appendix to Chapter 2, D. Allan for his help in the content of Chapter 9, Matthew A. Borrelli for proofreading several chapters, Nancy Foster for her excellent drawings and for having the patience of a saint, and Marianna Stewart for the excellent index. The help and support of the staff at Marcel Dekker, Inc., is also gratefully acknowledged, in particular, Rita Lazazzaro and Eric Stannard.

Nicholas F. Borrelli

Contents

MICROOPTICS TECHNOLOGY

1

Introduction

The emergence and rapid growth of the microelectronic industry and its concomitant drive for the miniaturization of electronic devices, together with the optical fiber telecommunication industry and its need to couple light in and out of single mode waveguides, have created in their wake, a new area of optics termed *microoptics*. What is meant by this term is optical elements of dimension of a millimeter or smaller. Primarily, this has meant lenses, or elements that act as lenses, as well as structures that redirect, polarize, and otherwise alter some state or direction of light. These would include elements such as mirrors, gratings, polarizers, and the like made in some ''micro'' form. If one includes optical waveguide structures as well, this broader classification can be considered under the name *integrated optics*. Another interesting aspect of the evolution of microoptics is that not only has the microelectronic industry supplied the need for tiny optical devices, but it has been able, in some cases, to supply the technology by which it can be fabricated. We will see in Chapters 4 and 7 that the very microfabrication technology used to fabricate microcircuits can be used to produce patterns that provide optical function through diffraction of light. This is but one way in which small optical elements, primarily lenses and lens arrays, can be fabricated.

One can find discussions of devices that contain microoptic elements, often done in a comprehensive but narrow way [1]. However, here the device and its performance are stressed, and the elements themselves are not discussed in any real detail. Moreover, the possible alternative methods are not compared or dis-

1

cussed. There are monographs dealing with particular technologies, like gradient index, and consequently they do not deal with the alternatives [2].

There are a few encyclopaedia-type publications that discuss optical elements but do not deal with the optical elements and the methods of fabrication in any detail [3,4]. In this book we try to stress the optical elements and how they are fabricated and try to give sufficient applications where the reader can appreciate the interplay and implications of a particular approach. We attempt to address some of these fabrication and application methods in some detail. The exposition is broken down into three parts. Initially we deal with the important optical element called the *microlens*. Microlenses are found in numerous optical devices and therefore will be given significant attention. Chapters 2 to 5 deal with the different ways that microlenses can be fabricated and provide a brief review of the underlying optic principles. Here we emphasize the properties and fabrication methods of microlens arrays. In the next part, Chapters 5 and 6, we deal with the variety of application of microlens arrays.

The structure of the first part of the book separates the lens fabrication techniques into types of lenses. For example, the categories refractive lenses, gradient index lenses, and diffractive lenses constitute individual chapters. For each of these lens types we will give a brief optics background, sufficient for the reader to appreciate the advantages, limitations, and problems that subsequently arise as a consequence of the optical imaging principles. This is particularly important when one is trying to compare, say, the efficiency of a gradient index lens to a diffractive lens. Although the function is the same the formulation of the imaging is done in a different way when gathering light from an object and directing it to an image; thus the reader must be familiar with the terminology if not the optical principles. Under each lens category, we further discuss the various fabrication methods. For example, under the refractive lens heading, we will discuss methods such as molded, photosensitive glass, etching, etc. The advantages and disadvantages of the fabrication method will be discussed. Because many of these fabrication methods are the property of commercial companies, we give the general approach to the fabrication rather than a recipe. Where we can, we refer specifically to the commercial vendors.

In Chapters 5 and 6 we try to cover the major application areas for microlens arrays. We cover the one-to-one imaging application of lens arrays in Chapter 5. The major applications will be as lens bars for scanning and reading documents. In Chapter 6, we cover the wide range of two-dimensional lens arrays that find application in many diverse areas. These two chapters clarify the distinction between applications where the arrays are used to image collectively, that is act like a conventional larger lens, and those where each microlens acts independently. In the fabrication of a microlens array, in addition to the attention that must be paid to develop the properties of the individual lens itself, equal attention must be paid to the manner in which exact positioning of the lenses relative to some fixed point can be accomplished. In a single lens, one can imagine a mounting fixture that permits the accurate positioning of the lens relative to the light source, or fiber. However, if a lens array is to be aligned to, for example, a laser diode array with a well-specified center-to-center distance as a consequence of its fabrication, alignment may never be possible, if the spacing between the lenses was not maintained during its fabrication. One is dealing with maintaining alignment of micrometers over centimeters. This adherence to dimensional stability over dimensions of many centimeters can make the defining difference in the choice of what method to use for any given application. This adds another aspect to the manufacture of lens arrays which brings in the temperature of the process and how it affects the geometric stability of the substrate. Since some common substrate materials, in particular glass, undergo some degree of irreversible volume change upon heating, the extent of which depends on the temperature relative to the fictive temperature, the maximum temperature achieved during the fabrication can be critical. Because of this more critical requirement, we will be mainly dealing in this book with the methods by which lens arrays are made.

Because the impetus for microoptic elements has come simultaneously from two different directions, microelectronics and optical telecommunications, the optical lens design and performance, as well as the size and layout, are different enough to influence the optimum fabrication method. This will be made clear in the subsequent discussions of the individual fabrication techniques as they relate to applications. Consider the following areas that have been of interest over the past few years.

- Compact optical devices requiring imaging optics to be confined to a small space. Examples of this are document readers, bar code readers, and scanners. These particular applications require erect one-to-one imaging. The advantage over a conventional lens of using a bar-shaped lens array for essentially a line-at-a-time imaging operation is the shortness of the working distance that can be achieved, and yet cover an 8.5 in. document. Total conjugate distances (distance from object to image) as short as 10 mm are achievable over paper-size distances. See Figure 1.1 for a schematic representation of this function.
- *Optical device interfaces with microelectronic structures, for example, CCD detector arrays.* The dimension of such structures requires small closely spaced lenses, registered precisely to the electronic elements. An example of this would be a LED printer bar where each pixel is imaged onto the detector (see Figure 1.1a).
- *Optical waveguide devices: lens to input and output light from single mode fibers, or arrays thereof.* There are a variety of applications, the most important of which is the efficient coupling of light from a laser diode to a single mode fiber (see Figure 1.1b). More recently, devices that are part of the WDM scheme (wavelength division multiplexing) are using microoptical elements.

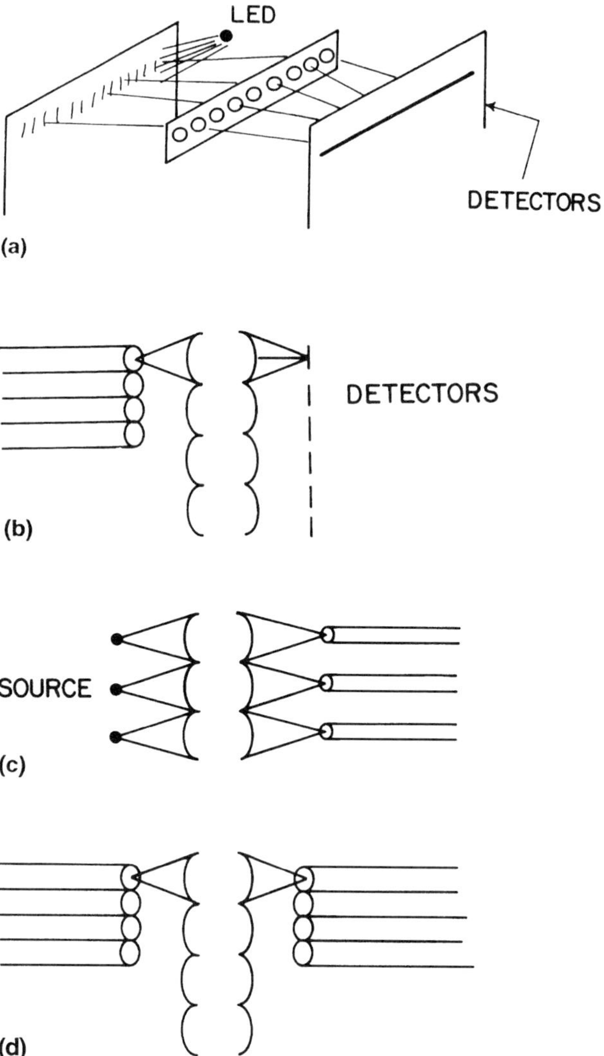

Figure 1.1 Representation of applications. (a) One-to-one imaging application where a document is imaged onto a detector array. Total distance between object and image plane is only 14 mm. (b) Microlens array imaging point onto a detector array. (c) Source to fiber array. (d) Microlens array collimating light from an optical fiber array, then reimaging onto the output fiber array.

As we go through and describe and discuss the techniques people have used to produce imaging structures, one will see that by virtue of the desired small dimensions, a wide variety of methods become viable which would not be otherwise practical on a larger scale. Examples are binary-optic structures, and to a large extent, gradient index structures. In the case of gradient index this is easily understood. The focal length of a lens with a parabolically shaped radial profile is proportional to $\Delta n/\Delta R$, not to Δn itself. Here $\Delta n/\Delta R$ represents the gradient of the refractive index change with the radius of the lens. Whereas it might be difficult to produce large index differences, it is not required if one can achieve small index differences over small distances. For diffractive lenses which are constructed from periodic structures of the order of wavelengths of light, the resolution capability limits one to 4 to 5 periods. For a typical case in the visible wavelength range this limits lens sizes to <100 μm.

Finally, in Part III of the book, we single out and discuss three new and important microoptic areas. Here one starts to get a feel for how the new devices utilize and integrate many different types of optical elements as well as share, to a large extent, similarity of fabrication methods. Chapters 7 to 9 deal with, respectively, the fabrication of microdiffraction gratings, the elements and properties of optical isolators, and finally a relatively new area called *photonic crystals.*

Microgratings and optical isolators have come to the fore, to a large extent, because of the emergence of optical communication systems. As an example, the importance of fiber gratings has come about from the application called *wavelength division multiplexing*, or WDM. This has to do with the optical communication scheme where more than one wavelength is propagating in the same optical fiber. The need to separate, and otherwise act independently on different propagating wavelengths, has produced the need for diffraction grating. What is new is that these gratings have to be implemented on a microscale, mated and integrated to an optical system. We cover all the important methods that have arisen over the last few years to satisfy this new need. We also cover the representative applications.

Chapter 8 is devoted to an optical device whose need has also arisen from the new *photonic* technology. This technology has to do with the next stage of the optical communication network. It deals with using light itself to perform many of the functions that are now done electronically. The initial example of this is the optical amplifier. Whatever the particular scheme, it will always need an optical element to prevent light from going backward in the train that would prematurely deexcite the inverted population. This optical element is the optical isolator. It is included not only to explain another important optical element but also because to provide an excellent example of how many different microoptic elements are brought together to produce an important new optical device.

The topic of photonic crystals covered in Chapter 9 is relatively new, even by the standards of the microoptics field in general. It is an excellent example of the way new innovative technology requires that the emergence and push of new methods mix with new theories, or new ways of thinking of old theories. In this case the new way of thinking about old theories was the realization that the mathematical formulation that is used to determine the behavior of electrons in solids, often called band theory, could also be applied to the way light propagates in periodic structures (5) The link to the technology of microoptics is that the length of the periodic scale of the structures for the important case of visible to near-infrared wavelengths corresponds to the wavelength itself. This means that fabrication methods appropriate for many of the elements discussed in previous chapters are again needed here. To make the cycle complete, the devices that one could imagine using this photonic crystal approach for are the very same optoelectronic applications discussed in the context of other approaches, which, in turn share the same fabrication techniques.

References

1. *Integrated Optical Circuits and Components*, L. D. Hutcheson, ed., Marcel Dekker, Inc., New York, 1987.
2. K. Iga, Y. Kokubun, and M. Oikawa, *Fundamentals of Microoptics*, Academic Press, New York, 1984.
3. *Handbook of Laser Science and Technology*, suppl. 2, Optical Materials, CRC Press, 1995.
4. *Handbook of Optics*, vol. 2, Devices, Measurements, and Properties, M. Bass, ed., McGraw-Hill, New York, 1995.
5. E. Yablonovitch, Photonic band-gap structures, *JOSA* B **10**(2), 283–295 (1993).

Part I
OPTICS AND FABRICATION

2

Refractive Elements

2.1 OPTICS REVIEW

2.1.1 Basics

In order to follow the development and performance of refractive microlens elements, some rudimentary understanding of geometric optics is required. A number of references are given for the reader to consider. What we need here, at the very least, is an understanding of the basic terminology. It might be useful to bear in mind that there is really no conceptual difference in the the optical principles relating to small refractive lenses as compared with large lenses; however, there is in the way one formulates the optical design. In the simplest case of paraxial ray tracing for large-diameter lenses, the so-called thin-lens approximation [1] is used, which simply means that the deviation of the rays through the lens thickness can be ignored. The paraxial specification means that Snell's law can be written as $\phi_i = n\phi_r$. Here ϕ_i and ϕ_r are the angles of incidence and refraction measured from the surface normal as shown in Figure 2.1. When the lens thickness is comparable, or exceeds, the lens diameter, one has to resort to what is often called the thick-lens formulation. Referring to Figure 2.2, one has a set of definitions that derive from the desire to maintain the simple lens maker's formula relating the focal length of the lens to the object and image distances.

$$\frac{1}{f} = \frac{1}{s_o} + \frac{1}{s_i} \tag{2.1}$$

9

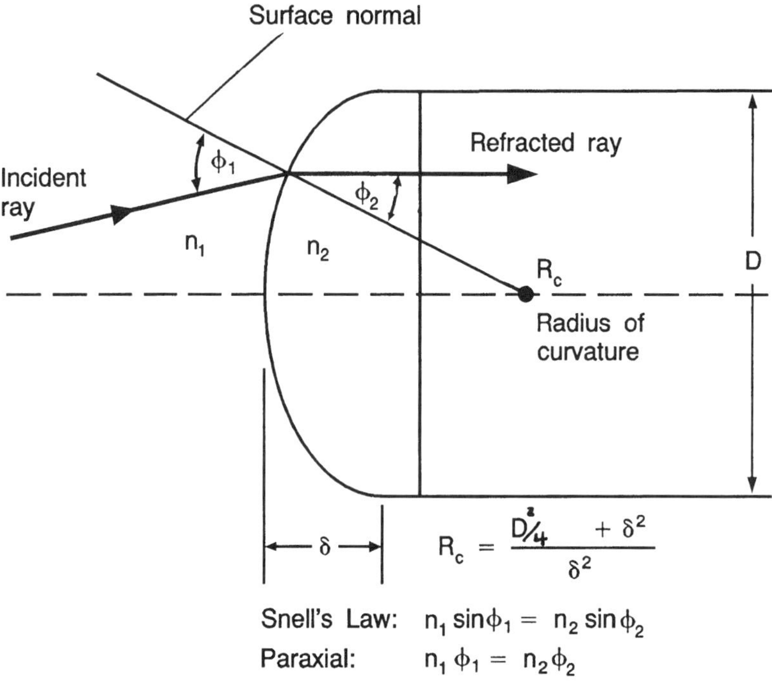

Figure 2.1 Definition of the salient terms to describe the ray path through a spherical lens. Ray incident from left in medium of index n_1 making an angle with respect to the normal at that point of ϕ_1, refracting to an angle ϕ_2 in the lens medium of refractive index n_2. The exact form of Snell's law is given relating the incident angle to the refracted angle, as well as the paraxial approximation.

Here, f represents the focal length of the lens, s_i is the image, and s_o is the object distance. In the thick-lens formulation one must use a new set of definitions of the various common lens terms. These definitions are summarized in Table 2.1 based on the description of Figures 2.1 and 2.2. One can see that the principal planes, P_1, P_2, are defined in such a way as to make Eq. (2.1) valid if the object and image distances are measured from them and not the lens boundaries. One should realize from the expression in Table 2.1 that the position of the principal planes can lie outside the physical lens. We can see from the drawing of Figure 2.2 how the principal plane is defined. The intersection of a parallel ray with the extension backward of a ray from the focus must lie on the principal plane.

With the aid of the Table 2.1 and Figure 2.2, the reader should be able to begin to understand the interplay of the various parameters as they impact a given

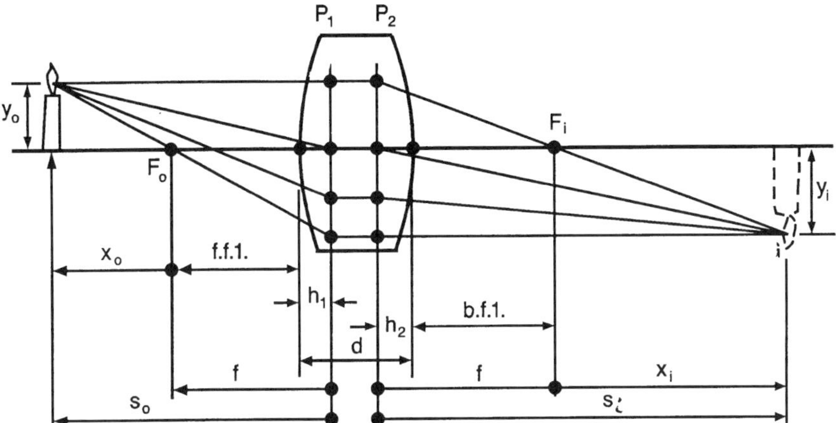

Figure 2.2 Diagram defining the various terms and distances relating to a thick lens which are included in Table 2.1.

desired optical application. For example, the distinction between the effective focal length (EFL), and the respective front and back focal distances is important. From a practical point of view, by way of an example of a trade-off, the distance most usually specified is the total conjugate distance. This is the distance from the object to the image, and usually represents a physical limitation in a design. Within this distance the lens thickness and working distances can be varied, but not independently from the EFL. One can see all this interdependence from the expression for TC given in the Table 2.1. We will refer back to this table at various times.

2.1.2 Performance Criteria

The literature is rich in describing the characterization of lens performance within the geometrical optics regime [1–3]. In general, the analysis is built up in terms of aberration theory which describes mathematically the distortion of a wavefront from the perfect spherical form. This is done in terms of an expansion of the difference between the emergent phase front and a perfect spherical wave, as shown in Figure 2.3a. The ray aberration as a consequence of the wavefront aberration is also indicated. One expands the difference in the expression for the wave from a perfect spherical wave in terms of the parameters described in the blowup of a portion of the wavefront shown in Figure 2.3b. The parameters are the height off the axis, ρ, and the polar coordinates, r and ϕ of the reference point on the wavefront. One expands the difference in a power series in ρ, ϕ,

Table 2.1 Definition of Optical Terms in Thick Lens Formulation (Referenced to Figure 2.1)

Term	Definition	Mathematical expression
R_1	Radius of curvature of first surface	
R_2	Radius of curvature of second surface	
s_O	Object distance (centerline distance from principle plane to Object)	
s_i	Image distance	
P_1	First principle plane	
h_1	Centerline distance from first lens surface to first principle plane	$h_1 = (n - 1)T_{\text{EFL}}/nR_2$
P_2	Second principle plane	
h_2	Centerline distance from second lens surface to second principle plane	$h_2 = (n - 1)T_{\text{EFL}}/nR_1$
EFL	Effective focal length	$[(n - 1)\{1/R_1 + 1/R_2 - (n - 1)/nR_1R_2]^{-1}$
NA	Numerical aperture	r_L/EFL
ffL	front focal length (front working distance)	$f - h_1$
bfL	back focal length (back working distance)	$f - h_2$
T	lens thickness measured at lens centerline)	
r_L	lens radius	
TC	Total conjugate (total distance from object to image)	$(s_O - h_1) + (s_i - h_2) + T$
$f\#$		$\dfrac{\text{EFL}}{2r_1} = 1/2\text{NA}$

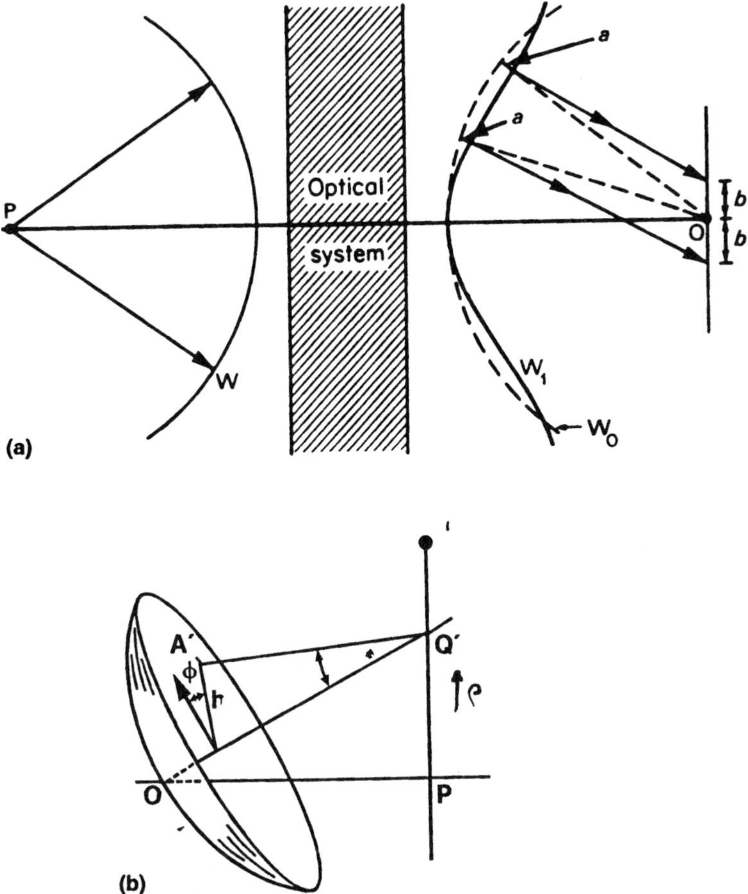

Figure 2.3 (a) Depiction of the wavefront deviation from the perfect spherical wavefront. Both the wavefront distortion, angle ϕ, and the ray distortion, b, are indicated. (b) Reference system for expansion of the deviation.

and r, the terms other than the pure r^2 term representing the deviation from the spherical wave, or aberrations. The convenient aspect of this approach is that one can relate certain terms in the expansion with specific types of common optical image distortions. For example, consider the following terms of the expansion:

$$Br^4 + Fr^3\rho \cos \phi - Cr^2\rho^2 \cos^2 \phi + Er\rho^2 \cos \phi + \cdots \tag{2.2}$$

The first term represents what is termed *spherical aberration*. Note that it is present on-axis ($\rho = 0$). The second term accounts for the optical distortion called *coma*. One can see that it is an off-axis effect and has an angular term. The third term is called *astigmatism*, and the fourth term is called *distortion* which can be positive or negative. All of these produce the commonly known distinct optical distortion effects. However, one can continue in the series to higher and higher order terms where the physical meaning becomes more obscure, but nonetheless represents terms of importance. The present analytical approach is to expand the difference in terms of Zernike polynomials, $Z_{nm}(r, \phi)$. One then can extend the analysis to as many terms as desired. In general the number used is 35. We will see this later on when we discuss the performance of specific microlenses.

One often encounters the term *paraxial approximation*. It essentially means that one is confined to on-axis rays, $\rho = 0$, making a very small angle with the axis; that is, r is very small. We shall see that this approximation is hardly ever met in microlens applications. As an obvious example, the assumption that ρ is 0 is difficult when one considers that the microlens radius is often not that much larger than the object, which could be an LED or an optical fiber.

Regardless of the size of the lens, one still has recourse to aberration treatment of lens to determine the lens performance. Clearly, diffraction ultimately defines the limiting performance of a lens, but before that limit is reached one must face all the limitations afforded from geometric optics. The geometric limitations for microlenses are more severe than for the conventional for the reason that with microlens applications only a single lens must supply the function rather than a multiple train where aberrations can be mitigated. For example, the effects of the geometric aberrations can be lessened by using many refracting surfaces. The other alternative is to use aspheric surfaces. This is the case where the lens figure is altered in such a way as to compensate for the various aberration terms in Eq. (2.2). For microlenses, the former is not practical because of space and alignment constraints, and the latter presents a challenging fabrication problem. As we will see, for single element microlenses, an aspheric molding operation can be used, but it is not practical for arrays. We will briefly mention the molded aspheric process in Section 2.3.3. Because most of the refractive microlens applications will be simple planoconvex or biconvex spherical lenses, we will be dealing in some detail on the performance of these cases.

2.1.3 Ray Tracing

One of the simplest and most useful ways to deal with the design, function, and to some extent, performance of a simple thick lens is to utilize a ray-trace analysis. Simply put, this means generating the algorithm to allow one to plot the ray trajectories. There are a number of relatively simple paraxial techniques, one of which we will describe below in detail, as well as the accurate ray-trace program with no assumptions. Today with computers it is relatively easy to generate one of your own, or to buy one of the commercial software packages. The simple paraxial approach is useful in determining the approximate relationships between the lens parameters for a given application. For example, one wants to collimate a source with a certain NA with a lens of diameter D. This will involve the interplay of the lens curvatures, lens thickness, and total conjugate distance. It gives aberration-free performance and thus cannot be used to evaluate actual optical performance. In the next sections a paraxial method will be described and an example of its use will be worked out. In the following section we will use a more accurate ray-trace program in which one can begin to estimate the actual performance of the lens. For this case, in contrast to the paraxial situation, we will see the onset and consequence of the various aberrations that derive from refraction from a purely spherical surface.

Paraxial Approximation

As mentioned above, there are a number of relatively simple ways to do paraxial ray tracing. For thick lenses, a particularly useful one makes use of what is called the transfer matrix [3]. A two-component vector is formed with the first component representing the distance above the optic axis y and the second the slope of the ray m through the point. See Figure 2.4.

(a)

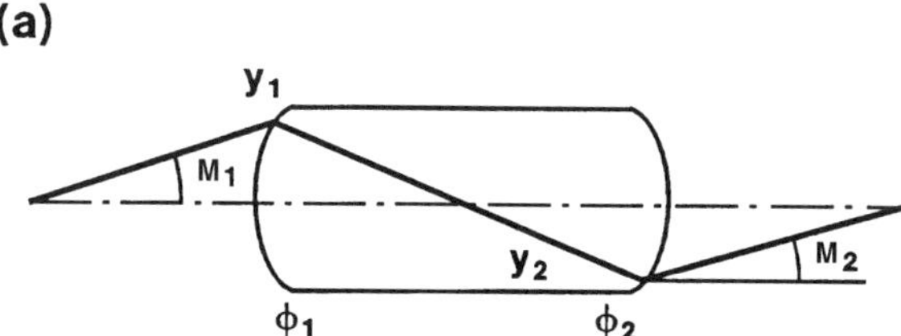

(b)

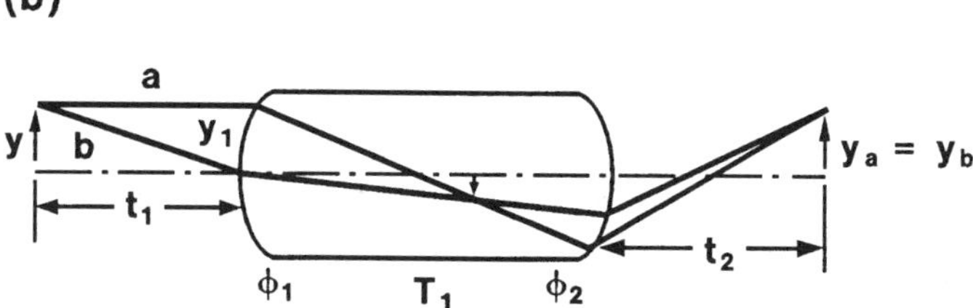

Figure 2.4 Definition of the ray parameters, the ray height h, and the ray slope m. In the lower portion of the figure are shown the test a and b rays.

$$\begin{pmatrix} y_2 \\ m_2 \end{pmatrix} = \mathbf{M} \begin{pmatrix} y_1 \\ m_1 \end{pmatrix} \tag{2.3}$$

where $\mathbf{M}$ is defined as

$$\mathbf{M} = \begin{pmatrix} 1 - T/nf_1 & T/n \\ T/nf_1f_2 - 1/f_1 - 1/f_2 & 1 - T/nf_2 \end{pmatrix} \tag{2.4}$$

Here T is the lens thickness, n is the refractive index of the lens material, and $f_i = R_i/(n-1)$; R_i is the radius of curvature of its surface. One can understand this from the point of view that $\mathbf{M}$ is made up of the product of three operations, the operation corresponding to the refraction of the first surface, followed by a translation operation through the thickness of the lens, followed by the refraction operation corresponding to the second surface.

$$\mathbf{M} = \mathbf{R}_2\mathbf{T}\mathbf{R}_1 = \begin{pmatrix} 1 & 0 \\ -1/nf_2 & 1/n \end{pmatrix} \begin{pmatrix} 1 & T \\ 0 & 1 \end{pmatrix} \begin{pmatrix} 1 & 0 \\ -1/nf_1 & 1/n \end{pmatrix} \tag{2.5}$$

In all cases Snell's law is taken with $\sin m = m$, that is, the paraxial approximation.

By way of a demonstration of how the simple ray trace aids in a evaluation of a microlens for a given example, consider the case where we want a microlens to collimate a source. Let the properties of the lens be given by the following set of values:

$$D = 250 \ \mu m$$
$$R_{c1} = 200 \ \mu m$$
$$R_{c2} = \text{infinite}$$
$$T = 589 \ \mu m$$

One sets up the matrix equation in the form

$$\begin{pmatrix} y_2 \\ 0 \end{pmatrix} = \mathbf{M} \begin{pmatrix} y_1 \\ m_1 \end{pmatrix} \quad \mathbf{M} = \begin{pmatrix} 1 - T/nf_1 & T/n \\ -1/f_1 & 1 \end{pmatrix} \tag{2.6}$$

where $f_i = R_{ci}/(n-1)$ and m_2 is taken to be zero since we want the light to be parallel. We want to obtain the value of m_1 since this will give us the distance we should place the source relative to the lens. Since y_1 is arbitrary, we will take $y_1 = D/2$, and $m_1 = D/2s_1$. Solving the algebraic equations yields the expected result that $s_1 = f_1 = R_{c1}/(n-1)$. If the flat surface of the lens faced the source, that is, R_{c1} is infinite and $R_{c2} = 200 \ \mu m$, then the object distance would by given by $s_1 = f_1 - T/n$. This suggests the interesting configuration of butting the source directly to the flat face by choosing the lens thickness to be $nR_c/(n-1)$. This is shown in Figure 2.5b.

For imaging an object that is a distance x_0 from the lens at a height y_0, one chooses two convenient rays as shown on Figure 2.4b. The *a*-ray has the initial coordinates $(y_0, 0)$ and the *b*-ray, $(0, m_1 = y_0/x_0)$. The simultaneous solution of the simple algebraic expressions will yield the position of the image.

Exact Method

In this case we use the exact ray trace where the refraction at the lens surface is properly taken into account without any approximations. The consequence of this accurate treatment is the appearance of the distortions of the image that are manifestations of the aberrations mentioned above. (In Appendix B of this chapter the algorithm is shown for a three-dimensional ray-trace program.) Here we reexamine the performance of the collimating lens treated above in the paraxial case.

We revisit the collimating example that we treated above in the paraxial approximation. We position the point source at the paraxially determined position of 0.4 mm. We will look at the resulting ray trajectory as a function of the NA of the source. In Figure 2.6a, the NA is only 0.05 and the collimation result is equivalent to what one would get from the paraxial approximation. As we let the NA increase to 0.10 (Figure 2.6b), we begin to see the consequence of spherical aberration. The marginal rays are no longer parallel. In the last case, Figure 2.6c,

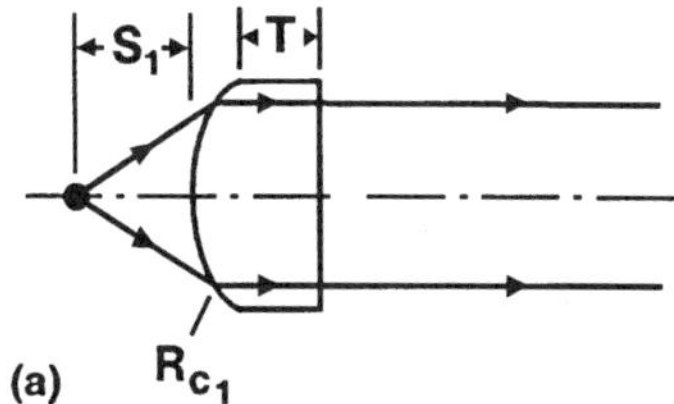

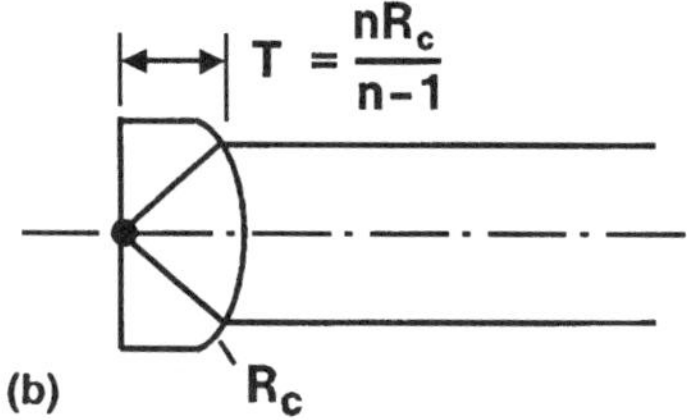

Figure 2.5 Two examples of the paraxial ray trace for the case of collimating a point source with a planoconvex lens. (See text.)

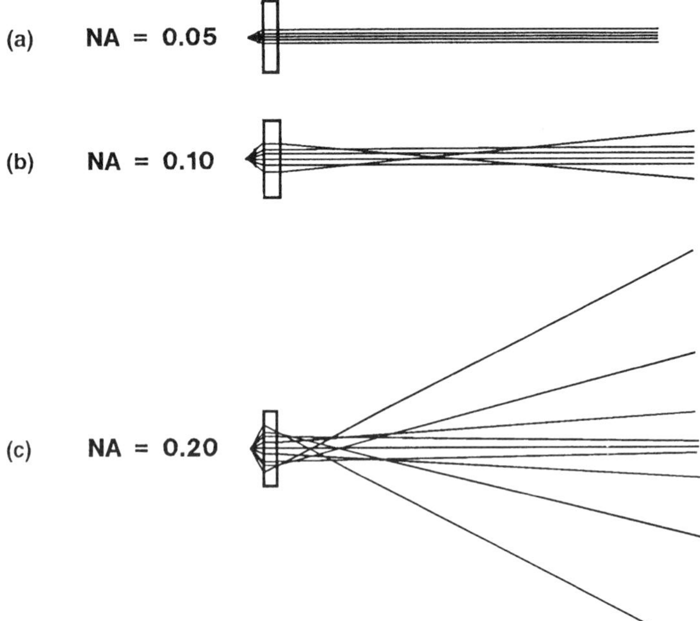

Figure 2.6 Exact ray trace of collimating function planoconvex lens specified in the text, for three different numerical apertures.

the NA is increased to 0.20 and the consequence of the spherical aberration is dramatically evident. Often in the face of this type of aberration, one moves the object closer than that indicated by the paraxial estimate. This has the effect of reducing the spread of the outermost rays at the expense of the inner rays no longer being parallel.

For the situation where one can use a biconvex lens, one can see the significant reduction in the aberration. We repeat the three situations given in Figure 2.6a–c for the case where $R_{c1} = R_{c2} = 500$ μm in Figure 2.7a–c. In this case, because the refraction is less at each surface, the overall spherical aberration is reduced for comparable NA.

The performance of certain designs, and the consequence of aberrations, will be dealt with in Chapter 5, when we deal with specific applications.

2.2 FABRICATION METHODS

2.2.1 Introductory Comments

A relative large number of ways to make small refractive-type lenses has been employed over the last 10 years. Each has had some perceived advantage. Some

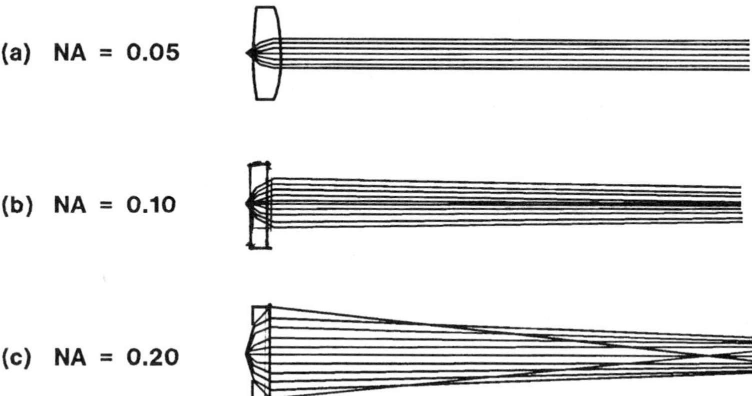

Figure 2.7 Exact ray trace of collimating function for biconvex lens with the same EFL as that shown in Figure 2.6.

are relatively inexpensive and utilize an existing technology, while others require new materials with special properties and unusual fabrication techniques. The particular application must determine the best material and fabrication technique according to the choices, glass/plastic, single element/array, aspheric/spherical. For microoptical applications involving single elements, that is, for lenses with diameter <1 mm, conventional grinding and polishing is impractical. The same is true for monolithic arrays of lenses. We will try to cover the important methods available for refractive lenses, keeping in mind that there are many variations. As we move to the next chapters, the reader will see more lens fabrication methods based on other types of lenses than refractive. In this section we will start with the most traditional and proceed to the more exotic.

2.2.2 Molding

Plastics

This is the classic method of preparation where a mold of the desired surface is made in a nonreactive material by some precise method. This provides the form which is translated to the end element by placing it in intimate contact with the mold. This is a quite straightforward process for the case of plastic elements where injection molding is used. See Figure 2.8 [4]. The technique, however has the disadvantage of the high initial cost of the mold, especially if single-point diamond turning technology is required. There have been plastic lens arrays made by molding, examples of which made by USPL [4] are shown in Figure 2.9.

There are somewhat simpler ways to produce plastic lenses, or lens arrays, for example, hot pressing. In this method a sheet of thermoplastic (polycarbonate,

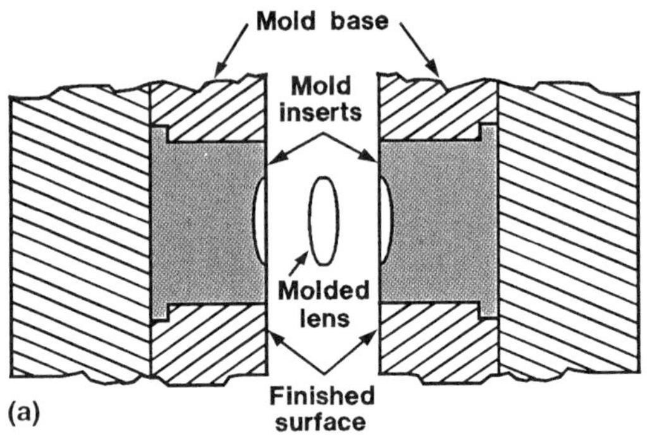

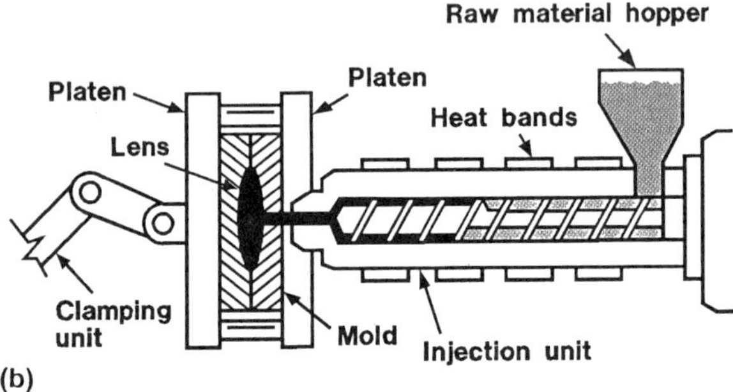

Figure 2.8 Diagram of extrusion molding process to make plastic lenses. The upper drawing shows a blow-up of the mold itself. (From Ref. 4.)

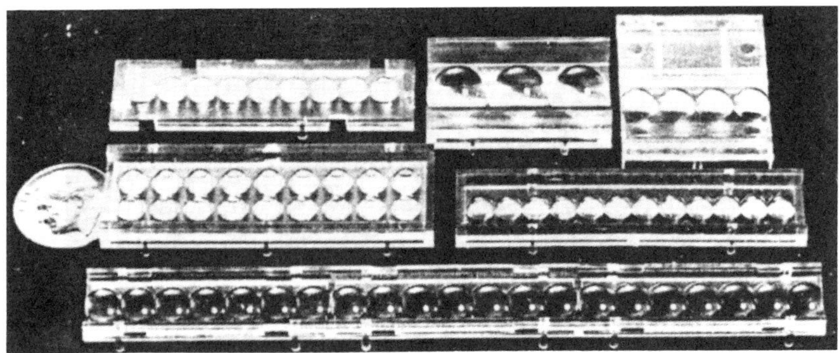

Figure 2.9 Photograph of various molded lens array configurations. (From Ref. 4.)

PMMA, or polystyrene) is heated so that it softens sufficiently to allow it to be permanently deformed. The heated sheet is then pressed against the mold. The softened lens material forms a protrusion at each aperture, as shown in Figure 2.10a. The mold may be made of stainless steel and the pattern formed by any convenient method such as chemical etching. This particular method has the advantage that the lens surface does not come in contact with the mold. A more complete drawing of the lens-forming operation is shown in Figure 2.10b.

The example results [5] indicate that lens arrays with individual lens diameters of the order of 1 mm are possible, with good spherical surfaces limited to the central region of the lens. The radius of curvature is in the range of 1.1–2.4 mm and controllable by the pressing time. The lens performance was estimated by a spot size test that was not well described. It general, this is a poor method to use to evaluate lens performance. It is not the size of the Airy disk that is important, but how much intensity is contained in the central ring.

Plastic on Glass

To avoid some of these problems there are hybrid methods, for example, plastic lenses on a glass substrate. Adaptive Optics Associates reports [6,7] fabricating lens arrays in this manner from a metal master. The master is made of a high-purity annealed and polished material. After the pattern is formed, a release agent is applied to the surface. Using standard replication techniques, a small amount of epoxy is placed on the surface and the optical substrate is placed on top. This is shown in Figure 2.10. Representative types of arrays are shown in Table 2.2.

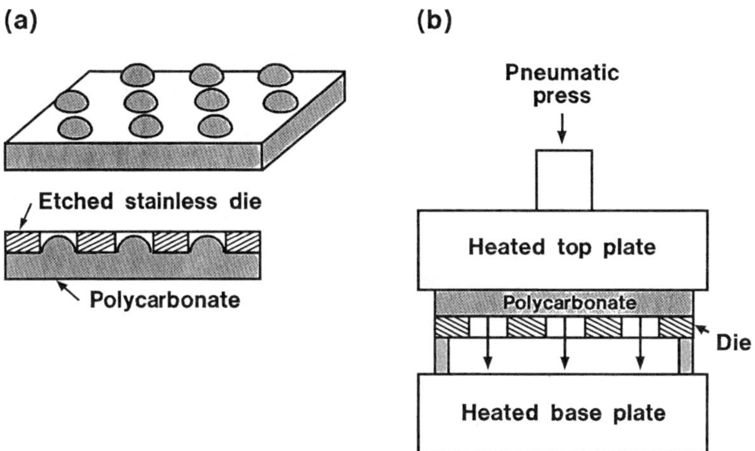

Figure 2.10 Schematic drawing of hot-pressed lens array in polycarbonate, also showing the heated press.

Table 2.2 Representative Lens Arrays Made by Molding

Aperture (μm)	Array size	f	FL (mm)	Format	Fill factor (%)
100	120 × 120	5	0.5	Square	>98
100	68 × 68	17	1.7	Square	>98
200	60 × 60	5	1.0	Square	>98
200	36 × 36	33	6.5	Square	>98
400	20 × 36	8	3.2	Square	>98
400	21 × 21	63	25	Square	>99
500	300 × 300	6.6	3.3	Square	>97
600	13 × 13	18	6.3	Square	>94
768	8 × 8	8	65	Circle	>79
1016	12 × 12	256	260	Square	>98

Source: Selected from data sheet of Adaptive Optics Associates (United Technologies), Cambridge, MA.

A SEM photograph of an array of 70-μm-diameter lenses made by this method is shown in Figure 2.11, along with the images produced by these $f/12$ lenses.

Using precisely controlled machines it is possible to make mold cavities with very high surface figure accuracy. As pointed out above, the application dictates what the material must be, glass or plastic, and the choice of material determines the difficulty of the subsequent molding operation. For example, for the high temperature process required for glass, the mold life, and in general, the difficulties accompanying the higher temperature process make the process quite different from what it would be for plastic. To make this point, we will describe a process reported for molding 5-mm-diameter aspherics as an example of what is involved. The fact that it is an aspheric surface is really not an issue in the problems encountered, save maybe in degree, because of the higher required accuracy of the replication process.

Glass

The molds are made to a surface accuracy of a few thousandths of a wave with a mirrorlike finish. To obtain such surfaces a technique called single-point diamond turning is used [8]. In its simplest description these are very precise lathelike machines that can control cut depths to 0.1 microinch. A typical diamond turning tool is represented in Figure 2.12. Machines such as those manufactured by Moore Specialty Tool Co. must be operated in a temperature-and humidity-controlled environment. The cost of such a machine, including the facility to place it in, can be as high as $1.5 million. An example of a complete molding assembly is shown in Figure 2.13. It consists of two molds (one for each lens surface), a

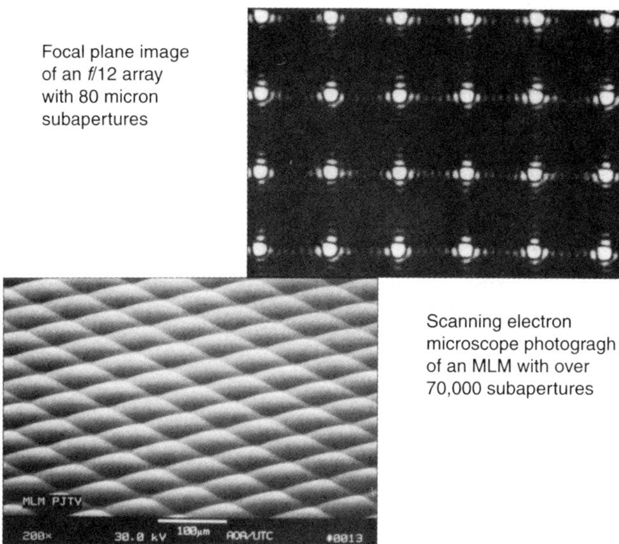

Figure 2.11 SEM photomicrograph of 80-μm-diameter molded plastic lenses and the images produced by the $f/12$ lenslets. (From Ref. 7.)

precision drilled alignment sleeve to control centration and tilt of the molds, a ring member to form the outside diameter of the lens, and finally a glass preform. In addition to the accurate tooling for high-numerical-aperture lens applications, there are two important factors in accurate replication: forming at high viscosity and maintaining an isothermal environment. Both of these conditions are required in order to limit distortion. There is a lower limit to the diameter of a lens that can be made by a single-point diamond turning mold and it stems from the radius of curvature of the cutting tool relative to the curve one is trying to cut. Typical curvatures are of the order of 300 μm.

The molding process described above might be more aptly called compression molding [9] of a glass preform. The glass preform is contained within the molding assembly which is heated to the molding temperature. Then pressing is accomplished by applying a load to one of the molds. Thereafter, the glass is cooled to below its transition temperature before removal.

The thermophysical properties of the glass play a crucial role in making the molding process a practical method. The glass has to have a lower transition temperature, $T_g < 400°C$, than typical glasses, so that the forming can be done at a sufficiently low enough temperature to make the tooling and molding cost effective. This primarily means what composition the mold material must be,

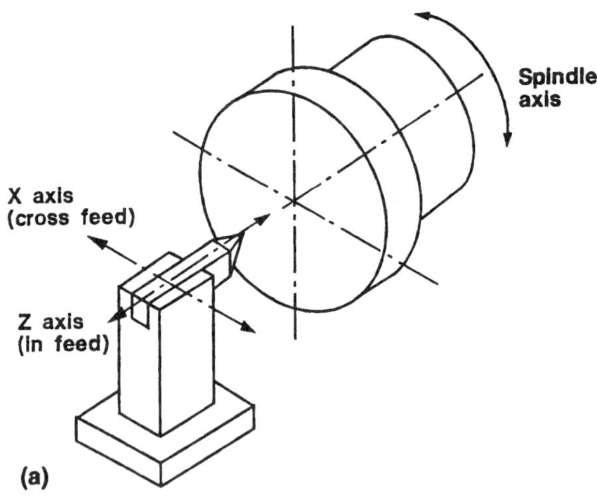

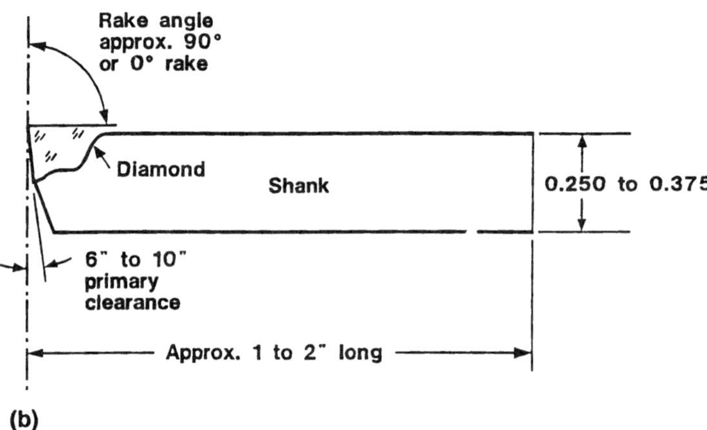

Figure 2.12   (a) Representation of a generic single-point diamond turning machine. (b) Typical cutting tool design. (From Ref. 8.)

and how many pressings it will allow. In general, in order to soften the glass, some sacrifice of the glass durability is made, which can be a problem. A glass composition that has been used to mold glass lenses is the Pb/Zn-flourophosphate system [10].

Another molding process has been reported for making lens arrays, termed

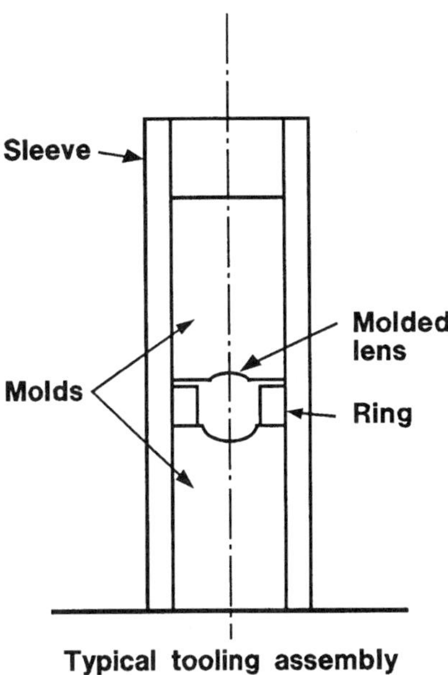

Figure 2.13 Schematic of the molding assembly. Consists of two molds, an alignment sleeve, and a glass preform as shown. (From Ref. 9.)

contactless molding [11]. The method is to heat the glass above T_g and then press into it with a mold made up in the form of a grid. The glass is forced up in the mold but does not come in contact with the mold. The glass is thus influenced by surface tension and forms a surface of minimum energy. In the simplest case, if the grid is made up of circles, then a spherical element is produced. In the case that the grid symmetry is something else, distortions away from a spherical shape occur at the boundaries. A picture of an array made in this way is shown in Figure 2.14. As in any molding process, the life of the mold is the major problem.

2.2.3 Glass vs. Plastic

In is probably worthwhile to generally compare the advantages and disadvantages of glass and plastic lenses for microoptics. A great deal of progress has been made over the years in making harder and tougher plastic materials with higher refractive indices. This has translated into improved scratch resistance and overall durability. But this is not the major issue with lens applications. The major drawback of high-performance plastic lenses has to do with the thermal stability of

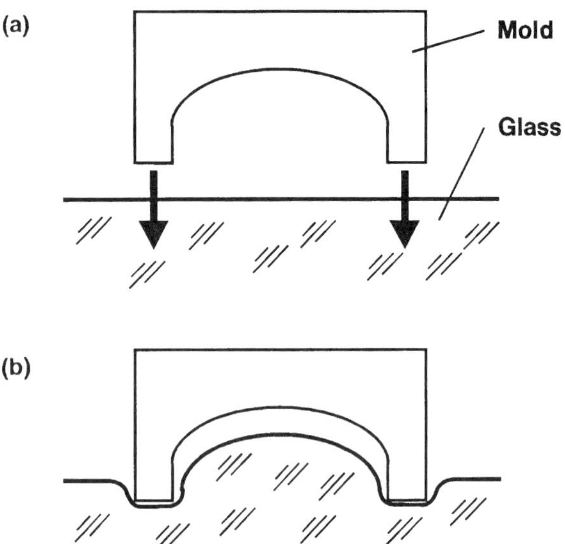

Figure 2.14 Schematic representation of contactless molding operation in glass. The ribs of the mold are pressed into the soft glass, which takes on a near-spherical surface because of surface tension. The soft glass never touches the interior of the mold. The mold is made of special alloys that do not react with the glass, and release easily. (See Ref. 11.)

the lens performance. One can quantify this property by defining the temperature coefficient of optical path, nL, through the expression

$$G = \frac{\Delta(nL)}{L\,\Delta T} = \frac{(n-1)\,\Delta L}{L\,\Delta T} + \frac{\Delta n}{\Delta T} = (n-1)\text{CTE} + \frac{dn}{dT} \tag{2.7}$$

where CTE is the coefficient of thermal expansion. One wants the value of G to be as small as possible. For glasses dn/dT is negative and partially cancels the CTE part, and the resulting value of G is relatively small. For plastics, the CTE is very large and dominates the value of G.

The overwhelming advantage of plastics is their low cost while maintaining a high performance level. Unfortunately, many of the applications of microlenses require ultra performance levels, not yet achievable by plastics. In less demanding applications, in particular for lens arrays, the plastics are more competitive. Different problems then arise which have to do with maintaining dimensions over long distances, or run-out as it is often called. In addition to being a less rigid material, the higher thermal expansion could pose a problem. Other aspects that have to be considered are how one bonds the plastic material to other materials. Plastics often absorb moisture and swell, creating problems with maintaining a secure bond. Using the plastic lenses on glass avoids all of these problems, leaving only the disadvantage of the softness of the lens surface with respect to abrasives.

2.2.4 Photoresist Based

There are a number of methods by which one can obtain reasonably good spherical microlenses that involve the use of patterned photoresist. This is the standard process where resist is spun on to a suitable substrate, then exposed to light through a mask to alter its solubility to a subsequent development solution. The convention is that a negative resist is one where after exposure it is resistant to the solvent, whereas a positive resist is one that dissolves where exposed [12]. In the following we will see how this can be used to produce microlenses.

Melted

The simplest and most ingenius way to create a microlens is to make a resist pattern in the shape of disks. In this case one exposes the resist of thickness L through a mask made up of clear circles with diameter D. Upon development, this turns into the disk shape shown in Figure 2.15 [13,14]. One now merely heats the substrate to melt the photoresist. The surface forces control the shape and a segment of a sphere represents the minimum energy surface. The first-order relationship between the circle diameter, the resist thickness, and the resulting radius of curvature of the surface, and ultimately the focal length of the lens so formed, is given by the expression

$$R_c = (n - 1)f = \frac{D^2}{4L} \tag{2.8}$$

It has been reported [14] that this method is effective in making lenses that are close to hemispherical, and correspondingly less so as one deviates to shallower lenses. The effective range of f-numbers (f/D) is in the range of 1–2.5. To increase the f-number one can replace the ambient medium with a fluid of intermediate index between the resist and air. 5 to 750 μm is the range of diameters of lenses that have been produced by this method. Another limitation is imposed by the thickness of resist that can be applied, around 50 μm. Yet another limitation is that one cannot place the lenses close enough to achieve high fill factors unless a different lens boundary shape is used.

Daly et al. [14] present a fairly comprehensive report on the fabrication process, including such facts as the actual resists that were used and some of the process steps. This is given in Table 2.3. They also describe some novel variations such as making a preform resist pattern before melting, as shown in Figure 2.16a,

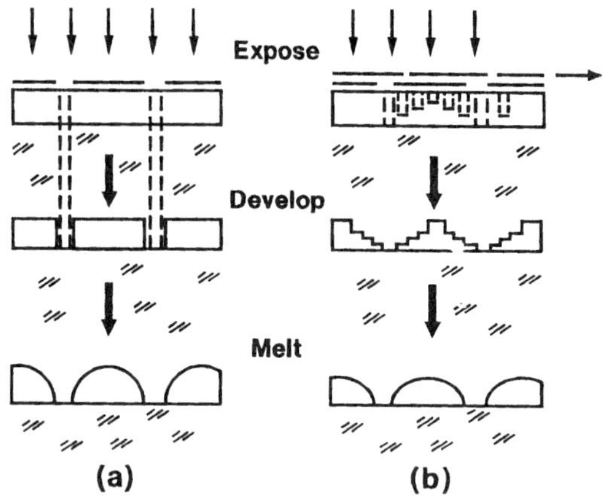

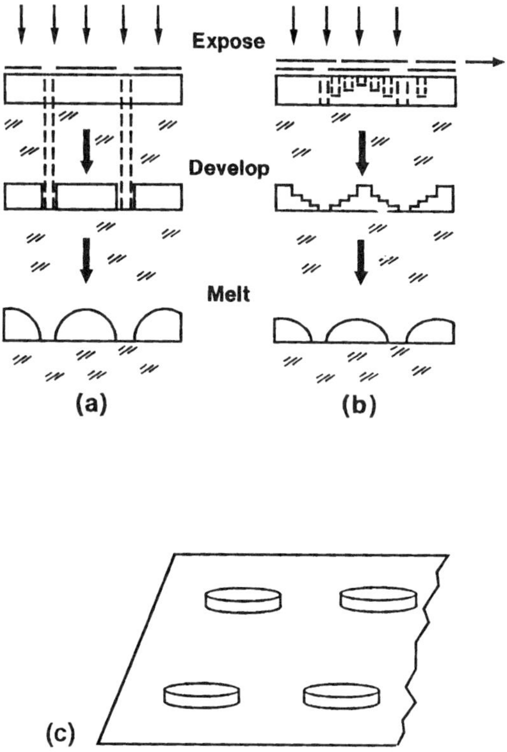

Figure 2.15 Representation of the melted photoresist method to form microlenses. Cylindrical-shaped columns or rows are patterned into the photoresist. Upon heating, the resist flows and spherical surfaces are produced. The resist can be prepatterned to obtain a better approximation to a spherical surface.

Table 2.3 Photoresists Used in the Melted Method

Type	Deposition (rpm/min)	Thickness (μm)
Shipley AZ 1400–37	1500/1	5
Hoechst AZ 4620-A	2500/0.16	8
	1200/.16	20
	800/0.1	40
	800/0.06	50

Taken from Ref. 14.

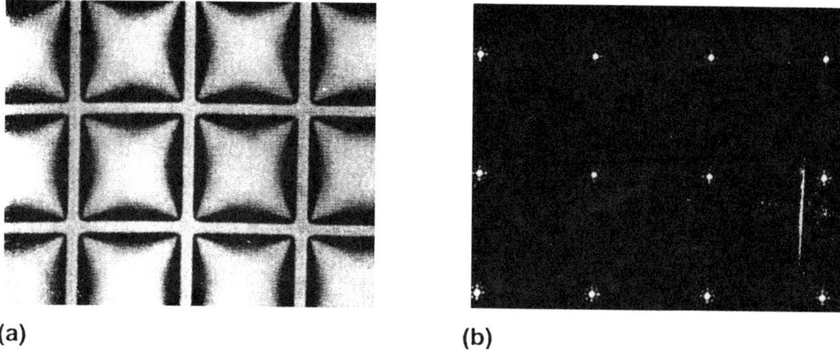
(a)　　　　　　　　　　　　　　　(b)

Figure 2.16 Square patterned lenses formed by orthogonal exposure. Images formed by the lenses are also shown.

to address the problem of making longer focal length lenses. Another interesting idea was to place the resist lenses on an existing metal apertured pattern to provide a dark matrix surround. Otherwise light would come through the nonlens portion of the array. They also showed lens with square cross section made by superimposing two cylindrical arrays, as shown in Figure 2.16a. They mention the expected poorer performance because of the edge effects. For close-packed designs, there appears to be a dependence on the ultimate shape and the shape before melting. For example, see the stepwise resist structure shown in Figure 2.16b. They report on a variety of lenses, both cylindrical and spherical, which we reproduce here as Table 2.4.

ETCHING

Following the procedure in the previous section, we can proceed to use the shaped resist to provide a way to etch the pattern into the substrate. For example, lenses etched in this way are shown schematically in Figure 2.17. Obviously, the major issue is to choose an etching process that is one to one in solubility. That is, the resist and glass must etch at the same rate. The etch of choice is RIE (reactive ion etching), where one can exercise some control over the relative etch rates by using a combination of oxygen-, and fluorocarbon-containing gases. Lenses with less than $\lambda/4$ waves of third-order spherical aberration have been made [15]; their diameters are comparable to those mentioned above.

Clearly, the substrate can be other than glass. There have been reports of lenses made in silicon and polyimide [15].

2.2.5 Microjet Fabrication

The use of the microjet technology, developed primarily for printing, has also been adapted to making microlenses [16]. The analogous microjet system used

Table 2.4 Representatives Sizes and Numerical
Aperture of Melted Resist Lenses

Focal length (μm)	Aperture (μm)	NA	Measured FWHH (μm)
Cylindrical			
78	88	0.56	1.0
105	95	0.45	1.1
147	100	0.34	1.4
200	102	0.25	1.95
300	102	0.17	5
500	100	0.1	10
Spherical			
630	750	0.6	1
570	280	0.31	3
180	125	0.34	1.7
110	65	0.3	1.8

Source: Ref: 14.

a piezoelectric ceramic with a microchannel machined in it with a nozzle on one end connected to a reservoir on the other. An electrical pulse bends the channel and forces a droplet through the aperture. The droplet is directed to a substrate which is mounted on an *xyz* micropositioner (Figure 2.18); the liquid drop solidifies on contact with the substrate and surface tension causes a spherical surface to form. The lens diameters are in the order of 80 μm and are roughly hemispherical. This method produces much the same result as the melted photoresist technique. One forms near hemispherical lenses and the packing fraction is limited because of the lenses running together. No description or characterization of the fluid used are given.

MicroFab Technology, Inc. reports being able to deliver spheres of fluid with diameters from 25 to 100 μm at rates of 6000 per second using this piezoelectric drop-on-demand ink jet printing technology [17].

2.2.6 Photosensitive Glass

Photothermal Process

One of the more interesting methods of producing refractive microlens arrays, at least as far as the phenomenon is concerned, is that which involves the use of a special photosensitive glass [18,19]. The basis of the effect is generated by

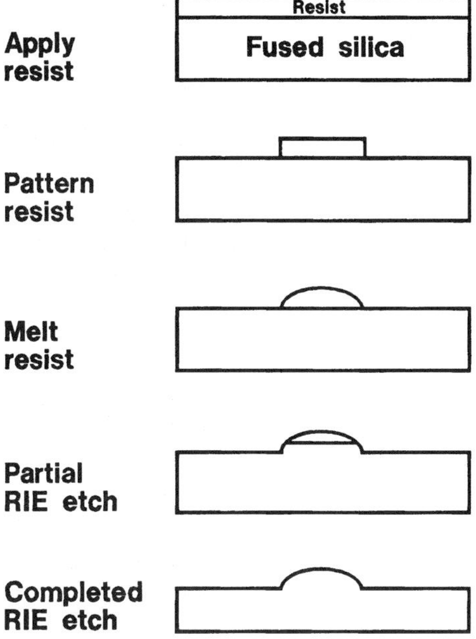

Figure 2.17 Etching of the lens into material using the melted photoresist as the graded mask. The initial steps are as shown above. It is important to maintain a one-to-one etch rate between the resist and the material.

a photonucleation of a phase crystallization (lithium meta-silicate) within the glass which produces a physical change in density. The total crystal content is of the order of 10–20% vol. Under the appropriate exposure pattern this density change can be used to produce surface features that ultimately act as lenses. A schematic of the process is shown in Figure 2.19. and a typical SEM result is shown in Figure 2.20. The photopatterning is done by conventional photolithographic masking techniques. After the glass is exposed, it is heated to about 600°C to effect the crystallization. What results are circular regions where the light is blocked, surrounded by crystallized regions of higher density. The effect of this is to squeeze the soft unexposed glass beyond the surface that then forms a minimum energy surface. The surface bumps are formed on both surfaces. We can estimate the height of the bump from a simple expression:

$$\frac{\delta}{T} = \left(\frac{2}{3}\right)\left(1 - \frac{\rho}{\rho_0}\right) \tag{2.9}$$

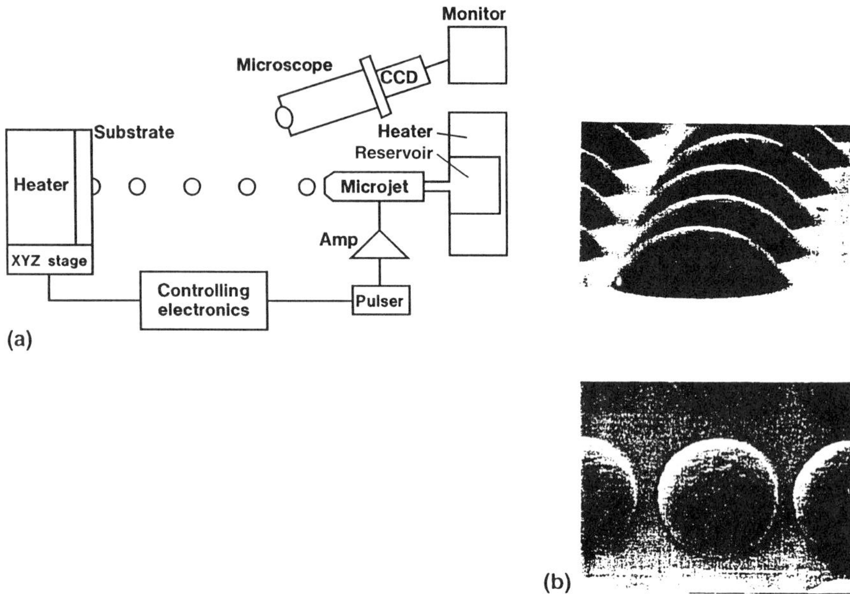

Figure 2.18 Schematic of microjet method of depositing microlenses. See text for description. Inset shows SEM photographs of lenslets on 100μm centers.

where T is the thickness, ρ is the unexposed glass density, and ρ_0 is the exposed crystallized glass density. The maximum density change observed is in the order of 1.5%. The effective depth is determined by the absorption coefficient at the activating wavelength, which is 310 nm. In general, using conventional Hg-Xe exposure lamps, exposure times of 10–100 s are required and the depth of subsequent development is 3 mm. The samples can also be simultaneously exposed from both sides by extending the maximum thickness to 6 mm and by stacking configurations [20] as shown in Figure 2.21.

Nonspherical lenses can also be formed. The shape of the lens is determined by the geometry of the exposed region. For example, lenses that represent segments of ellipsoids of revolution are easily produced by exposure through an elliptical mask. Thus it is possible to form an anamorphic lens by using an elliptically shaped mask. For dense packing even hexagonal-shaped lenses have been produced. A variety of shapes that have been formed are shown in Figure 2.22. In Appendix A to this chapter, we will derive the minimum energy shape that evolved from an arbitrary closed-boundary curve. It turns out that a reasonably uniform surface evolves except near the boundaries that contain edges.

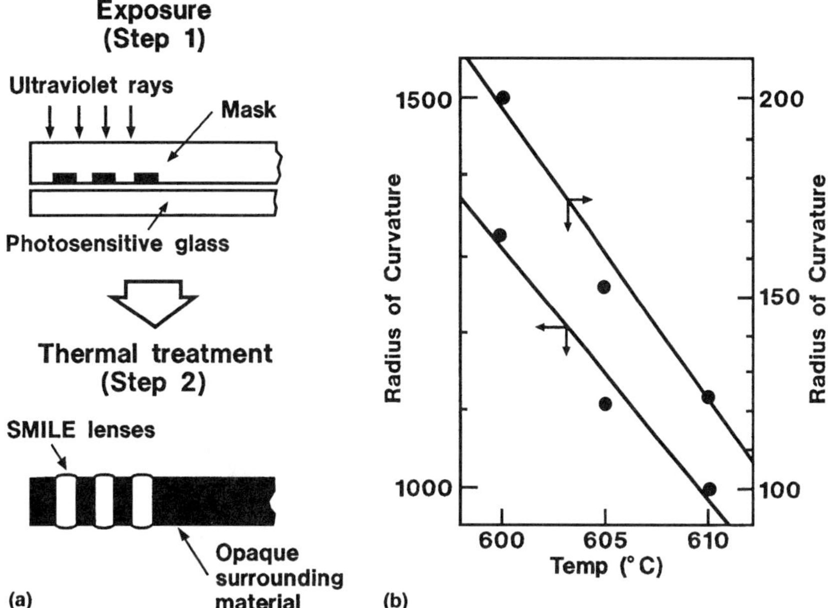

Figure 2.19 On the left, a schematic of the photosensitive process. Special glass is exposed through a mask, then heated. Exposed area densifies and squeezes soft unexposed glass to form spherical protuberances. On the right is a graph of the radius of curvature of the induced lens as a function of the top temperature of the thermal schedule. It is shown for two different lens diameters.

Ion-Exchange Stuffing

Another way was found to make lenses in this photosensitive glass using an ion-exchange stuffing mechanism [20]. It is shown that the ion-exchange of Na or K for Li occurred only in the unexposed glass. The substitution of the larger ion Na or K for the Li in the unexposed glass caused the glass to expand. In the exposed and developed region the ion exchange did not occur, and moreover at the temperature of the ion-exchange the material is rigid. As a consequence the soft unexposed glass is squeezed beyond the surface and spherical bumps are produced. A representative result is shown in Figure 2.23 where the radius of curvature of the lens is plotted vs. the square root of the exchange time in a KNO_3 at 550°C.

There are practical advantages of the ion-exchange method of producing lenses over the purely thermal method. The major one is the temperature unifor-

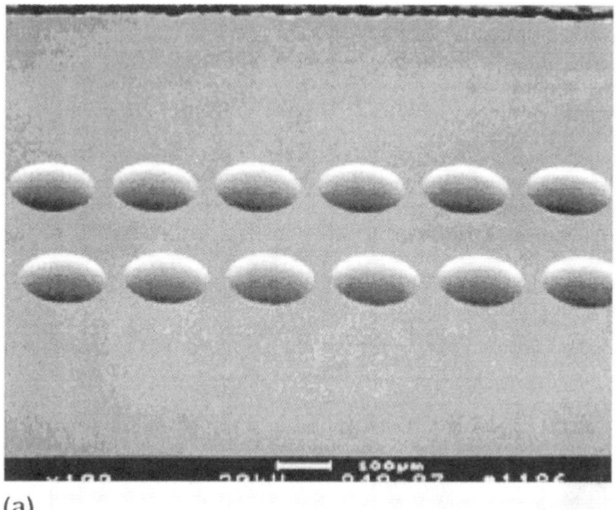

(a)

(b)

Figure 2.20   (a) Scanning electron micrograph of lenses formed by SMILE. In this case the lens diameter was 160 μm on 195 μm centers with a focal length of 0.4 mm. (b) A cross-sectional view showing the crystallized intervening material.

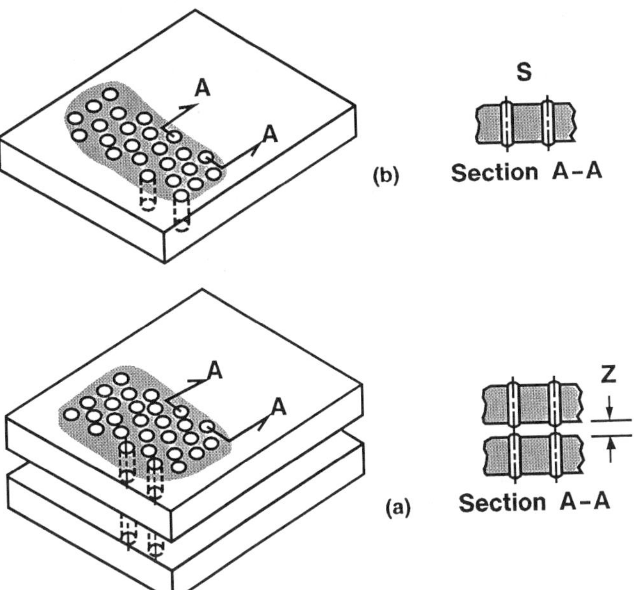

Figure 2.21 Schematic drawing of result of double side exposure of the photosensitive glass to produce a biconvex lens. Also shown is the stacking of two arrays.

mity afforded by the immersion of the sample in a constant temperature bath over that achievable in a furnace. This is particularly true when processing large two-dimensional arrays. Another advantage is for applications where compound structures are required. One can finish the exposed and developed patterned array to its final thickness before the ion-exchange development of the lenses. One has two pristine lens surfaces which can then be attached together as previously shown in Figure 2.20. In the normal thermal method one side of the sample is in contact with a substrate during the heating and this produces small defects on the lens.

Fabricated Lenses

The lens diameters range from 80 to 1000 μm and the corresponding focal lengths is greater than 100 μm. Since the process can yield double convex lenses, the effective focal length range is extendable via the compound lens effect. The effective focal length, EFL, of a biconvex lens is given by

$$\frac{1}{\text{EFL}} = \frac{1}{f_1} + \frac{1}{f_2} - \frac{T/n}{f_1 f_2} \tag{2.10}$$

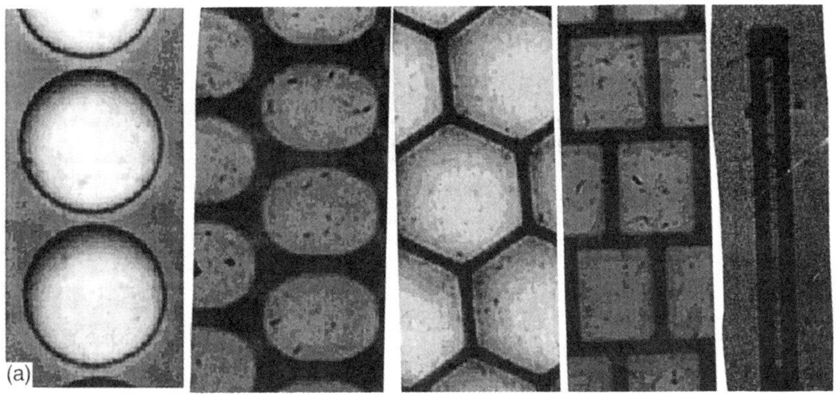

(a)

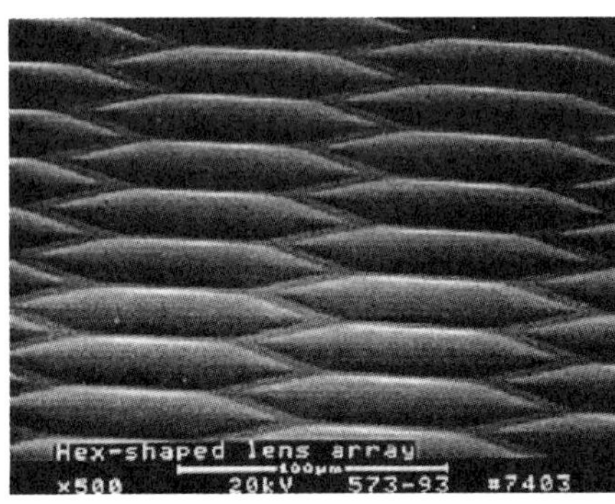

(b)

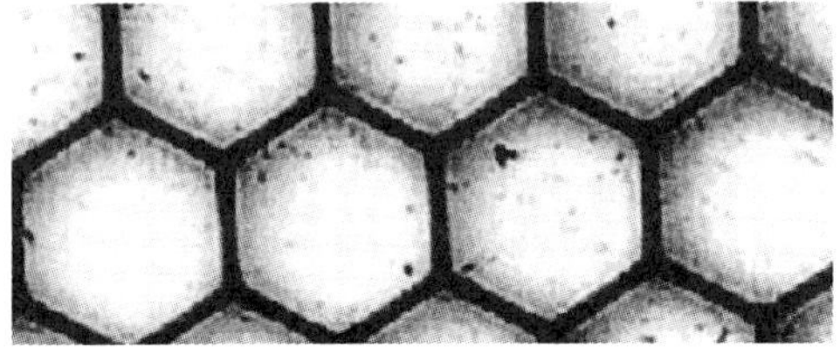

(c)

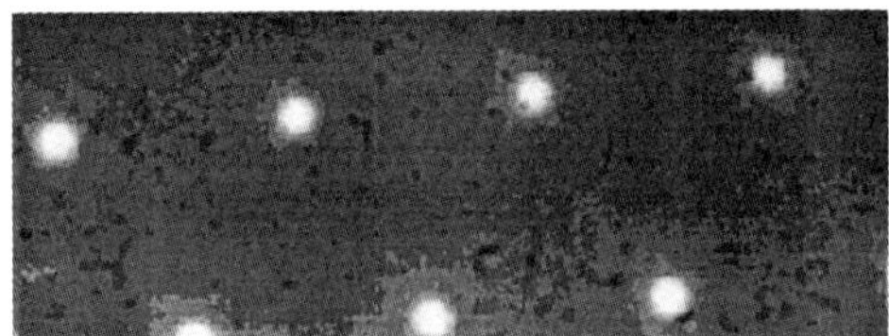

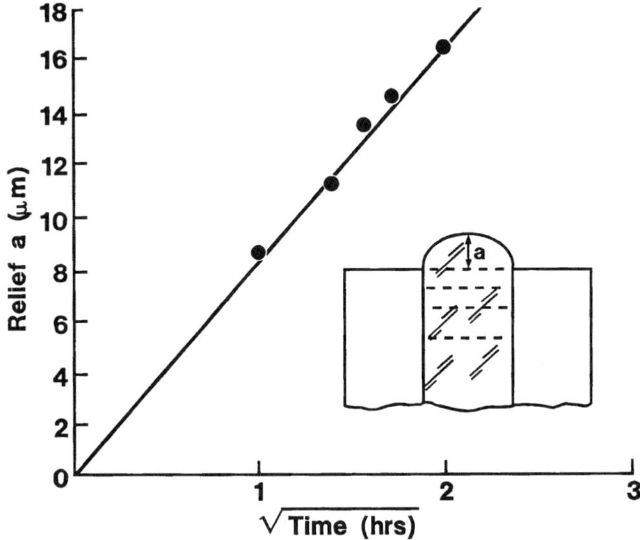

Figure 2.23 Ion-exchange stuffing to produce microlenses in photosensitive glass. Ion exchange proceeds more completely in the unexposed glass. Graph shows height of relief as a function of diffusion time at a fixed exchange temperature.

where T is the lens thickness, and f_1 and f_2 would be the focal lengths of the individual front and back lenses. Table 2.5 lists the range of lens diameters and configurations that have been fabricated. The lens arrays can be arranged in any two-dimensional pattern limited only by the minimum separation distance of 10 μm.

This photothermal method lends itself to the fabrication of erect one-to-one imaging arrays because the natural formation is a biconvex structure [21]. If the lenses are strong enough, a relay image can be formed midway between the two lenses. In Chapter 5, the applications of such lens bars will be explained and their performance discussed.

Figure 2.22 (a) Photomicrographs of various lens shapes and arrangements made from the photosensitive glass method under trade name SMILE™. (b) Regular hexagonal pattern, upper photo shows electron microscope view and the lower the geometric layout. (c) Images formed by the lens array shown in (b).

Table 2.5 Lenses Made by Photothermal Method

Type	Diameter (mm)	Spacing (mm)	FL (mm)	Shape/pattern
Plano	0.08–1.0	>0.01	0.2–	Round/any
Biconvex	0.12–1.0	>0.01	0.2–	Round/any
Plano	0.08–0.2	>0.01	>1	Hex, square/any
Stacks	0.12–1.0	>0.01	0.2–	Round/any

2.2.7 Miscellaneous Methods

Laser Heating

There have been other methods that have been described as leading to lenslike structures. One method makes use of a laser whose emission wavelength is strongly absorbed such as to produce local melting at the focal spot [22a,b]. The explanation put forward for the formation of the bump is that the glass is initially melted and then resolidified. Since the soft glass with the higher temperature has a lower density, the excess volume wells up out of the colder solid substrate glass and reforms upon solidification into a lenslike shape.

A more recent study [23] indicates the mechanism shown in Figure 2.24. The circular depression around the raised lens which was caused by the local heating was measured and is shown in Figure 2.25. Typical exposure conditions leading to lens of various diameters are shown in Figure 2.26. These data are all taken from Ref. 23.

Another intriguing method recently described by Lawandy et al. [24] is where the chemical etchability of certain borosilicate glasses can be altered by preexposure to high intensity light. The mechanism is not really understood but is attributed to an atomic level of damage induced by the irradiation that subsequently affects the way the material reacts with typical glass etchants like HF. In the data reported, the glass etches preferentially where exposed.

2.3 COMPARISON AND ANALYSIS

The intention of this section is to give the reader some way to compare the various ways that microlenses can be prepared as outlined above. Clearly, the specific application defines and orders the important performance criteria; there is no overall ''best way.'' When cost is included, the choice is even more difficult. However, we can strive to set up a method of comparison where one can get some idea of what kind of property and performance one can expect from the

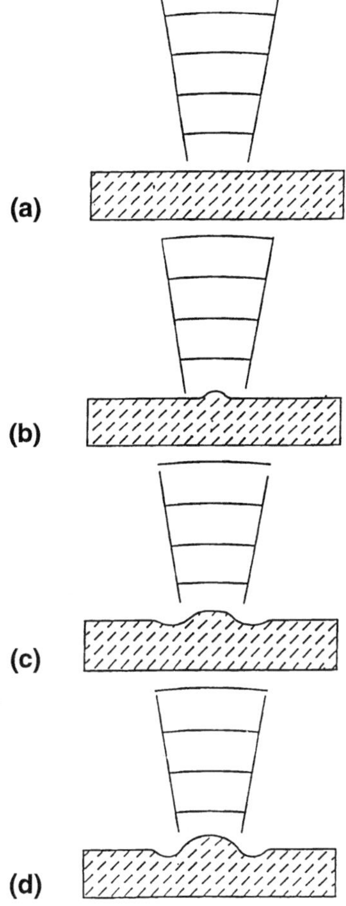

Figure 2.24 Schematic drawing of how lens forms by local laser heating. In (b) the temperature is sufficient to soften the central region; in (c) with higher temperature being achieved, the onset of the large deformation is seen. (After Ref. 23.)

various ways lenses and lens arrays are made. This will be done in two ways. The first will be to summarize from the above methods what can and what cannot be done by each of the fabrication methods. We will compare the range of achievable focal lengths, lens diameters, etc., in a chart form so we can readily see what is possible. The second comparison will be more detailed and will try to put together what data there is pertaining to the actual lens performance. Ideally,

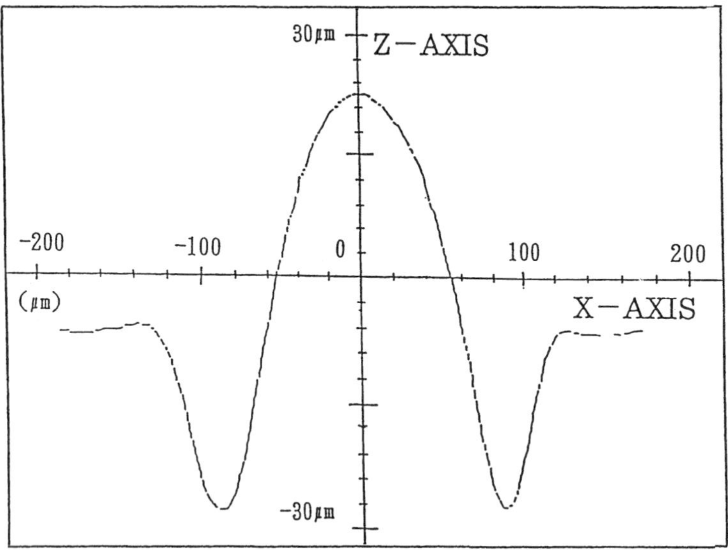

Figure 2.25 Profilometer trace over heated region. (After Ref. 23.)

this would include a common method of characterization such as interferometry, but unfortunately little real lens characterization has been reported. Nonetheless, what has been published will be summarized.

2.3.1 Comparison of Capabilities

As mentioned above, this section deals with the type of lenses that can be made and roughly the range of parameters available. We will do this in a tabular form to permit a quick reference. We list the fabrication method vertically and the property or capability horizontally.

By Asphere, we mean that such a surface can be prepared by this method. Under Shapes is whether such arbitrary boundary shapes as ellipses or squares can be made. Array means whether the method lends itself to making arrays of lenses. The focal length, FL, and the Diameter refer to the limiting values producible by the method. Packing means how close the lenses can be spaced, ultimately expressed as a percentage fill factor. Geometry refers to how the lenses can be arranged in a two-dimensional pattern. The size refers to the dimensions of a piece that can be processed.

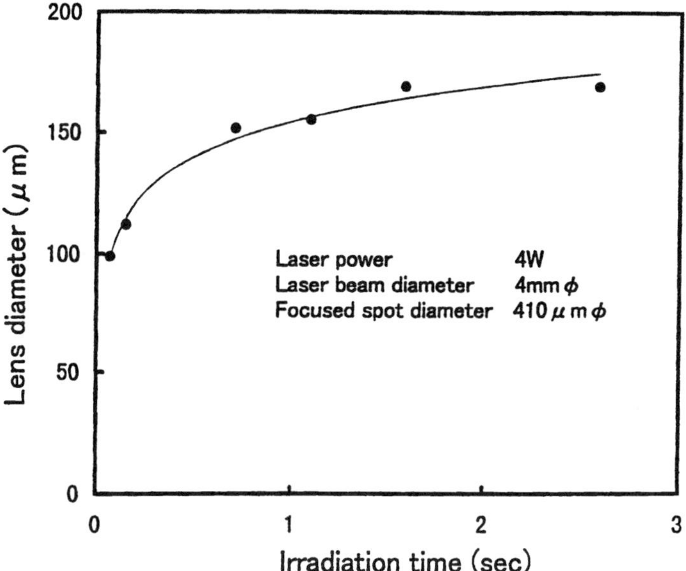

Figure 2.26 Diameter of lens as a function of irradiation time, for parameters as listed. (From Ref. 23.)

Table 2.6 Comparison Table

Fabrication	Asphere	Shapes	Array	FL (µm)	Diameter (mm)	Packing/ Geometry
Molded						
Plastic	Y	Any	Y	>100	>0.1	100%/any
Glass	Y	Any	N	>500	>2*	—
Contactless	N	Any†	Y		>0.1	80%/any
Resist						
As is	N	Circle	Y			
Etched	N	Circle	Y			
Microjet	N	Circle	Y		0.08–1	80%/any
Photosens glass	N	Any†	Y	>200	0.08–1	80%/any
Laser	N	Circle	Y	>100	>0.1	?/any

*Depends on radius of curvature of cutting tool.
†The surface figure is dependent on the boundary curve.

2.3.2 Methods of Analysis of Lens Performance

It does not matter how ingenious and easy the lens fabrication method is if the lens performance is not adequate for the application. The question is then how to characterize lenses in such a way as to be able to obtain a reasonable estimate of how they will perform. The most common performance characterization that is given is that the lens is diffraction limited. What is usually meant is that the spot size in the image plane was somehow measured and compared to the diffraction limited number which is either $1.22\lambda/\mathrm{NA}$, if the intensity across the field is flat, or $(2/\pi)/\mathrm{NA}$ if the intensity is gaussian. Unfortunately, the measured spot size, as interpreted as the central portion of the Airy profile, does not change even in the presence of significant spherical aberration. As we will show below, even if the surface curve is not perfectly spherical, the first ring diameter is not significantly altered.

The irradiance distribution in the focal plane in the presence of a phase aberration, $\phi(\rho)$, is given by the following expression [25].

$$\frac{I(r)}{I_0} = \left(\frac{k}{R}\right)^2 \left| \int \exp[i\phi(\rho)] J_0\left(\frac{kr\rho}{R}\right) \rho \, d\rho \right|^2 \tag{2.11}$$

where $k = 2\pi/\lambda$, R is the distance from the pupil plane to the image plane (focal length of the lens), a is the radius of the aperture in the pupil plane (lens radius), and ρ is the radial position in the image plane. This rotationally symmetric phase aberration term can be expressed in terms of the classical Zernike polynomials representing the fourth, sixth and eighth order spherical aberrations. For $\phi(\rho) = 1$ corresponding to the perfect lens, one can integrate Eq. (2.11), and obtain the well-known Airy function.

$$\frac{I(r)}{I_0} = \left| \frac{2J_1(\mathrm{kar})/R}{\mathrm{kar}/R} \right|^2 \tag{2.12}$$

Mahajan [25] investigated a number of informative cases with this formulation which could form a basis of an ultimate performance comparison for microlenses. He used the Strehl ratio as the parameter measuring the magnitude of the aberration. The Strehl ratio is defined as the value of Eq. (2.11) at $r = 0$, that is, $I(0)$. One can show that the Strehl ratio is very well approximated by the expression

$$S = \exp\left(-\sigma_\phi^2\right) \tag{2.13}$$

where σ_ϕ is the standard deviation of the spherical aberration. In a typical Zernike analysis of the interferometric data for a given lens, this number would correspond to RMS or peak-to-valley number under the appropriate aberration. The

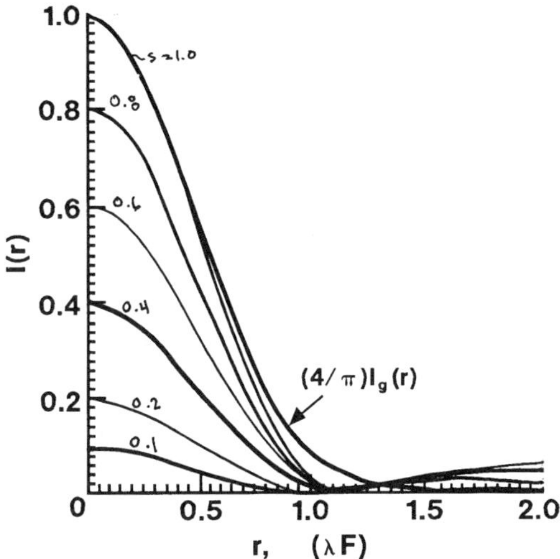

Figure 2.27 Computed point spread function vs. radius(normalized to the product of the wavelength λ and the focal ratio F). The various curves are for different degrees of aberration, as indicated by the Strehl ratio S. Also drawn is the Gaussian profile. (Graph is replotted from Ref. 24.)

results are shown in Figure 2.27. The interesting aspect of this calculation is that the central ring diameter is not changing with the degree of aberration. The main effect of the spherical aberration is the reduction of the intensity of the central ring and the corresponding increase in the outer ring.

2.3.3 Data Comparison

What the analysis in the above section points out is that for a true comparison of the microlens performance by any of the methods described above, either the point spread function (PSR) be measured or a Zernike analysis of interferometric data should be done. The reader is referred to Table 2.7 for a listing of the Zernike terms.

Zernike polynomial analysis of the molded glass asphere has been reported. These are relatively large lenses in the microoptic sense. These lenses are produced to provide high resolution at high numerical aperture. A typical result for a molded bisphere is shown schematically in Figure 2.28 representing an overall rms 0.05 λ distortion from a perfect spherical wavefront. The numerical aperture for this lens was 0.45 [26].

Table 2.7　Zernike Poynomial Notation

			Z_{nm}	
n	*m*		Polynomial	Characterization
0	0	0	1	Piston
1	+1	1	$\rho \cos \phi$	*X* tilt
	−1	2	$\rho \sin \phi$	*Y* tilt
	0	3	$2\rho^2 - 1$	Focus
2	+2	4	$\rho^2 \cos 2\phi$	Astigmatism 0° or 90°
	−2	5	$\rho^2 \sin 2\phi$	Astigmatism ±45°
	+1	6	$(3\rho^2 - 2)\rho \cos \phi$	*X* coma and tilt
	−1	7	$(3\rho^2 - 2)\rho \sin\phi$	*Y* coma and tilt
	0	8	$6\rho^4 - 6\rho^2 + 1$	Spherical and focus
3	+3	9	$\rho^3 \cos 3\phi$	
	−3	10	$\rho^3 \sin 3\phi$	
	+2	11	$(4\rho^2 - 3)\rho^2 \cos 2\phi$	
	−2	12	$(4\rho^2 - 3)\rho^2 \sin 2\phi$	
	+1	13	$(10\rho^4 - 12\rho^2 + 3)\rho \cos\phi$	
	−1	14	$(10\rho^4 - 12\rho^2 + 3)\rho \sin \phi$	
	0	15	$20\rho^6 - 30\rho^4 + 12\rho^2 - 1$	
4	+4	16	$\rho^4 \cos 4\phi$	
	−4	17	$\rho^4 \sin 4\phi$	
	+3	18	$(5\rho^2 - 4)\rho^3 \cos 3\phi$	
	−3	19	$(5\rho^2 - 4)\rho^3 \sin 3\phi$	
	+2	20	$(15\rho^4 - 20\rho^2 + 6)\rho^2 \cos 2\phi$	
	−2	21	$(15\rho^4 - 20\rho^2 + 6)\rho^2 \sin 2\phi$	
	+1	22	$(35\rho^6 - 60\rho^4 + 30\rho^2 - 4)\rho \cos \phi$	
	−1	23	$(35\rho^6 - 60\rho^4 + 30\rho^2 - 4)\rho \sin \phi$	
	0	24	$70\rho^8 - 140\rho^6 + 90\rho^4 - 20\rho^2 + 1$	
5	+5	25	$\rho^5 \cos 5\phi$	
	−5	26	$\rho^5 \sin 5\phi$	
	+4	27	$(6\rho^2 - 5)\rho^4 \cos 4\phi$	
	−4	28	$(6\rho^2 - 5)\rho^4 \sin 4\phi$	
	+3	29	$(21\rho^4 - 30\rho^2 + 10)\rho^3 \cos 3\phi$	
	−3	30	$(21\rho^4 - 30\rho^2 + 10)\rho^3 \sin 3\phi$	
	+2	31	$(56\rho^6 - 105\rho^4 + 60\rho^2 - 10)\rho^2 \cos 2\phi$	
	−2	32	$(56\rho^6 - 105\rho^4 + 60\rho^2 - 10)\rho^2 \sin 2\phi$	
	+1	33	$(126\rho^8 - 280\rho^6 + 210\rho^4 - 60\rho^2 + 5)\rho \cos\phi$	
	−1	34	$(126\rho^8 - 280\rho^6 + 210\rho^4 - 60\rho^2 + 5)\rho \sin\phi$	
	0	35	$252\rho^{10} - 630\rho^8 + 560\rho^6 - 210\rho^4 + 30\rho^2 - 1$	

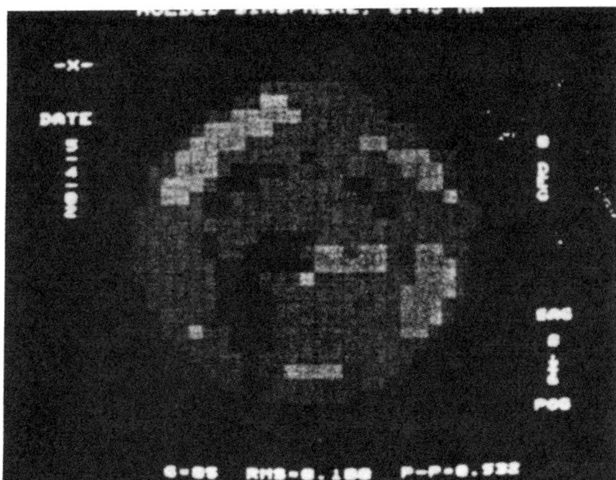

Figure 2.28 Interferometrically measured optical phase difference map of a molded aspheric bisphere. (From Ref. 25.)

Aperture limitations imposed on interferometry make the measurement of very small lenses a problem. Nonetheless interferometric data has been reported for microlenses made by the photosensitive glass process, SMILE. The complete tabulation of the Zernike interferometric analysis is given for a 0.4-mm diameter plano-convex lens with a radius of curvature of 1.25 mm (f/6.25). The coefficients for the 36 polynomial fit are listed in the left-hand column. In the Table 2.8, on the right, a synopsis of the contributions are made in terms of the conventional aberration descriptions such as spherical, coma, astigmatism, etc. At the very bottom is a row which gives the overall least square deviations, listed as P–V, peak-to-valley, RMS, root-mean-square, and Strehl ratio. The RMS characterization is 0.125 λ.

There are no interferometric results reported for the other microlens types, likely due to their small size. Adaptive Optics Associates claims 0.25 λ deviation for lenses greater than f/4, although how this was obtained was not specified.

The point-spread function is a more accessible characterization method, at least in principle. There are a number of experimental techniques by which one can obtain fairly accurate 2D profiles of the intensity pattern in the focal plane of the lens. These encompass scanning slit systems, pyroelectric-based systems, and the newer CCD-based 2-D arrays used in conjunction with a microscope.

Measured PSFs for three SMILE[tm] lenses are shown in Figure 2.29 [20]. They are compared to the calculated Airy curves. In all cases one can see the wings are larger, particularly for the larger lens diameter. This clearly indicates aberrations. Unfortunately, the method compared the normalized curves that did not permit

Table 2.8 Zernike Analysis of SMILE Lens

(Diameter = 0.4 mm, focal length = 1.25 mm)

TERM	COEFF	RMS (0.0001W)
1 TILTX	0.0335	167
2 TILTY	1.6721	8360
3 FOCUS	0.2701	1559
4 AST30	0.0046	19
5 AST31	−0.0154	63
6 COM30	0.0777	275
7 COM31	−0.0785	278
8 SPH3	0.2583	1155
9 3TH50	−0.0016	6
10 3TH51	−0.0064	23
11 AST50	−0.0004	1

RMS ERROR (0.0001 WAVES)

ORD	SPHE	COMA	ASTI	3THE	4THE	5THE	SUBT
3	1155	391	66				1221
5	37	13	13	23			47
7	10	24	13	2	16		33
9	15	43	26	29	9	7	61
11	13						13
SUBT	1156	394	73	37	18	7	1224
					RMS RESIDUE		112
					TOTAL RMS		1229

12 AST51	0.0040	13
13 COM50	−0.0035	10
14 COM51	−0.0029	8
15 SPHS	0.0098	37
16 4TH70	0.0048	15
17 4TH71	−0.0016	5
18 3TH70	0.0007	2
19 3TH71	−0.0003	1
20 AST70	0.0023	6
21 AST71	−0.0045	12
22 COM70	−0.0004	1
23 COM71	0.0094	24
24 SPH7	−0.0030	10

P−V ERROR (0.0001 WAVES)

ORD	SPHE	COMA	ASTI	3THE	4THE	5THE
3	3875	2210	321			
5	196	91	80	132		
7	43	189	101	16	102	
9	98	388	223	230	66	45
11	66					

Overall average values (in waves)

PEAK	VALLEY	P−V	RMS	STREHL RATIO
0.414	−0.246	0.660	0.125	0.540

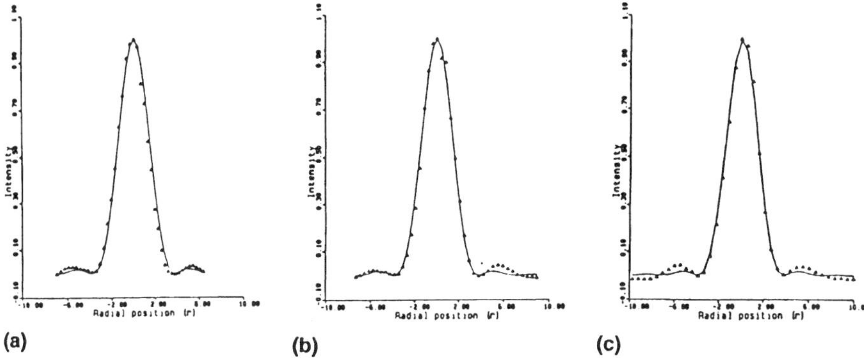

Figure 2.29   Measured point spread functions for SMILE lenses compared to best-fit Airy pattern. The lens diameter was 0.4 mm and the EFLs were 0.14, 0.19, and 0.36.

a quantitative estimate of the extent of aberrated behavior. Daly et al. [14] report PSF data for the melted photoresist lenses over a range of diameters which is shown in Table 2.4. They make a comparison to the Airy curve based on the full width at half maximum. The results are not particularly consistent. In addition, it is not clear what the FWHH really means, since Majahan has shown that the normalized patterns over a wide range of aberrations, have essentially the half width.

APPENDIX A

A.1   LENSES DERIVED FROM SURFACE TENSION

Three of the fabrication methods described above relied on surface tension to produce the surface figure. The methods were the melted photoresist, the contactless molding, and the photosensitive glass. In many applications (see Chapter 5) it is important for arrays of lenses to have a high fill-factor, that is, to have the lenses occupy nearly 100 percent of the area. This requirement is a problem for the lens fabrication methods mentioned above because the lens figure is not independent of the shape of the lens. In this section we will calculate the minimum energy surface figures for different boundary conditions which will be relevant for applying these techniques for making lenses with noncircular boundaries.

The reader is referred to Figure A.1 for a statement of the problem. We want to determine the equation of the surface $z = f(x, y)$ bounded by the curve C, where a uniform pressure p is applied to the surface whose material has a surface tension coefficient α. This is entirely equivalent to the variational problem of finding the minimum (extremum) of the integral area:

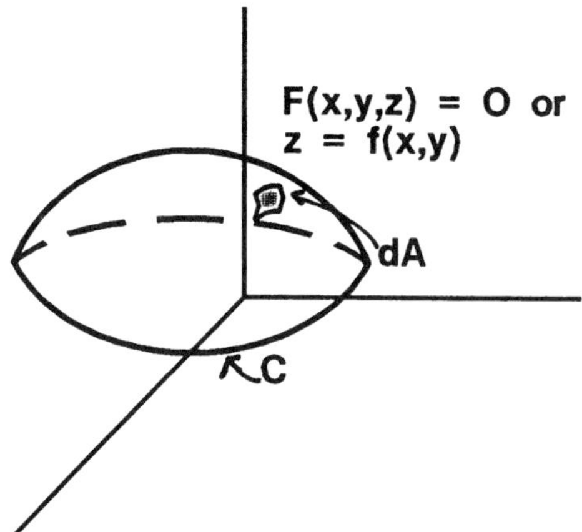

Figure A.1 Representation of the surface described by the equation $F(x, y, z) = 0$ bounded by the curve C.

$$A[f] = dA \sqrt{1 + \left(\frac{\delta f}{\delta x}\right)^2 + \left(\frac{\delta f}{\delta y}\right)^2}\, dx\,dy \tag{A.1}$$

under the constraint of the constant volume. The variational problem is expressed as $\delta(A + \lambda V)$ where λ is the Lagragian mutiplier and is determined by the condition $V(\lambda) = V_0$ The Euler Lagrange solution of Eq. A.1 is

$$\nabla \cdot \left\{\frac{\nabla f}{\sqrt{1 + |\nabla f|^2}}\right\} - \frac{p}{\alpha} = 0 \tag{A.2}$$

For the well-known case where C is a circle, one can obtain an analytical solution to Eq. (A.2) in cylindrical coordinates assuming axial symmetry. The solution is an equation of a portion of a spherical surface, viz.

$$\frac{f(\rho)}{R_0} = \sqrt{\left(\frac{2}{P}\right)^2 - \left(\frac{\rho}{R_0}\right)^2} - \sqrt{\left(\frac{2}{P}\right)^2 - 1} \tag{A.3}$$

where R_0 is the radius of the boundary circle, and $P = R_0 p / \alpha$.

For other shapes of the boundary curves one has to resort to computer numerical solutions as available from such software packages as PDE/Protran. We calculated the square and the hexagon since both of these would lead to high fill factors. The data is presented as isodeviation lines, where the deviation is from a perfect sphere. For the square geometry, because of the symmetry, only one

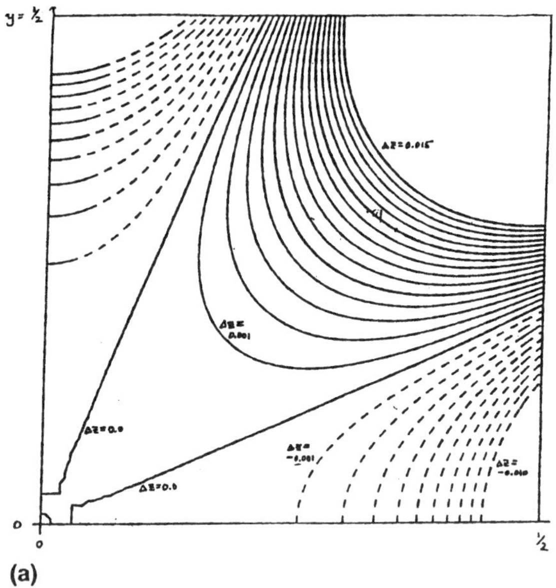

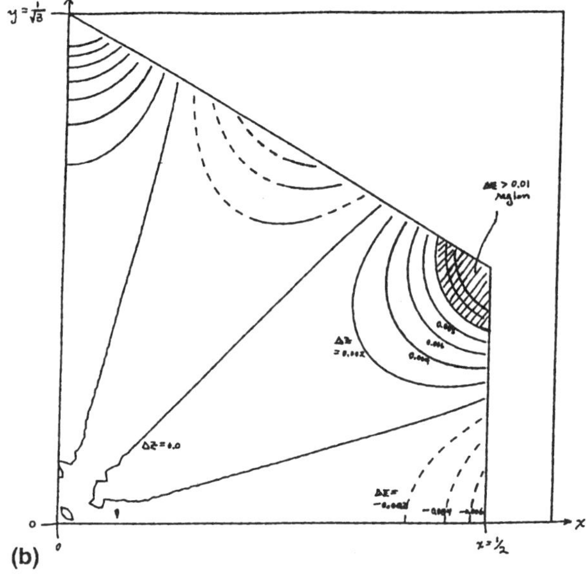

Figure A.2   (a) Quadrant map of isodeviation of surface height relative to a segment of a spherical surface for the case of a lens with a square cross section of width L. All quantities x, y, Δz are normalized to L. The normalized radius of curvature is $R_c/L = 1.4$. (b) Same as in (a) except the cross section is hexagonal with characteristic length L.

quadrant is shown in Figure A.2a. The length of the square is L and all the coordinates are normalized to L, including the deviation term Δz. One can clearly see that the distortion cuts deeply into the lens. For $L = 400$ μm, a Δz of 0.001 corresponds to roughly one wave at 550 nm.

The situation is much improved for the hexagonal geometry situation where again we have shown the result for the smallest repeated section in Figure A.2b. The distortion corresponding to $\Delta z = 0.001$ is only around the apices of the hexagon, and the major portion of the lens conforms to a sphere within one wave at 550 nm.

There is an application that relates to this concern for the nature of the curvature of microlenses. This is the application where it is desired to circularize a beam. Often the emitting regions of solid-state sources are not regular, resulting in a different beam spread in the x and y directions. If one collimates this output in the conventional way, one obtains an ellipse for the shape of the beam. It is frequently desirable to have the beam circular. One approach is to use two cylindrical lenses, as shown in Figure A.3. By choosing the correct ratio of the focal lengths f_1/f_2, one can compensate for the different beam-spreading angles. Where space is an issue, there is a way to use a microoptic lens that has the appropriate difference in radii of curvature in the x and y directions. It turns out that the shape required is that from a section of a torus, shown in Figure A.4. The equation for sag of the lens cut from such a section is

$$x^2 + y^2 + z^2 = a^2 + R^2 + 2R(a^2 - z^2)^{1/2} \tag{A.3}$$

Here R is the radius of the torus and $2a$ is the circular thickness.

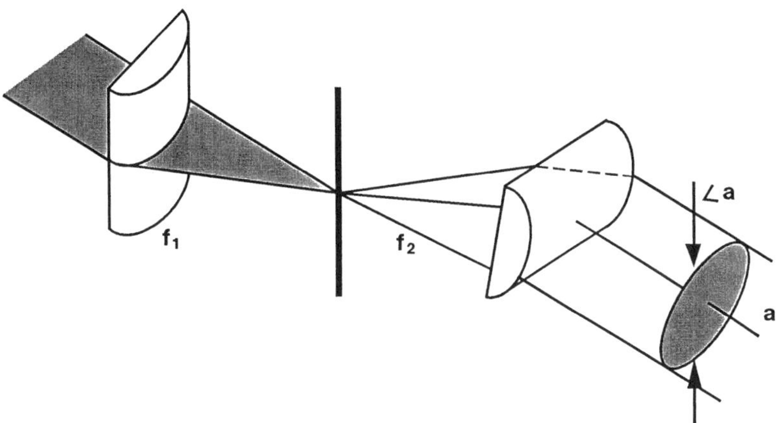

Figure A.3 Method of circularizing a beam with the use of two cylindrical lenses. The ratio of the focal lengths determines the correction.

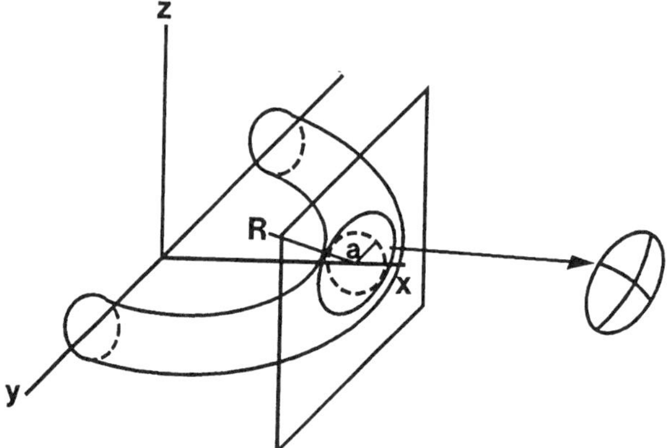

Figure A.4 Sketch of the slice of a toroid to produce an anamorphic lens with two spherical surfaces. The one radius of curvature corresponds to the radius of the toroid and the other to the radius of the thickness.

APPENDIX B

B.1 RAY-TRACE ALGORITHM

B.1.1 Analysis

The reader is referred to Figure B.1. for the definition of the symbols used. In addition, we define the following relationships:

$$x_1 = -\left(\frac{T}{2} + t_1\right)$$

$$x_4 = \frac{T}{2} + t_2$$

$$(x_2 + a_1)^2 + y^2 + z^2 = R_1^2$$

$$(x_3 - a_2)^2 + y_3^2 + z_3^2 = R_2^2 \tag{B.1}$$

$$a_1 = \frac{T}{2} - R_1$$

$$a_2 = \frac{T}{2} - R_2$$

where t_1, t_2, a_1, a_2, R_1, R_2, and T are all greater than 0. Starting at an arbitrary object point (x_1, y_1, z_1) the direction cosines of the ray to a point on the lens (x_2, y, z) are given by

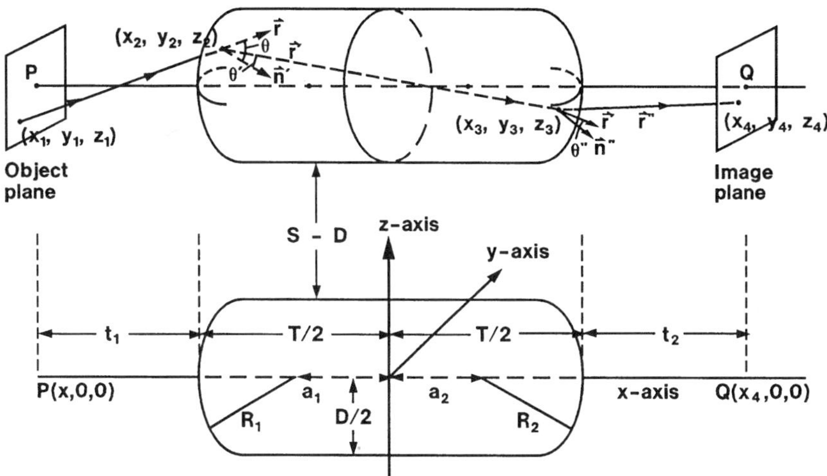

Figure B.1 Drawings showing the symbols and notation used in the ray-trace algorithm. The upper figure shows the 3D perspective showing the coordinate points at each surface consistent with the ray-trace algorithm. The lower curve indicates the dimensional variables.

$$l_1 = \frac{x_2 - x_1}{B}$$

$$m_1 = \frac{y - y_1}{B} \tag{B.2}$$

$$n_1 = \frac{z - z_1}{B}$$

$$B = \sqrt{(x_2 - x_1)^2 + (y - y_1)^2 + (z - z_1)^2}$$

where

$$x_2 = -\sqrt{R_1^2 - y^2 - z^2} + a_1 \tag{B.2a}$$

At the point (x_2, y, z) on the lens, the direction cosines of the normal to the surface are

$$l_2 = -\frac{(a_1 + x_2)}{C}$$

$$m_2 = \frac{-y}{c} \tag{B.3}$$

$$n_2 = \frac{-z}{c}$$

with

$$C = \sqrt{(a_1 + x_2)^2 + y^2 + z^2}$$

The angle of incidence θ is obtained from the scalar product of the unit vector $\mathbf{r}(l_1, m_1, n_1)$ of the incident ray with the unit vector $\mathbf{n}(l_2, m_2, n_2)$ of the normal, where both $\mathbf{r}$ and $\mathbf{n}$ point toward the right, as do all other unit vectors,

$$\cos\theta = l_1 l_2 + m_1 m_2 + n_1 n_2 \tag{B.4}$$

The refracted ray with unit vector $\mathbf{r'}$ must be in the plane of incidence, that is, the plane determined by $\mathbf{r}$ and $\mathbf{n}$.

Mathematically this can be expressed as

$$\mathbf{r'} = a\mathbf{r} + b\mathbf{n}$$

The value of a and b can be obtained by subtracting the following equations

$$\mathbf{r'} \cdot \mathbf{r'} = 1 = a^2 + b^2 + 2ab\cos\theta \tag{B.5}$$
$$(\mathbf{r'} \cdot \mathbf{n})^2 = \cos^2\theta' = a^2\cos^2\theta + b^2 + 2ab\cos\theta$$

which yields

$$1 - \cos^2\theta' = a^2(1 - \cos^2\theta)$$

or

$$\sin^2\theta' = a^2\sin^2\theta$$

where θ' is the angle that the refracted ray makes with the normal. Applying Snell's law one obtains

$$a = \frac{1}{n} \tag{B.6a}$$

where n is the refractive index of the glass. To obtain b one substitutes the value of a into the above equation and solves for b:

$$b = \frac{1}{n}\{\sqrt{\cos^2\theta - 1 + n^2} - \cos\theta\} \tag{B.6b}$$

The direction cosines of the refracted ray can now be obtained using the Eqs. (B.7):

$$l_3 = al_1 + bl_2$$
$$m_3 = am_1 + bm_2 \tag{B.7}$$
$$n_3 = an_1 + bn_2$$

The equation of the refracted ray is given by

$$y_3 - y = \frac{m_3}{l_3(x_3 - x_2)} \tag{B.8}$$

$$z_3 - z = \frac{n_3}{l_3(x_3 - x_2)}$$

where the point (x_3, y_3, z_3) is located on the back spherical surface whose equation is

$$(x_3 - a_2)^2 + y_3^2 + z_3^2 = R_2^2 \tag{B.9}$$

Solving these two equations simultaneously yields the value of x_3.

$$x_3 = D \pm \sqrt{D^2 - E} \tag{B.10}$$

where

$$D = (m_3^2 + n_3^2)x_2 + a_2 l_3^2 - m_3 y l_3 - n_3 z l_3$$
$$E = (m_3^2 + n_3^2) x_2^2 - 2 x_2 l_3(m_3 y + n_3 z) + l_3^2(y^2 + z^2 - R_2^2 + a_2^2)$$

Substituting the value of x_3 (the positive root) into Eq. (B.8) yields y_3 and z_3.

The direction cosines of the unit vector $\mathbf{r}''$ normal to the surface at (x_3, y_3, z_3) are now determined to be

$$l_4 = \frac{x_3 - a_2}{F}$$

$$m_4 = \frac{y_3}{F} \tag{B.11}$$

$$n_4 = \frac{z_3}{F}$$

where

$$F = \sqrt{(x_3 - a_2)^2 + y_3^2 + x_3^2}$$

The incident angle on the back surface can now be obtained from

$$\cos \theta'' = l_3 l_4 + m_3 m_4 + n_3 n_4 \tag{B.12}$$

The exit ray's direction cosines are determined as before by writing

$$\mathbf{r}'' = e\mathbf{r}' + f\mathbf{n}'' \tag{B.13}$$

and following the same procedure as done on the front surface, we find that

$$e = n \quad \text{and} \quad f = -n \cos \theta'' + \sqrt{n^2 \cos^2 \theta'' - n^2 + 1}$$

The direction cosines of the exit ray are thus

$$l_5 = el_3 + fl_4$$
$$m_5 = em_3 + fm_4 \tag{B.14}$$
$$n_5 = en_3 + fn_4$$

If the image plane is located at $x_4 = {}^T/_2 + t_2$, the exit ray forms an image spot with coordinates (y_4, z_4) given by

$$y_4 = y_3 + \frac{(x_4 - x_3)m_5}{l_5}$$

$$z_4 = z_3 + \frac{(x_4 - x_3)n_5}{l_5} \tag{B.15}$$

REFERENCES

1. M. Born and E. Wolfe, *Principles of Optics*, Pergamon Press, London.
2. R. W. Ditchburn, *Light*, Blackie and Son Ltd, Glasgow, Scotland, 1963.
3. C. S. Williams and O. A. Becklund, *Optics*, Wiley Interscience, New York, 1972.
4. *Handbook of Plastic Optics*, 2nd ed., U.S. Precision Lens Inc., Cincinnati, OH, 1983.
5. P. Pantelis and D. J. McCartney, Third Int. Conf. Electrical, Optical, and Acoustic Properties of Polymers, conference proceedings, Plastics and Rubber Institute, Sept. 16–18, (1992), London.
6. D. D'Amato, S. Berletta, P. Cone, J. Hizny, R. Martinsen, and L. Schumtz, Fabrication and testing of monolithic lenslet module (MLM) arrays, Opt. Fab./Testing, Tech. Digest, vol. 11, Opt. Soc. Am, Monterey, CA, June 12–14, 1990.
7. Data sheet from United Technologies, Adaptive Optics Associates, Cambridge, MA.
8. P. Hannah and R. Rhorer, Basics of diamond turning, tutorial, Am. Soc. Precision Eng., Atlanta, GA, October 24, 1988.
9. M. A. Fitch, Molded optics, *Photonics Spectra*, October 1991.
10. A. R. Olszewski, assigned to Corning Inc., U.S. Patent 4,362,819, 1982.
11. T. Dannoux, P. Calderini, G. Pujol, and J.-P. Themont, Procede et dispsitif de fabrication de reseaux de microlentilles optiques, French Patent 94-08420, July 1974.
12. S. M. Sze, *VLSI Technology*, chap. 4, McGraw-Hill, New York, 1988.
13. C. D. Popovic, R. A. Sprague, and G. A. Neville Connell, Techniques for monolithic fabrication of microlens arrays, *Appl. Opt. 27*, 1281–1284 (1988).
14. D. Daly, R. F. Stevens, and M. C. Hutley, Manufacture of microlenses by melting photoresist, IOP short meeting series 30, Institute of of Physics, May 1, Teddington, UK, 1991.
15. K. O. Mersereau, C. R. Nijander, A. Y. Feldblum, and W. P, Townsend, Fabrication and measurement of fused silica microlens arrays, SPIE Int. Sym. Opt. Appl. Sci. Eng., July 19–24, San Diego, CA (1992).
16. D. L. MacFarlane, V. Narayan, W. R. Cox, T. Chen, and D. J. Hayes, Microjet

fabrication of microlens arrays, *IEEE Photonics Techn. Lett. 6*(9), 1112–1114 (1994).

17. D. B. Wallace, D. J. Hayes, and C. J. Frederickson, High Throughput Array for Fluid Microdispensing in DNA Diagnostics, product information, MicroFab Technologies, Inc., Plano, TX (1994).

18. N. F. Borrelli, D. L. Morse, R. H. Bellman, and W. L. Morgan, Photolytic technique for producing microlenses in photosensitive glass, *J. Appl. Phys. 42*, 2520–2525 (1985).

19. N. F. Borrelli and D. L. Morse, Microlens arrays produced by a photilytic technique, *Appl. Opt. 27*, 476–479 (1988).

20. N. F. Borrelli, Generation of lens arrays using photothermal techniques, IOP short meeting, series 10, Institute of Physics, May 1, Teddington, UK, 1991.

21. N. F. Borrelli, R. H. Bellman, J. A. Durbin, and W. Lama, Imaging and radiometeric properties of microlens arrays, *Appl. Opt. 30*(25), 3633–3642 (1991).

22a. V. P. Veiko, E. B. Yakovlev, V. V. Frolov, V. A. Chujko, A. K. Kromin, M. O. Abbakamov, A. T. Shakola, and P. A. Fomichov, Laser heating and evaporation of glass and glass-boring materials and its application for creating, *MOC, SPIE 1544*, 152–163 (1991).

22b. V. P. Veiko, Essence and comparison of different laser technologies for microoptics fabrication, SPIE Int. Symp. Optical Applied Science/Eng. Miniature/Microoptics,'' July 19–24, San Diego, CA, 1992.

23. M. Wakaki, Y. Komachi, and G. Kanai, Microlens and microlens array formed on a glass plate using a CO_2 laser, *Appl. Opt. 37*(4), 627–31 (1998).

24. E. Sauvain, J. H. Kyung, and N. M. Lawandy, Multiphoton micrometer-scale photetching in silicate-based glasses, *Opt. Lett. 20*(3), 243–245 (1995).

25. V. N. Mahajan, Aberrated point-spread functions for rotationally symmetric aberrations, *Appl. Opt. 22*(19), 3035–3041 (1983).

26. R. O, Maschmeyer, R. M. Hujar, L. L. Carpenter, B. W. Nicholson, and E. F. Vozenilek, Optical performance of a diffraction-limited molded-glass biaspheric lens, *Appl. Opt. 22*(16), 2413–2145 (1983).

3

Gradient Index

3.1 INTRODUCTION

Although the binary-optic method of making microlenses may be the source of the current excitement in the microoptics field, as we will see in Chapter 4, it was the gradient index method pioneered by Nippon Sheet Glass that paved the way, and in many ways still dominates the field. We will see in succeeding sections of this chapter how the ingenious use of an ion-exchange process in glass provided the first practical way to make microlenses. This method of making an imaging device is generically called GRIN, for *gr*adient *in*dex, so you will see people use the description *GRIN lens* to mean any lens formed through refractive index profile.

The idea that an index profile could produce an image similar to that produced by a curved refracting surface goes back to Wood [1]. Intuitively, it would seem that one should mimic the phase front that occurs through a spherical lens, as a starting point. In other words, if the optical phase difference $\Delta(nL)$ is equal to $n\,\Delta L(r)$ for a refracting lens, then reproducing $L\,\Delta n(r)$ might produce the same effect. We shall see that this turns out to be essentially true, although we will derive it from a more rigorous basis.

Marchand [2] presents an excellent and complete exposition of the phenomenon related to gradient index optics. We utilize his approach in our brief optics review given below. Iga et al. [3] is another valuable and extensive reference. The scope is more restricted, but it also contains applications of devices other than lenses.

What we are concerned with is the demonstration that there is an index profile $n(x, y, z)$ that images light to a given point in the manner similar to that of a refractive spherical lens. Moreover, we want to show what sort of limitations or approximations are necessarily imposed to achieve this end. We begin with the equation that governs the direction that a ray of light will take through a medium with an arbitrary index profile $n(x, y, z)$. We then look at the possible solutions and condition that lead to imaging under the assumption of a purely radial refractive index profile. We outline a simple ray-trace procedure that is useful in the paraxial approximation. The planar version of this process will be discussed separately because it presents a more difficult problem where the index is both a function of the radial distance and the axial distance. From there we proceed to the description of actual methods of fabrication, and finally to an analysis of the GRIN imaging performance.

3.2 OPTICAL THEORY

3.2.1 Path in a Rod with Gradient Index

The path a beam of light will travel in a medium with a refractive index distribution given by $n(x, y, z)$ can be treated in one of two ways. One way is to solve the wave optics problem as contained in the scalar form of Maxwell's equations with a spatially varying refractive index [4]. A second approach is to treat it from the standpoint of geometric or ray optics using the ''least action'' principle [2]. The former is more useful for the general case of an arbitrary profile, but for the kind and dimension of the devices that we will be discussing, it is more instructive to take the ray approach.

The least action principle dictates that light will travel a path in such a way as to minimize the time of flight. The integral representation of this statement, Fermat's principle, is that the time of flight (distance divided by the velocity)

$$\delta t = \int \frac{ds}{v} \tag{3.1}$$

should be a minimum. One can put this in the classic differential arc element, $ds = \sqrt{dx^2 + dy^2 + dz^2}$, replacing the velocity with the variational calculus form by substituting for the expression $v = c/n(x, y, z)$. The variational problem can now be stated: For what function $n(x, y, z)$ is this integral a minimum?

At this point we anticipate the practical situation to come, which deals with rod lenses, and use cylindrical coordinates. Further, in the same vein we initially assume that n is a function of r alone consistent with their ion-exchange fabrication. The differential arc element in cylindrical coordinates is $ds = \sqrt{dr^2 + r^2\, d\theta^2 + dz^2}$, so Eq. (3.1) can be written as

$$\delta t = L\left(r, \theta, z, \frac{dr}{dz}, \frac{d\theta}{dz}\right) dz \tag{3.2}$$

$$= \left[\frac{n(r, \theta)}{c}\right] \sqrt{1 + (dr/dz)^2 + r^2(d\theta/dz)^2} = dz$$

The solution proceeds via the use of the Euler-Lagrange differential equations of variational calculus [5], which are formally expressed as

$$\frac{\partial L}{\partial r} = \frac{d}{dz}\left(\frac{\partial L}{\partial r'}\right) \tag{3.3a}$$

$$\frac{\partial L}{\partial \theta} = \frac{d}{dz}\left(\frac{\partial L}{\partial \theta'}\right) \tag{3.3b}$$

where L is the integrand of Eq. (3.2) and $r' = dr/dz$ and $\theta' = d\theta/dz$. For the sake of simplicity in our development, we suppress the θ dependence. The reader is referred to Marchand [2] for the full derivation. As a consequence of our 2D approach we will be looking at what would constitute meridional rays in a cylindrical coordinate system (see Figure 3.1). These are rays that contain the z axis. We will point out the consequence of this as we discuss skew rays.

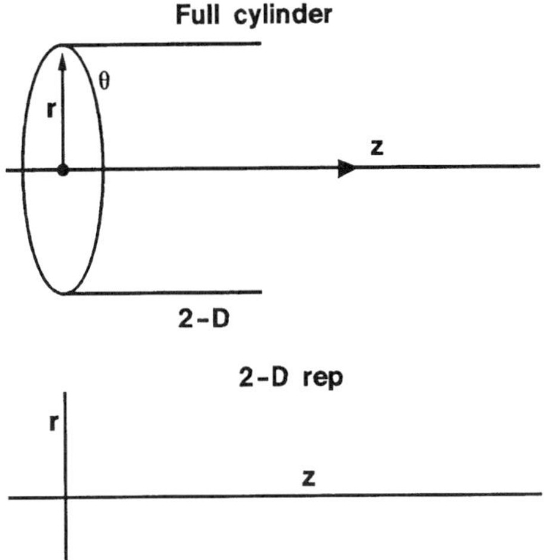

Figure 3.1 Full cylindrical coordinate system used for the complete development of ray paths; (*bottom*) reduced 2D coordinate system used in the case of assumed axial symmetry.

Applying Eq. (3.3a) to L as defined by Eq. (3.2) yields

$$\sqrt{l + r'^2}\left(\frac{dn}{dr}\right) = \left(\frac{d}{dz}\right)\left\{\frac{nr'}{\sqrt{l + r'^2}}\right\} \tag{3.4}$$

where we use r' to signify dr/dz. Multiplying both sides by n, Eq. (3.4) can be written in the form

$$\frac{d\{n^2\}}{dz} = \frac{nr'}{\sqrt{l + r'^2}}\frac{d}{dz}\left(\frac{nr'}{\sqrt{l + r'^2}}\right) = \frac{d}{dz}\left\{\frac{n^2 r'^2}{l + r'^2}\right\} \tag{3.5}$$

where we used the relation $dn/dr = (l/r')dn/dz$. We can now equate the terms in the curly brackets to obtain

$$\frac{(nr')^2}{l + r'^2} = n^2 + l^2 \tag{3.6}$$

or upon rearranging and taking the square root,

$$\frac{n}{\sqrt{l + r'^2}} = + l \tag{3.7}$$

where l is a constant independent of z. Solving for r' in Eq. (3.7) and then integrating with respect to z yields the equation of the ray,

$$z - z(0) = \pm\int k\frac{dr}{\sqrt{n^2 - l^2}} \tag{3.8}$$

This equation defines the ray $r(z)$ once we put in a specific form for $n(r)$. We have taken the long way around to obtain the ray Eq. (3.8) to show the more general approach to the solution of the problem. There is, however, a much quicker route to Eq. (3.8) in view of the restrictions we put on the problem; viz., n is only a function of r with no θ dependence. One could have expressed the Euler-Lagrange equations in parametric form. For example, Eq. (3.4) could have been written as

$$\frac{\partial n}{\partial r} = \frac{d}{ds}\left\{\frac{n\,dr}{ds}\right\} \tag{3.9a}$$

using the identity $d/ds = (l + r'^2)^{-1/2}\,d/dz$. Similarly, the other two equations for θ and z would have been

$$\frac{\partial n}{\partial\theta} = \frac{d}{ds}\left\{\frac{n\,d\theta}{ds}\right\} \tag{3.9b}$$

and

$$\frac{\partial n}{\partial z} = \frac{d}{ds}\left\{ \frac{n\,dz}{ds} \right\}$$

(3.9c)

Since n is not a function of z, then one could have integrated (3.9c) directly:

$$\frac{n\,dz}{ds} = \text{constant}$$

(3.10)

This is nothing more than the statement that the z direction cosine is invariant. This equation is exactly that expressed above in (3.7), so we can immediately identify the constant as the z direction cosine which because of its invariance is identically equal to its value at $z = 0$, l_0.

3.2.2 Imaging Forms of the Radial Profile

For a specific $n(r)$, one can use Eq. (3.8) to determine the ray path. Our concern is with those $n(r)$ that lead to a formation of an image. What this means is that from a fan of rays from a given point z_0, they will pass through a given point at some other z. Consider the radial profile given by the expression [2,3,6],

$$n^2 = N^2 \left(1 - Ar^2\right)$$

(3.11)

Substituting this into (3.8) and integrating yields the solution

$$(z - z_0)\left(\frac{z}{l_0}\right) = \left\{ \begin{array}{c} \sin^{-1}\left(\dfrac{r}{r_0}\right) \\[2ex] \cos^{-1}\left(\dfrac{r}{r_0}\right) \end{array} \right\}$$

(3.12)

or, putting the boundary conditions at $z_0 = 0$, one obtains

$$r = r(0)\cos\left(\frac{N\{A\}^{1/2}z}{l_0}\right) + \left\{\frac{(dr/dz)_0}{N\sqrt{A}}\right\}\sin\left(\frac{N\{A\}^{1/2}z}{l_0}\right)$$

(3.13)

Clearly this represents ray paths that lead to an imaging condition for meridional rays as long as l_0 is constant for all rays involved. Consider the situation

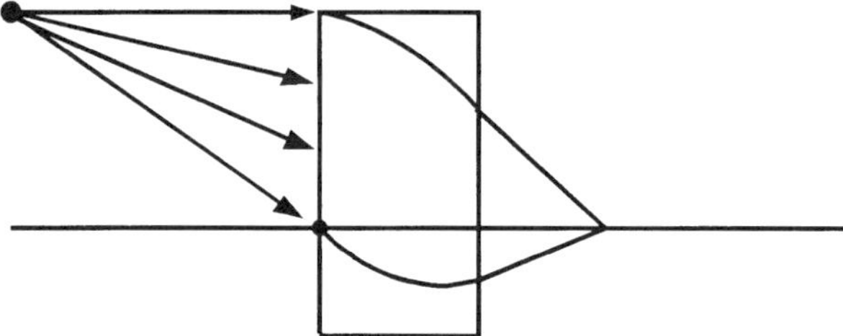

Figure 3.2 Ray trace through a body with parabolic index profile [see Eq. (3.11)] showing the imaging property.

depicted in Figure 3.2. In this case one can easily see that all rays emanating from point P will intersect at the point z defined by

$$\tan z_{\text{int}} = \frac{r_i - r_j}{m_j - m_i} \tag{3.14}$$

We saw from above that l_0 is a constant for each ray. We are interested in the question, what would make it a constant for all rays, irrespective of initial position and slope. Considering the definition,

$$l_0 = \frac{n(r)}{\sqrt{l + r'^2}} \tag{3.15}$$

it is clear that it would remain relatively constant if $(dr/dz)^2$ is small compared to unity, and the change in n itself with r is small over the region of interest. This is tantamount to a paraxial approximation.

There is a radial profile that images meridional rays without regard to any approximation [7]. The form of the radial profile is

$$n(r) = N_0 \operatorname{sech}(az) \tag{3.16}$$

If one substitutes this into Eq. (3.8), one obtains the expression

$$\sinh(ar) = A \sin\left\{\sin^{-1}\left[\frac{\sinh(ar_0)}{A + az}\right]\right\} \tag{3.17a}$$

or,

$$\sinh(ar) = A \cos\left\{\cos^{-1}\left[\frac{\sinh(ar_0)}{A + az}\right]\right\} \tag{3.17b}$$

where $A = \sqrt{(N_0/1_0)^2 - 1}$. Equations (3.17) represent periodic functions of z and correspond to a sharply imaged condition for meridional rays which is independent of the ray or the starting point chosen [2].

In a real sense (3.17) constitutes the ideal profile for a gradient index lens with respect to meridional rays. By ideal is meant that all rays are sharply focused irrespective of the initial ray position or slope. It is like an aspheric surface is to a refractive lens. However, if we consider the possibility of skew rays as well, then we find that there are problems. The consideration of skew rays involves maintaining the θ dependence as indicated by Eq. (3.3b) throughout the analysis. Because of the lengthy derivation, we will be content to reproduce the profile that images helical rays derived by Rawson et al. [8] of the form

$$\frac{n(r)}{N_0} = [1 + (Ar)^2]^{-1/2} \tag{3.18}$$

The importance of the skew rays depends upon the particular lens application. We see that the best approximation with respect to all kinds of rays is the following. Consider the expansions of all three of the index profiles considered viz., Eqs. (3.11), (3.16), (3.18). A comparison of the sech to the parabolic profile is shown in Figure 3.3. They are all expressible in the form where the lead terms are identical:

$$n(r) = N_0\left[1 - \left(\frac{A}{2}\right)r^2\right] \tag{3.19}$$

We can see, as pointed out above, that this profile will only be good to the extent of the paraxial approximation. Inclusion of higher order terms leads to what would be equivalent to aberrations, as we will see later. The ray solutions for this profile are the same as that obtained above; see Eq. (3.13)

3.2.3 Matrix Ray Trace for Rod Lenses

Kapron [9] first suggested a matrix-based ray-trace formalism for the paraxial case. This was subsequently augmented by NSG [10] and used in their product **a** description/performance publications. This has proven to be very convenient and useful for lens design. Others have published more general ray-trace methods [11], but in most cases of interest the paraxial approximation is sufficient.

The method follows the same matrix format as for the thick-lens ray tracing that was done for refractive elements in the paraxial approximation in Chapter 2 (see Section 2.1.3). With the use of Eq. (3.13), one can write matrix expression relating the initial ray position and slope for a lens of thickness D.

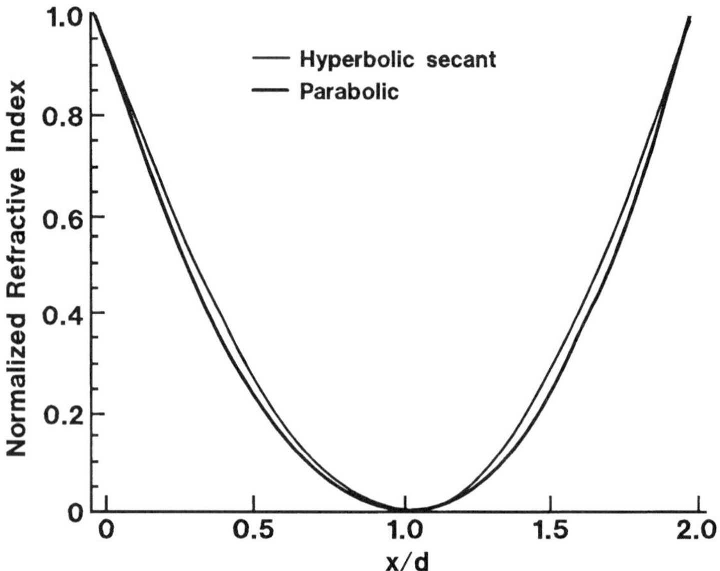

Figure 3.3 Comparison of the radial refractive index profile derived from the exact imaging sech (Ar) to that of the paraxially imaging parabolic profile, $n = n_0[1 - (A/2)r^2]$.

$$\begin{pmatrix} r_1 \\ m_2 \end{pmatrix} = \begin{pmatrix} \cos(\sqrt{A}\,D) & (1/N_0\sqrt{A})\,\sin(\sqrt{A}\,D) \\ -N_0\sqrt{A}\,\sin(\sqrt{A}\,D) & \cos(\sqrt{A}\,D) \end{pmatrix} \begin{pmatrix} r_0 \\ m_0 \end{pmatrix} \qquad (3.20)$$

From this one can write down the relationships for the focal length, working distance, and location of the principal planes. These results are shown in Figure 3.4. The computed ray diagram for a simple case of parallel light being focused is shown in Figure 3.5.

The simple lens maker's formula can be used in the same way as it was for the conventional thick lens.

$$\frac{l}{L_1} + \frac{l}{L_2} = \frac{l}{f} \qquad (3.21)$$

where the L's are measured from the principal planes. Another interesting aspect of a GRIN lens with a parabolic profile is that its numerical aperture is a function of the radial position as shown in Figure 3.6.:

$$\mathrm{NA} = \sqrt{n(r)^2 - n_0^2} \qquad (3.22)$$

These expressions will be useful when we deal with the applications.

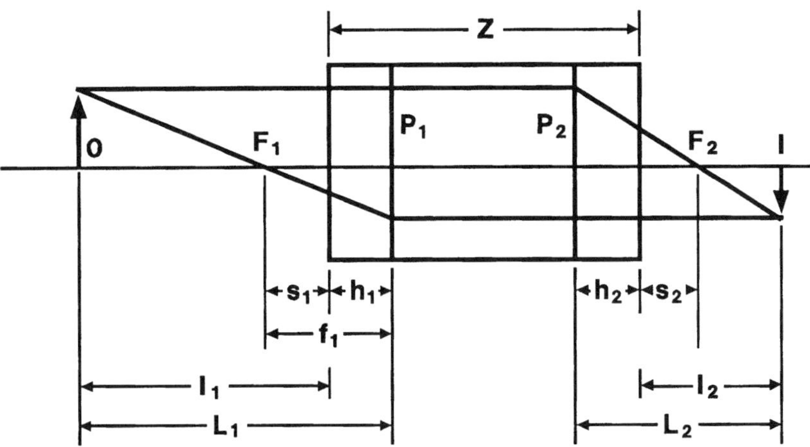

Figure 3.4 Drawing relating the commonly used parameters to describe the imaging of a GRIN lens of thickness Z. The parameters are formally defined in the same manner as shown in Figure 2.1 and listed in Table 2.1 for conventional refracting lens. However the actual relationships are defined through the use of Eq. (3.20). Unfortunately, the convention for the symbols are not all the same. In particular, the symbol most commonly used for the object and images distances in the GRIN notation is l rather than s.

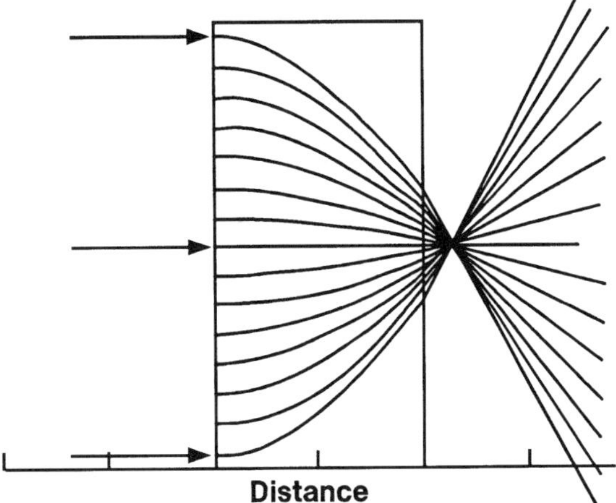

Figure 3.5 Computed ray trace through a GRIN lens with a parabolic refractive index profile. Approximate path obtained from Eq. (3.20) with $m_0 = 0$. Positions on the exit face are $r_{in} \cos (\sqrt{A}\, D)$ with slopes of $-N_0 \sqrt{A} \sin (\sqrt{A}\, D)$; thus the equation of exit ray is $r = r_{ex} - m_{ex}z$, or $r = r_{in} \cos (\sqrt{A}\, D) - zN_0 \sqrt{A} \sin (\sqrt{A}\, D)$.

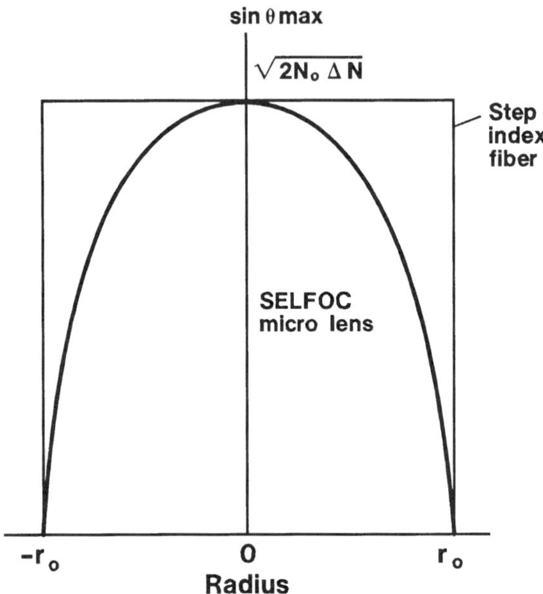

Figure 3.6 Variation of numerical aperture as a function of radial position of a GRIN lens with a parabolic profile.

3.3 PLANAR LENS

The analysis of the imaging behavior of the so-called planar version of GRIN lenses [3,12,13] is somewhat more difficult to mathematically describe than that of the rod lens geometry discussed above. The way the index profile is produced is quite different, and consequently some of the simplifying assumptions made in the treatment above are not valid. In the next section we go into the fabrication methods in somewhat more detail, but for the present it is instructive to compare how the index profile is derived with a simple diagram. In Figure 3.7a we show how the rod lens is made. In the ion-exchange method, for example, the refractive index is altered uniformly in the radial direction and constant along the axial direction. One would thus expect the refractive index profile to look as sketched in Figure 3.7b. It is with this situation in mind, since it is the preferred way to make such lenses, that the above derivation of the ray trace was done. In the planar version of GRIN, the index is altered by ion exchange through a small opening in a mask on the surface, as shown in Figure 3.8a. The expected index profile is shown in Figure 3.8b. Clearly the index is a function of both r and z. From the shape of the profile Iga et al. [3] proposed the function

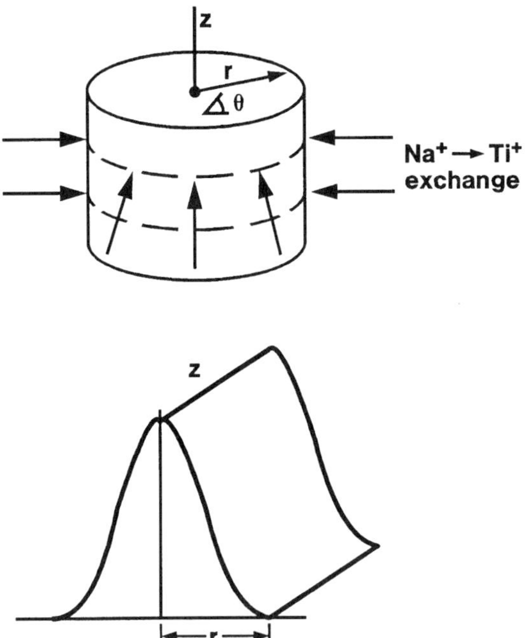

Figure 3.7 Schematic representation of the molten salt bath ion-exchange process into a glass rod to produce a radial refractive index gradient. The lower figure is a sketch of the index profile in terms of the radial and axial position.

$$n(r, z)^2 = N_0^2(R(r) + Z(z)) = N_0^2\{1 - Ar^2 - v_n Az^n\} \tag{3.23}$$

One can easily see that this would constitute an approximation to the profile shape of Figure 3.8b.

3.3.1 Planar Ray Trace

In this section we present a shortened approach to the development of the equation of the ray for the planar case. Iga et al. [3,13] have given complete derivations and the reader is referred to these publications for the details. We will try to follow the same general form used above for the rod case. One starts with Euler-Lagrange equations in parametric form; see (3.9a–c), with a change of independent parameter from the differential arc length ds to it normalized to the index $n(r, z)$.

$$d\tau = \frac{ds}{n} \tag{3.24}$$

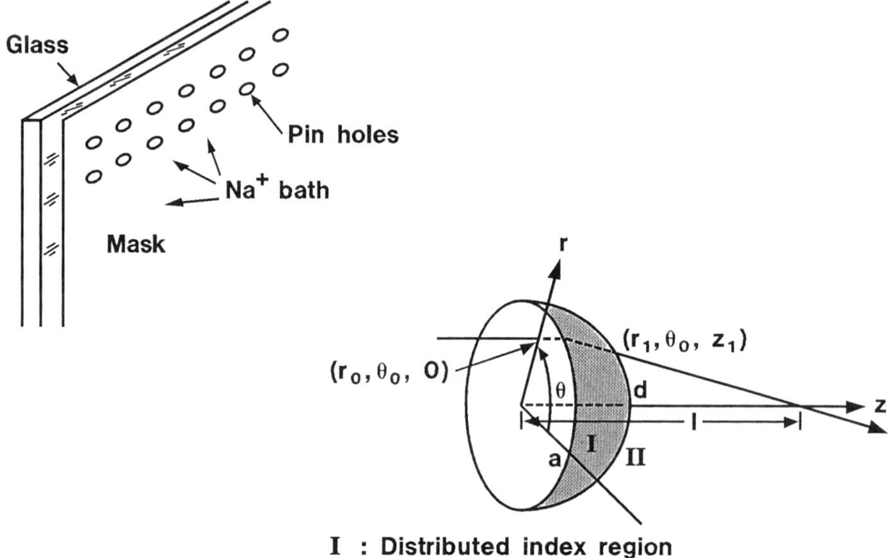

Figure 3.8 Schematic representation of planar GRIN process where molten salt bath ion-exchange is done through a mask in contact with the surface, as shown. The lower figure shows the coordinate system by which the refractive index profile is characterized. One notes that the major difference produced by the planar process is that the profile is also a function of z, as well as r and θ.

Equations (3.9) now can be written as the following:

$$\frac{d^2 r}{d\tau^2} - r\left(\frac{d\theta}{d\tau}\right)^2 = \left(\frac{d}{dr}\right)\left(\frac{n^2}{2}\right) \tag{3.25a}$$

$$\left(\frac{d}{d\tau}\right)\left(\frac{r^2\, d\theta}{d\tau}\right) = \left(\frac{d}{d\theta}\right)\left(\frac{n^2}{2}\right) \tag{3.25b}$$

$$\frac{d^2 z}{d\tau^2} = \left(\frac{d}{dz}\right)\left(\frac{n^2}{2}\right) \tag{3.25c}$$

To simplify matters, we again assume axial symmetry, and remove the θ dependence and return to considering meridional rays only. Using the form for $n(r, z)$ given in Eq. (3.23), one has

$$\frac{d^2r}{d\tau^2} = \left(\frac{N_0^2}{2}\right)\left(\frac{dR}{dr}\right) \tag{3.26}$$

and

$$\frac{d^2z}{d\tau^2} = \left(\frac{N_0^2}{2}\right)\left(\frac{dZ}{dz}\right) \tag{3.27}$$

In Eq. (3.26), multiply both sides by $dr/d\tau$, followed by integrating with respect to τ, and finally taking the square root, to obtain

$$\frac{dr}{d\tau} = \pm N_0\sqrt{R(r)} + C_1 \tag{3.28}$$

where C_1 is a constant of integration which can be evaluated in terms of the incident r direction cosine of the ray at $z = 0$. In a similar manner, Eq. (3.27) becomes

$$\frac{dz}{d\tau} = \pm N_0\sqrt{Z(z)} + C_2 \tag{3.29}$$

with C_2 defined in a similar way to C_1, viz., the z direction cosine at $z = 0$. Dividing (3.29) by (3.28) and performing the appropriate algebraic manipulations yields the expression

$$dz = \pm\sqrt{Z(z) + l_0^2 R(r_0) + (l_0^2 - 1)Z(0)} \tag{3.30}$$
$$dr = \pm\sqrt{R(r) - l_0^2 R(r_0) + (1 - l_0^2)Z(0)}$$

where l_0 is the incident z direction cosine, and the initial condition is given by

$$\left(\frac{dz}{d\tau}\right)_{z=0} = n(r_0, 0)l_0 N_0\sqrt{R(r_0)} + Z(0) \tag{3.31}$$

One can then cast (3.30) as two integrals, whose equality defines the desired relation between r and z when particular forms are assumed for $R(r)$ and $Z(z)$.

$$\int\frac{dr}{\sqrt{R(r) - l_0^2 R(r_0) + (1 - l_0^2)Z(0)}} = \pm\int\frac{dz}{\sqrt{Z(z) + l_0^2 R(r_0) + (l_0^2 - 1)Z(0)}} \cdot \tag{3.32}$$

We will show solutions for special cases in the next section.

3.3.2 Imaging Condition

Unlike the rod lens situation, the ideal profile is not known. What one can do is to look at the imaging from the solutions of (3.31), which is not at all an obvious procedure. A simplification can be made in order to investigate the imaging behavior. This is to consider only rays which are parallel to the z axis. For this case $l_0 = 1$. Further, one can assume the profile suggested by Iga expressed in (3.22), $R(r) = (1 - Ar^2)$ and $Z(z) = -v_n A z^n$. For the parabolic case, $n = 2$, a closed form solution (3.31) was obtained by Iga [3,13] of the form

$$\frac{r}{r_0} = \cos \frac{1}{\sqrt{v_2}} \sin^{-1} \frac{\sqrt{(n_2 A)z}}{(\sqrt{1 - AR_0^2})} \tag{3.32a}$$

(A closed-form solution was also obtained for the $n = 1$ case.) Suffice to say at this time that focusing does occur as a consequence [Eq. (3.32a)] over a reasonable portion of the lens. In the limit of the argument of the $\sin^{-1}$ term being small, one can replace it with the argument itself. Also, the cosine could then be written as $\sqrt{1 - x^2}$, could reduce to the simple form:

$$\frac{r}{r_0} = \sqrt{\frac{1 - Az^2}{1 - AR_0^2}} \tag{3.32b}$$

3.4 METHODS OF FABRICATION

The fabrication of GRIN lenses has been dominated by the ion-exchange method. In this case a solid material, glass or polymer, is immersed in a bath containing an ion that will diffuse into the body, exchanging for a corresponding ion that diffuses out of the glass into the bath [14]. The net effect of the exchange of ions is to modify the refractive index. We will try to cover the details of this process. See Figure 3.9.

An older method was the idea of building up a series of tubes of glass inside of tubes each with a prescribed different refractice index. The nest of tubes was

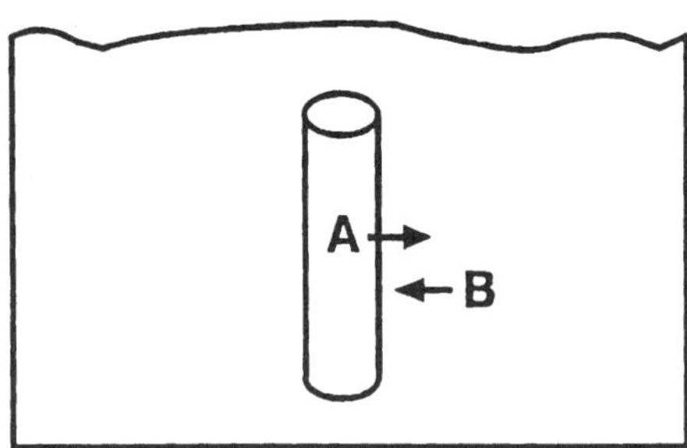

Figure 3.9 Generic representation of the ion exchange into a rod to produce a radial refractive index gradient. The bath can be a molten salt or solution. The rod will contain the counter ion. The exchange occurs through normal diffusion processes.

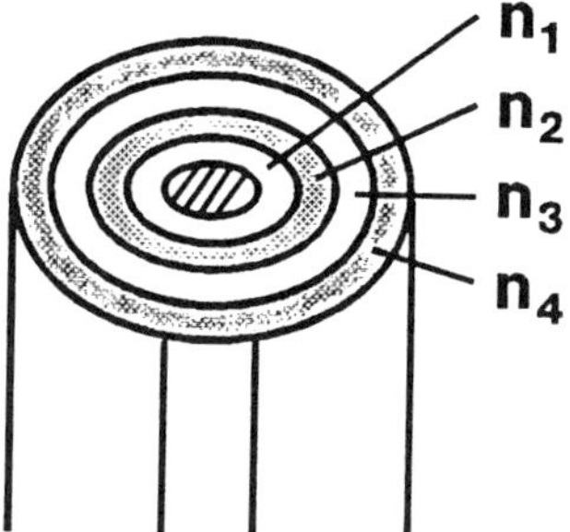

Figure 3.10 Nested rod and tube structure intended to produce a radial refractive index profile. The tubes each are of a slightly different refractive index. The idea is to heat the nested tubes until they fuse.

then heated to fuse them. See Figure 3.10. A higher temperature variation on this method was to use rings of powdered glasses [15].

Finally, there are a number of photosensitive processes where the action of light on a special glass [16,17], or polymer [18], produces a chemical or structural change that leads to a refractive index change. By exposing with the appropriate intensity distribution one can produce a desired $n(r)$.

We proceed through each one of these methods in turn.

3.4.1 Ion Exchange

Ion Exchange is a well-known phenomenon used in glass processing going back many years. For example, ion exchange provides strengthening of glass [19]. The idea is to utilize the diffusion of ions from a bath rich in component A into a glass poor in that component. The process is treated as a classic diffusion problem, where we have chosen the equation in cylindrical coordinates:

$$\left(\frac{1}{r}\right)\left\{\frac{d}{dr}\right\}\left\{rD_a\left(\frac{dA}{dr}\right)\right\} = \frac{dA}{dt} \tag{3.34}$$

with the following boundary conditions:

$$A = 0 \qquad 0 < r < r_0,\ t = 0$$

$$\frac{dA}{dr} = 0 \qquad r = 0,\ t > 0$$

$$A = A_0 \qquad r = r_0,\ t > 0$$

For the case where D is a constant, the solution for the concentration of A as a function of the radius and time is [20].

$$\frac{A}{A_0} - 1 = -2\sum\{\exp(-\beta_n^2 T)\}\frac{J_0(\beta_n r/r_0)}{\beta_n Jl(\beta_n)} \tag{3.35}$$

where β_n are the zeros of the zero-order Bessel function J_0 and $T = Dt/r_0^2$. The solution is shown in graphical form in Figure 3.11.

There, of course, will be another equation involving species B, with diffusion coefficient D_b, which diffuses out of the glass. The process will be controlled by the slower of the two reactions and our diffusion constant will implicitly mean D_{ab}. The distinction being made is between the interdiffusion process governed by $D(A, B)$ and the self-diffusion processes stipulated by $D(A)$ and $D(B)$. The diffusion constants are themselves concentration dependent over some range of concentration because of mutual interactions. For a good discussion of this see Ref. 21. There is a straightforward way to measure the concentration dependence of the diffusion coefficient, usually referred to as the Boltzman method [22]. It is based on the measurement of the diffusion profile in a semi-infinite slab after

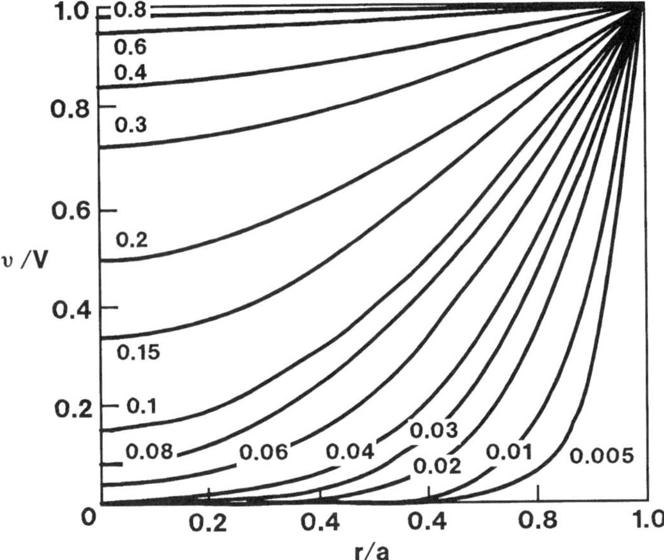

Figure 3.11 Classical solution of the diffusion equation in cylindrical coordinates, for the normalized concentration of the ion diffusing into the rod. Zero represents the center of the rod. The parameter is Dt/r_0^2, where D is the diffusion coefficient, r_0 is the radius of the rod, and t is the time.

some time t. From this single measurement one can estimate $D(c)$. One can reduce the partial differential equation

$$\left(\frac{d}{dx}\right)\left(\frac{D\ dc}{dx}\right) = \frac{dc}{dt} \tag{3.36}$$

to an ordinary differential equation by the change of variable, $y = x/\sqrt{t}$. One then obtains the equation

$$\left(\frac{d}{dy}\right)\left(\frac{D\ dc}{dy}\right) = \left(\frac{-y}{2}\right)\left(\frac{dc}{dy}\right) \tag{3.37}$$

One can integrate this expression with respect to y to obtain an explicit expression for $D(c)$, after changing back to the original variables,

$$D(c) = \left(\frac{1}{2t}\right)\left(\frac{dx}{dc}\right)x\ dc \tag{3.38}$$

Referring to Figure 3.12, one can see how the method is used. The profile $c(x)$ is measured at some time t. The integral represented by the shaded area is numerically determined. Then for each value of c, as shown for an arbitrary point on the figure, the derivative is estimated (dashed line) and the value of $D(c_1)$ is

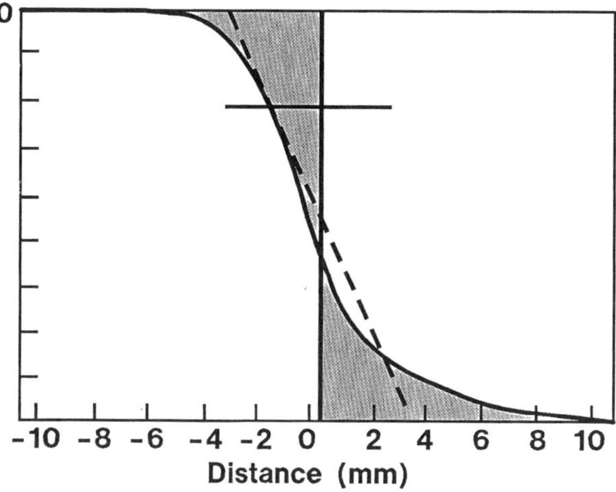

Figure 3.12 Method used to estimate the concentration dependence of the diffusion coefficient. The measured concentration of a given species is plotted vs. distance. See text and Eq. (3.38) for explanation of method.

obtained from (3.38). One then proceeds to do this for any value of c to generate the full $D(c)$ relationship. We will come back to this relationship in the next section to show how it might be used to an advantage in the design of a refractive index profile.

Relation of Refractive Index to Concentration

Explicitly, one assumes that the refractive index is proportional to the concentration of species A, or more precisely, to the difference between A and B. There are equations that relate the refractive index to the concentration of various constituents with some success. One is the Lorentz-Lorenz equation which relates the refractive index to the atomic polarizability α of the constituent ions that make up the glass structure:

$$\frac{n^2 - 1}{n^2 + 2} = \left(\frac{4\pi}{3}\right)\Sigma N_i\alpha_i \tag{3.39a}$$

or, an often-used simpler form,

$$n^2 - 1 = \left(\frac{4\pi}{3}\right)\Sigma N_i\alpha_i \tag{3.39b}$$

The polarizabilities are a numerical measure of the ease of which an external electric field can influence the electron distribution about the ion. For oxide glasses, certain ions are known to dominate the sum written in Eqs. (3.39). We have listed the polarizabilities of some common ions in Table 3.1. The additional problem encountered is the ease with which these ions diffuse, which is related to how the ions are bonded in the structure. Presently, most have followed the simple rule that the singly charged ions are the most ionic and thus are the most mobile. For applications where a large index change is desired, the ions of choice have been Ag^{+1} or Tl^{+1} for alkali.

For the sake of simplicity in the single ion-exchange process, it is convenient to express the refractive index as [21]

$$n^2 = a_1c + a_2 \tag{3.40}$$

where c is the concentration of the exchanged constituent, and a_1 and a_2 are empirical constants.

Control of Index Profile

In the last section we described the ion-exchange process as it pertains to a way to spatially change the refractive index of a rod. In light of the mathematical description of the required refractive index given above, it is important to see how people have used the diffusion process to achieve the desired concentration gradient.

Table 3.1 Polarizabilities of
Various Ions (cubic angstroms)

Ion	α
Li^{+1}	0.03
Na^{+1}	0.41
K^{-1}	1.33
Rb^{+1}	1.98
Cs^{+1}	3.34
Ag^{+1}	2.4
Tl^{+1}	5.2
Ca^{+2}	1.1
S^{-2}	4.8–5.9
O^{-2}	0.5–3.2
Cl^{-1}	2.96
I^{-1}	6.43
F^{-1}	0.64
Pb^{+2}	4.9

Source J. R. Tessman, A. H. Kahn, and
Wm. Schockley, Electronic polarizabilities of ions in crystals, *Phys. Rev. 92*(4).

Consider a simpler form of Eq. (3.38b) where we consider only the contributions of the ions that are to be exchanged:

$$n^2 = c_0 + c_1 A + c_2 B \tag{3.41}$$

which is true at all values of r. Further, since $A + B = 1$, one can write this as

$$n^2(r) = (c_0 - c^2) + (c_1 + c_2)A = C_0 + C_1 A \tag{3.42}$$

For constituent A one substitutes the expression (3.35). Consider the expansion
of $J_0(\beta_n r/r_0)$ for small values of the argument:

$$J_0\left(\frac{\beta_n r}{r_0}\right) = 1 - \left(\frac{\beta_n}{2r_0}\right)^2 r^2 + \frac{(\beta_n/2r_0)^4 r^4}{4} \tag{3.43}$$

One can see that for each term of the series of (3.35) there will be an r^2, r^4,
r^{2k}, component, resulting in an overall form of (3.43) of

$$n^2(r) = C_0 + C_1(b_1 + b_2 r^2 + b_4 r^4 + \cdots) \tag{3.44}$$

where $b_n = F_n(t, D, r_0\beta_n)$, This is the desired form of the profile. The constants
relate to the constants of the material and of the process. So, what one has to work
with are the conditions that best allow the form of (3.44) to achieve whatever level

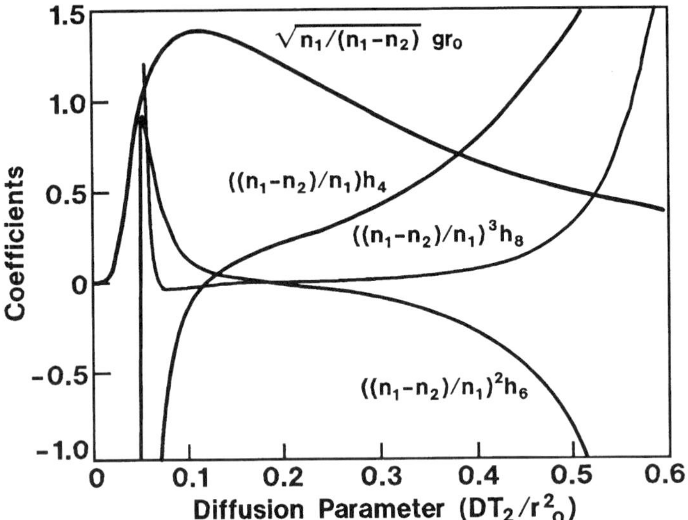

Figure 3.13 Refractive index coefficients as defined from Eq. (3.44) as a function of the diffusion parameter DT/r_0^2. Each coefficient represents a different order in the radial expansion and the idea is to choose the appropriate value of the parameter to produce as minimum an aberration as possible. The n_1 and n_2 refer to the maximum values obtainable with constituent 1 or 2. (After Iga [3].)

of approximation is desired. For example, in the parabolic case one would choose the parameters such that the coefficients $>b_2$ were as small as possible. Iga [3] shows in a graph (Figure 3.13) how these coefficients vary with the parameter DT/r_0^2, where D is the diffusion coefficient, T the time of the diffusion, and r_0 the rod radius. One sees that one would choose a value of the parameter in the vicinity of 0.15. One can also use this graph to see if it were possible to achieve an approximation to the hyperbolic secant profile if one could find a value of the parameter where $b_4 = {}^2\!/_3 b_2^2$ and $b_6 = {}^{17}\!/_{45} b_3^2$.

Another approach taken to obtain the desired parabolic or secant hyperbolic profile is that taken by Messerschmidt et al. [21]. They propose to use the glass composition through the ionic interaction energy to modify the concentration dependence of the diffusion coefficient. It is shown that it is this interaction between the two diffusing species that accounts for the concentration dependence of the diffusion constant. They show that for a parabolic profile to ensue, using the Boltzman method described above, the diffusion parameter Dt/r_{02} would have to vary with concentration in the following way:

$$\frac{D(A)t}{r_0^2} = \frac{\sqrt{A}}{4} - \frac{A}{6} \tag{3.45}$$

The proposal is to study the concentration dependence of the diffusion coefficient in different glass systems and seek out the parameters that would allow one to approach the form indicated by Eq. (3.45).

Process and Material Considerations

There are a number of important material and process properties that bear on the performance and practicality of ion-exchange fabrication. Since these are often complex subjects in themselves, we will only briefly touch on them here to give the reader an idea of what is involved.

The first set of properties pertain to the glass composition. The specific glass composition impacts the magnitude of Δn that can be achieved through the solubility of the exchangeable ion [23], thus influencing the diffusion coefficient, the activation energy for diffusion, the dependence of the diffusion coefficient on concentration [24,25], and the chemical durability to the ion-exchange bath. All these properties of the glass can play an important role in the performance for certain applications, and therefore understanding the origins and mechanism of the particular property can lead to a better or more versatile lens.

The second set is that which involves the ion-exchange process itself, in particular that which has to do with the composition and properties of the exchange bath. In Table 3.2 we have listed the commonly used bath compositions with comments in terms of the problems, if any.

Methods of Fabrication

Rod Lenses The manufacture of GRIN rod lenses by the ion-exchange technique is accomplished by a large batch process [14]. Meter-length or longer millimeter-diameter rods are bundled together and placed into large ion-exchange

Table 3.2 Compositions and Properties of Ion-Exchange Bath

Ion	Radius in pm	Coordination number	Refractive index change ΔM	LD_{50} in mg/kg	Salt	Remarks
Li^+	59	4	0.02	710	Li_2CO_3	high tensile stresses
	76	6				
Na^+	99	4	-0.02 to 0.002	1955	$NaNO_3$	tensile stresses
	102	6				
K^+	138	6	0.009	1894	KNO_3	compressive stresses
Rb^+	152	6	0.01	1200	$RbCl$	high price
Cs^+	167	6	0.04	1200	$CsNO_3$	slow diffusion
Ag^+	126	6	0.1 (0.22)	2820	Ag_2O	low thermal stability
Tl^+	150	6	0.1	25	Tl_2SO_4	additonal expenditure for safety

LD_{50} refers to toxicity level.

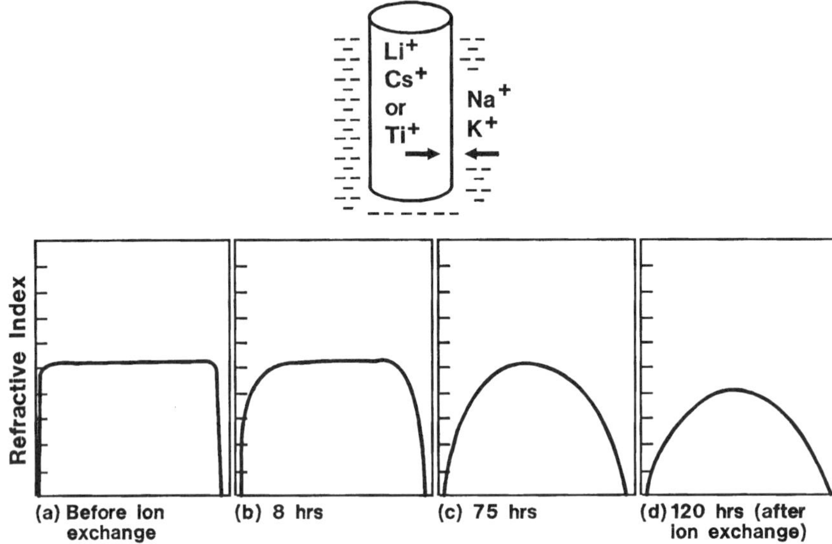

Figure 3.14   Evolution of radial profile as a function of time after ion exchange into a rod. (From Ref. 26.)

baths for times as long as 120 h. They are cut and polished to length to produce the desired optical characteristic. NSG reports a typical sequence of profiles taken at various times for a Cs-exchange done at 570°C, as shown in Figure 3.14 [26]. The particular bath was not specified. The optical characterization parameter is the pitch which means when the argument of the expressions in Eq. (3.20), viz., $\sqrt{A}D$, is equal to 2π. The ray diagram for a 0.25, 0.50, and 0.75 pitch lens is shown in Figure 3.15. An example listing of the types of SELFOC SML (Selfoc Micro Lens) lenses that are available is shown in Table 3.3.

As we will see in Chapter 5 when we discuss applications, one of the major device areas has to do with lens bars that produce one-to-one erect imaging. We show this schematically in Figure 3.16a. These lens bars are made by laying out the rods into rows. Then one row is placed on top of the other, as shown in Figure 3.16b. We will discuss the application of such devices in Chapter 5.

It is not clear what limitation, if any, this process imposes on possible rod lens diameters that can be practically made. It may involve such cost features as rods of glass that are much smaller in diameter and have a greater propensity to break during the process. We will see in the next chapter when we deal with the radiometric aspects of one-to-one imaging bars that the lens diameter does play a role.

There is a similar process available in plastic rods where the ion exchange in the anion [3,18].

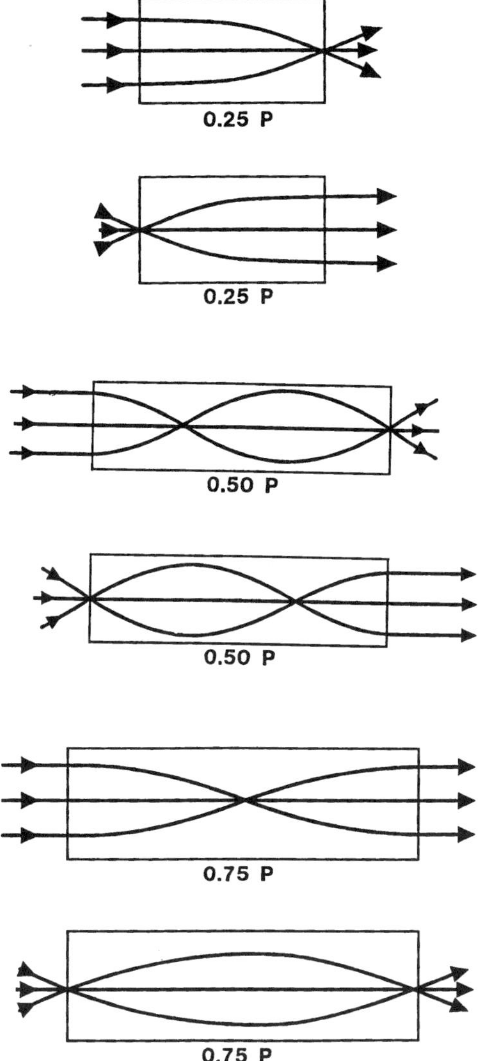

Figure 3.15 Schematic ray trace of rod lens with parabolic profile in typical use configurations as a function of pitch P which is defined is when $(\sqrt{A}\,D) = \pi/2$ in Eq. (3.20).

Table 3.3 Range of SELFOC GRIN Microlenses

NA	Pitch[a]	Working distance[b]	Diameter	Length
0.46	0.23	0.21 mm	1.8 mm	4.26 mm
0.46	0.25	0.21	1.0	2.58
0.60	0.25	0.21	1.8	3.65
0.46	0.29	—	1.8	5.37

[a]See text, equals $A^{1/2}D = 2\pi$.
[b]Distance from output surface to image.
Source Newport Catalog.

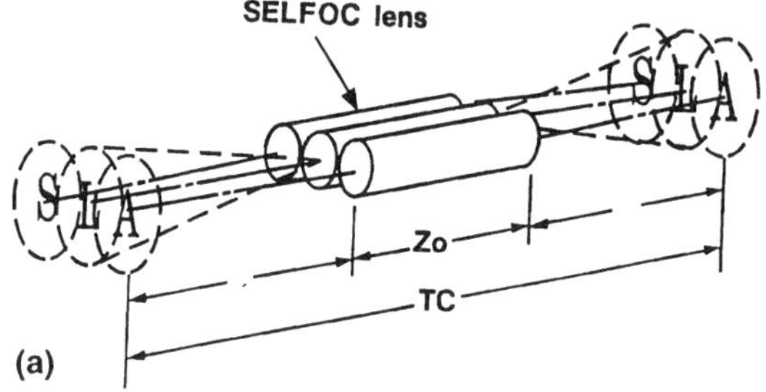

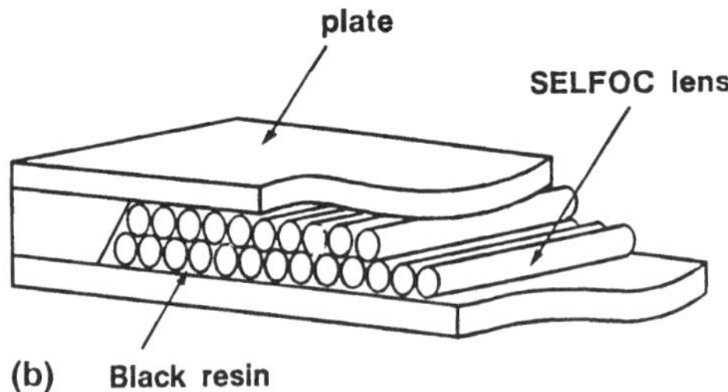

Figure 3.16 Schematic representation of an array of rod lenses used to image in an erect one-to-one manner. Lower figure shows the way the rod lenses are arranged to form the two row bar lens. See Chapter 5 for the applications of such arrays.

Planar The fabrication technique is outlined in Figure 3.17a. A metal mask with open holes of radius r_m is photopatterned onto the glass that is then exposed to an ion-exchange bath. The distributed index region $n(r, z)$ results from the concentration profile indicated in Figure 3.17b. An interferometric picture of an exchanged region is shown in Figure 3.18. NSG [26] has modeled the concentration profile and found that good fits can be obtained by taking into account the concentration dependence of the diffusion coefficient through the expression

$$D(C_a) = D_0 \exp \frac{KC_a}{C_0} \tag{3.46}$$

The effect this has on the profile, as measured by the value of the parameter K, is shown in Figure 3.19. Values of K between 5 and 7 seem to provide a reasonable fit to the experimental measurement of the profile.

Often the ion-exchange is done in a field-assisted fashion where an electrode is attached to the back side to direct the flow normal to the surface. In this case the diffusion equation is modified to account for the bias produced by the field.

$$\frac{dC}{dt} = \frac{Dd^2C}{dx^2} - \frac{\mu EdC}{dx} \tag{3.47}$$

The right-hand portion of Figure 3.17a indicates an additional process whereby one obtains a refractive contribution from the swelling effect produced by the ion-exchange and the index gradient [26a]. As noted, this swelling occurs in general but is slight when the ion exchange in not carried out to a large extent. In this case, the slight raised bumps are polished off to obtain the pure GRIN case of the left-hand depiction. Alternatively, when the exchange is carried out to a much greater depth as shown, the exchanged regions nearly overlap and the volume expansion is more significant. The advantage of this swelling feature is that it approximates a spherical shape and thus increases the effective NA of the lens.

A typical set of properties for a single planar lens and a coupled pair of the conventional plano-plano type are shown in Table 3.4 [3]. Lens diameters that can be produced by this method range from the order of 100 μm to 1 mm. As in some of the data that follows, the spatial characterization is displayed in terms of the parameter r/r_m, where r_m is the radius of the hole through which the ion exchange is carried out. This implies equivalent performance of the lenses over this size range. We will see examples of this range in the next Chapter 4 where we will deal with the applications of such arrays.

3.4.2 Thermal Mixing/Diffusion

The ideas here are ways to somehow provide an index profile by bringing mixtures of different materials into radial contact in a controlled way such as to

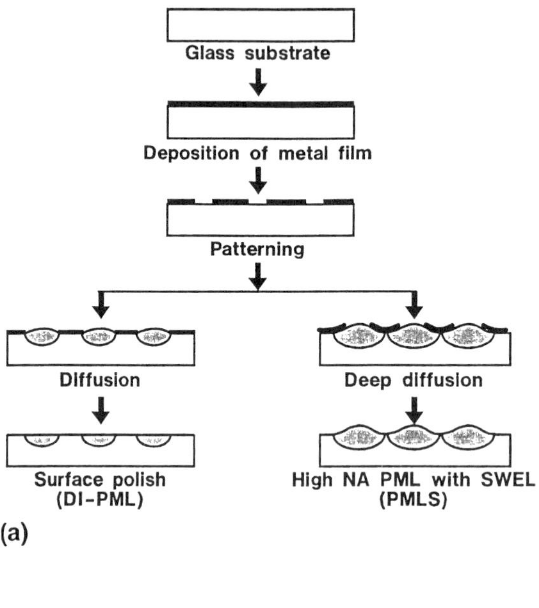

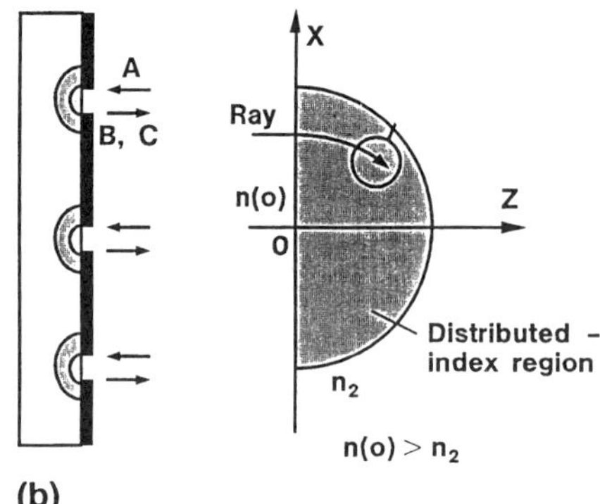

Figure 3.17 Upper figure is a representation of the planar ion-exchange process. The first stage provides a patterned mask for the subsequent exchange, As a consequence of the volume difference upon exchange, a swelling occurs as shown. This provides a refractive component to the lens, if desired. The lower figure shows an indication of the shape of the index alteration produced by this type of ion exchange.

Figure 3.18 Interferogram of a thin axial slice of a plane GRIN lens. (From Ref. 3.)

subsequently be able to produce a given refractive index profile. As usual, there are high-temperature requirements for inorganic glasses, and low-temperature methods for polymeric materials. We will cover these separately.

Glass

The classical way of producing a gradient index pattern in ordinary glass is to make a cylindrical nest of glasses, as shown in Figure 3.20, and then to heat the set to above the softening temperature. The idea is to blend the glasses together to smooth out the abrupt index changes. Ideally one would hope to smooth the profile into something like a continuous profile. In practice, a lot of problems can and do occur. The blending at the interfaces does not extend very far in the tubular regions. This makes it difficult to achieve a reasonably continuous profile unless one uses a large number of very thin shells. It is also difficult to obtain a continuous range of compatible glass compositions within a large refractive index interval. It is doubtful this method could make very small lenses and it is even harder to imagine how it could be applied to an array.

Lightpath, Inc. [15] has suggested a way around the blending issue by a process combining glasses with differing refractive indices at a high temperature where the glass can more easily flow together. It appears at present that the process has been used only to produce axial gradients, and no radial gradient results have been presented.

A third method proposes to use a flame-deposition technique which is used in the manufacture of optical waveguide blanks [27]. Here a silica precursor and a dopant precursor are burned in a flame to form a mixed composition soot which deposits onto a rotating mandrel. The mandrel is removed and the soot blank is consolidated at 1500°C. What results is a glass with almost theoretical density. This process is shown schematically in Figure 3.21. By controlling the proportion

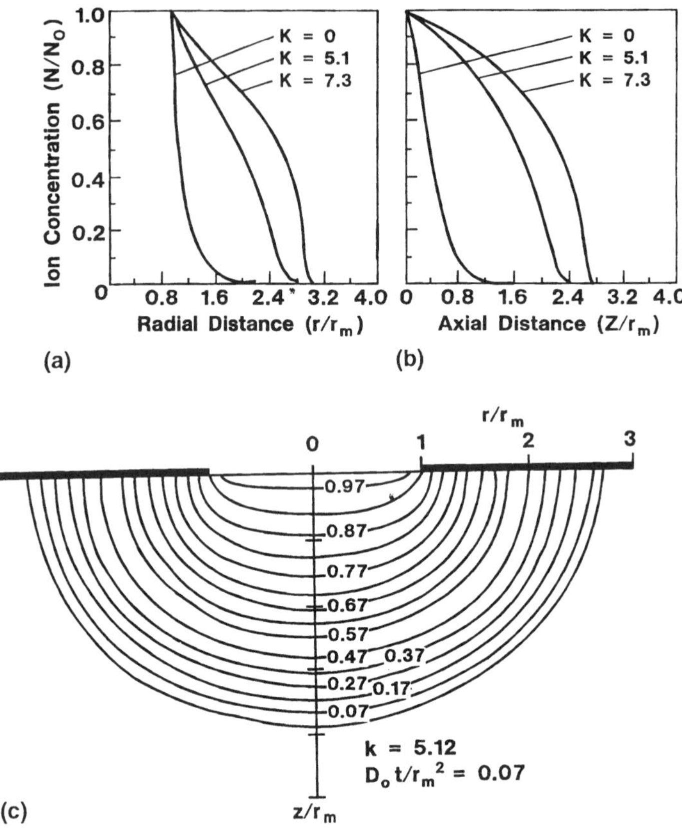

Figure 3.19 Upper set of curves: spatial concentration profiles calculated with concentration dependent diffusion coefficients from Eq. (3.46). The magnitude of K indicates the degree of deviation from ideal diffusion. Lower curve is the computed version of the refractive index profile shown in Figure 3.18. (From Ref. 26.)

of the two constituents as a function of time, and therefore composition as a function of radial thickness, one can produce any desired refractive index profile [27].

There are inherent problems in this process if it is to be utilized for the fabrication of GRIN lenses. For example, there is a layered nature to the deposition which survives the consolidation. The net effect of this is to produce what amounts to a staircase approximation to the desired profile. This structure produces undesirable features for a GRIN lens application; thus the problem is to

Table 3.4 Typical Data for Planar Microlens

	Single lens	Coupled
Lens diameter	0.9 mm	0.9 mm
Depth of index	0.45	0.45
Focal length (in air)	2.9 mm	—
(in glass)	4.0–4.5	2.5
Numerical aperture	0.15	0.25
Index of substrate	1.52	

Source Ref. 34.

reduce the preform size sufficiently so as to make the discontinuous structure effectively disappear. Even after a reduction of 50× one can still make out the ringlike structure as seen in the photograph in Figure 3.21.

A corresponding problem relates to the center opening of the preform where the mandrel was. Consolidation does close the hole but there is still some remnant of the hole. If one reduces the size sufficiently by redrawing the blank at an appropriate temperature, then eventually both these problems would disappear.

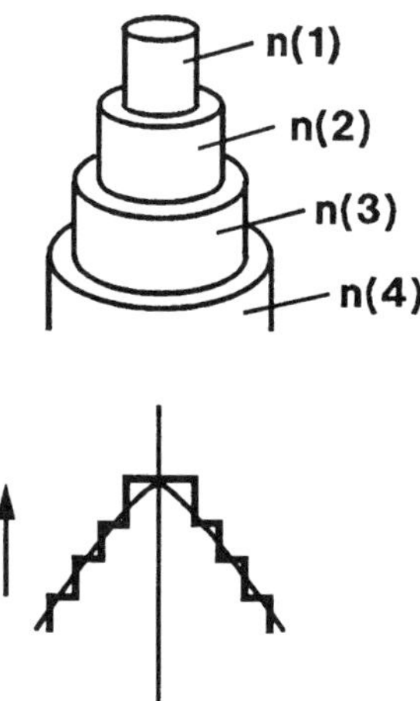

Figure 3.20 Sketch of the nested tube approach, producing a radial index gradient. As indicated, the profile will be stepwise unless the thickness of each tube is thin and there is some degree of interdiffusion at the boundaries.

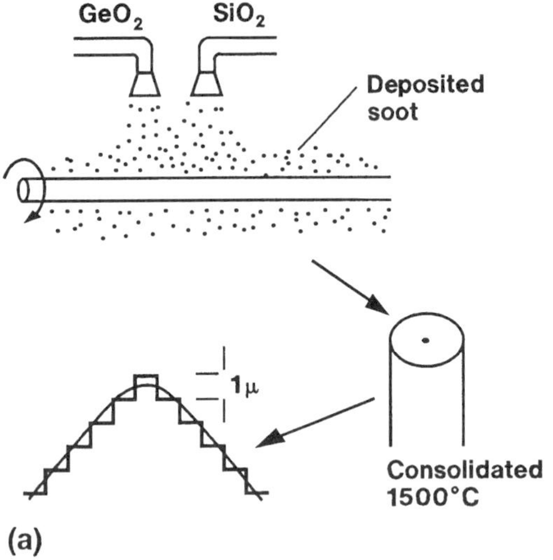

Figure 3.21 Flame deposition method of producing radial gradient index profile. Mixture of the precursors are burned to form the soot in proportion to the desired index change with radial distance. The layer thickness per revolution of the bait rod yields the step dimension as shown in the lower sketch. A photomicrograph (100×) of a pattern produced by this method is shown. Note the resolvability of the laydown layers.

Polymers

The polymers present a way to do gradient index at a lower temperature. This can be done either by mixing different materials or by the use of the degree of polymerization. It should be kept in mind that the intention in most studies to date has been to produce optical fiber with a radial profile to increase the system bandwidth, not to make microlenses. It should be clear from the analysis presented above that the lens application is much more demanding of the accuracy of the profile. As a consequence, one does not see much lens characterization, but rather results pertaining to light propagation in such fibers. Because of this we will describe only the methods where accurate control of the index pattern is likely.

The most controlled method is based on a method described in a Kodak patent [28] which we show in Figure 3.22. A mixture of monomers is injected into a centrifugal mold, the composition of the mixture being varied according to that resulting in the desired profile by essentially creating a plurality of concentric cylinders of differing refractive index. The mixture is then polymerized. Another method, sketched in Figure 3.23, uses the idea of a rotating tube. A zoned polymerization is effected by means of illumination from the side through a mask [18].

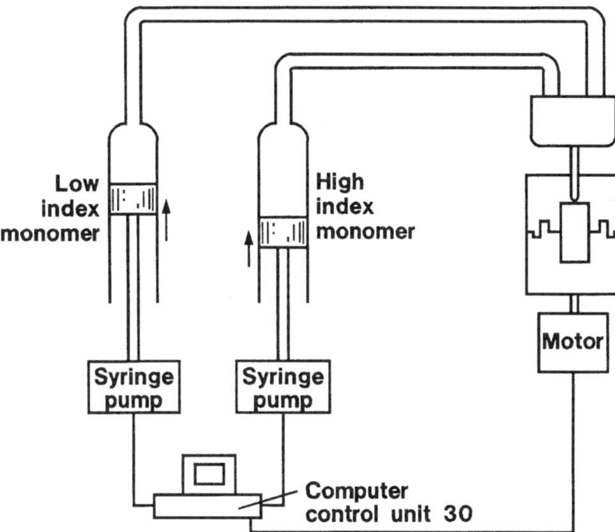

Figure 3.22 Drawing of the method used to produce a radial refractive index pattern in a polymer-based system. Each of the pumps control the relative amount of the high and low index monomers being delivered to the rotating cell.

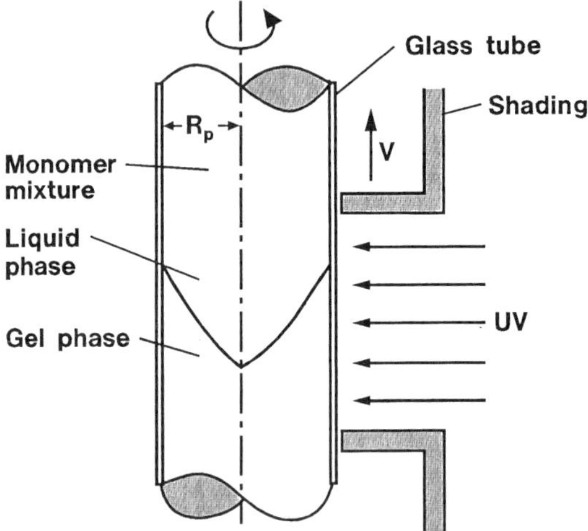

Figure 3.23 Photoinitiated method of producing radial reactive index gradient. UV exposure from the side of the rotating mixture controls the polymerization.

3.4.3 Photochemical Methods

Ideally one would like to permanently alter the refractive index of a solid body by direct exposure to light through the initiation of some photochemical reaction. The index profile would then result as a consequence of the intensity profile of the beam. As we will see, this latter condition is not so simple to achieve. Although one can imagine a number of examples of how a reaction initiated by light might lead to a refractive index change, for example, photopolymerization, surprisingly only a few have been reported.

Photochemistry in Porous Glass

It has been reported that the introduction of photosensitive organometallic compounds into a fine porous structure of a leached borosilicate glass (Vycor, trade name of Corning, Inc., also known as Corning Code 7920) followed by exposure to light, produced sufficiently large refractive index changes that small microlenses could be made [16]. The porous glass is characterized as having about 30% porosity with an average pore size of 6 nm. The glass is essentially transparent above 250 nm.

A schematic of the process steps to form a lens is shown in Figure 3.24. The porous sample is initially dipped into a solution containing the photosensitive agent for sufficient time to allow the solution to infiltrate the entire thickness. The sample is then air-dried to allow the solvent to evaporate. At this stage the photosensitive material is considered to be entrapped but not bound to the surface

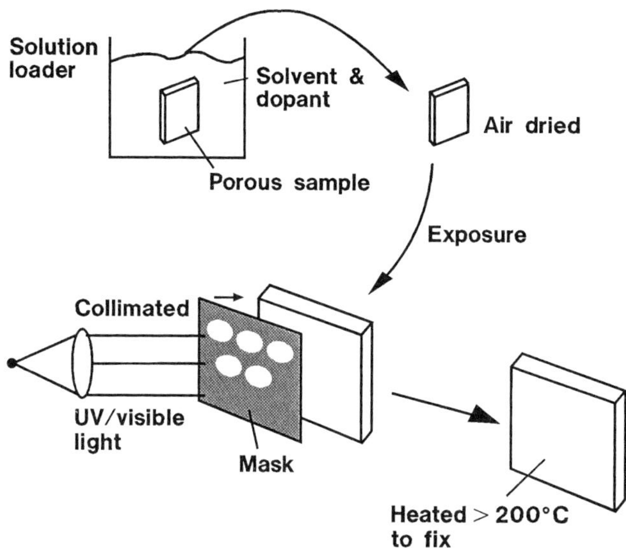

Figure 3.24 Flowchart of the steps to produce a radial refractive index pattern in porous glass. The glass is loaded with a photosensitive material and then exposed through a mask. The light produces irreversible chemical changes that lead to permanent incorporation of the dopant.

of the pores. A representative sampling of the compounds used is contained in Table 3.5 [17]. The sample is then exposed to collimated light through a mask made up of clear holes on an otherwise opaque background. The wavelength of the exposure light is determined by the absorption spectrum of the compound. In the exposed region a photoreaction proceeds, allowing the compound to bind to the glass surface. Before the exposure the organometallic can be removed by solvent extraction; after exposure it cannot. A proposed mechanism for the photolysis reaction is shown in Figure 3.25 [29]. At this stage there is a color change in the exposed regions indicating that a photoreaction has taken place. The last step involves heating the sample to oxidize, or otherwise stabilize the metal ion, which ultimately produces the desired refractive pure index change in the absence of any absorption. The explanation is that one considers the exposed region as doped with a small concentration of metal oxide species. The image formed from lenses made by this process using $(CH_3)_3SnI$ as the photosensitive agent is shown in Figure 3.26. In the lower photo is shown the interferogram of the lenses. The lens diameter was 150 μm.

A red-orange color is reported upon photolysis, which is attributed to the formation of I_2 from the photoreaction involving $(CH_3)SnI$. Thereupon, heating to

Table 3.5 Photosensitive Organometallic Compounds

Compound	Exposure wavelength
$Fe(CO)_5$	<500 nm
$Fe_3(CO)_{12}$	<700 nm
$M(CO)_{10}$	<500 nm
M = Mn, Re	
$M(CO)_6$	<400 nm
M = Cr, W, Mo	
$Ru_3(CO)_{12}$	No effect
Ti(cycopentadiene)$_2$Cl$_2$	<600 nm
$(CH_3)_3SnI$	<700 nm

Source: Ref. 17.

150°C returns the exposed region to a colorless state with the refractive index permanently altered. The index change has been reported to remain even into the consolidated state [29].

What is interesting about this process, and not explained in the papers dealing with the results, is how the near-parabolic profile arises. If it were simply a photoprocess, then the index profile should be relatively flat-topped across the exposed circular aperture with some rounding at the boundary from a combination of lack of parallelism of the exposure light and the near-field diffraction from the mask openings. The fit to a parabolic profile obtained from the interferometric data is shown in Figure 3.27. It is reasonably good out to a radius of 60 μm, compared to the radius of the exposure of 75 μm. The index change across this radius was typically of the order of 2 to 3 $\times$ 10^{-3}. Using the profile form given above,

$$\frac{Dn}{n_0} = \left(\frac{A}{2}\right)r^2 \tag{3.48}$$

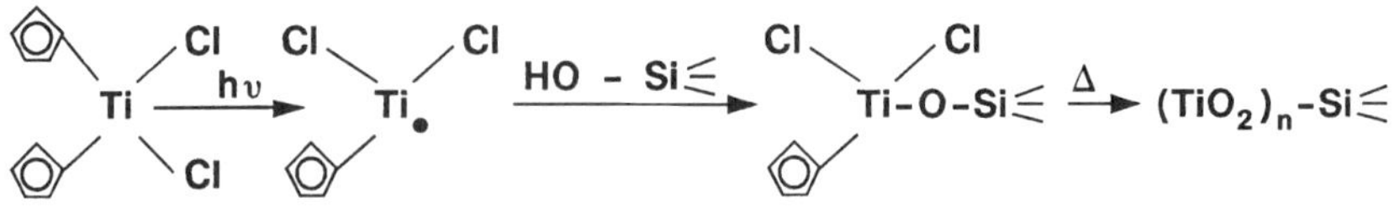

Figure 3.25 Proposed photoreaction in titanocene-doped porous glass that leads ultimately to incorporation of TiO_2 species in exposed regions. The extent of photoreaction is proportional to the intensity.

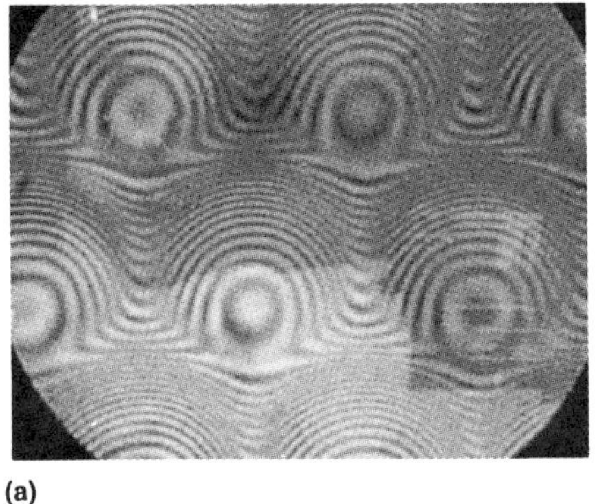

(a)

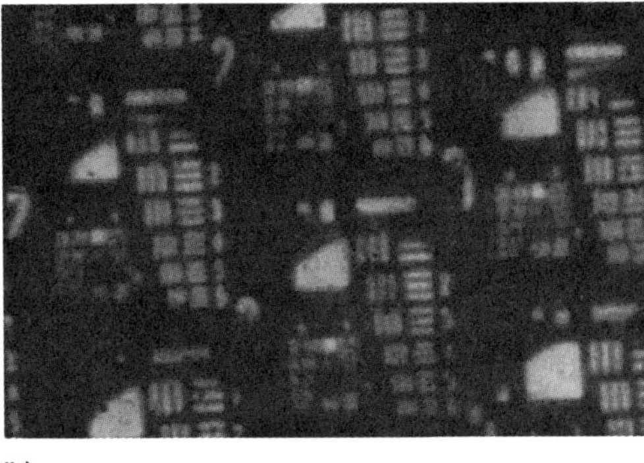

(b)

Figure 3.26 Upper photo is an interferometric trace of the induced refractive index pattern produced in a $(CH_3)_3$ SnI-doped sample exposed through a mask containing 150-μm-diameter clear circles. The lower photo shows the images produced by the induced radial profile.

one obtains a value of $\sqrt{A}$ equal to 0.8 mm^{-1}. SELFOC GRIN lenses have values that range from 0.2–0.6 mm^{-1}. As a point of comparison, recall that the focal length of a GRIN lens is expressed as

$$f = [n_0\sqrt{A}\ \sin\ (\sqrt{A}\,Z)]^{-1} \tag{3.49}$$

where Z is the lens thickness.

There are two likely reasons that the profile is not flat. The first is that the index change proceeds layer by layer so that the initial exposed region acts on the light for deeper layer exposure. In a way, it can be thought as a weak focusing

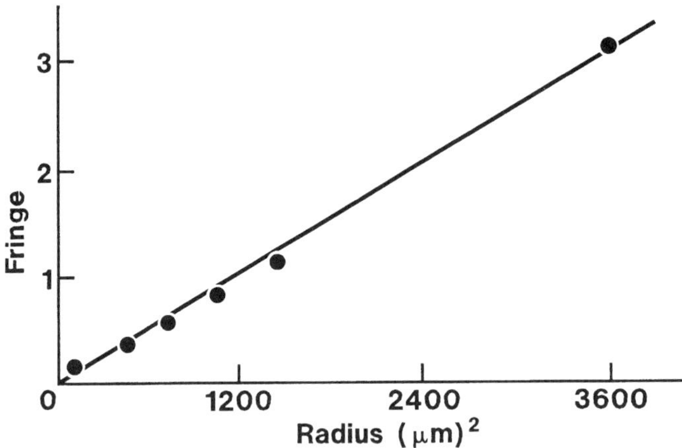

Figure 3.27 From the interferometric trace of Figure 3.26a, the fringe displacement is plotted against the square of the radial position to ascertain the extent of a parabolic profile.

effect. One can imagine the effect of this with the aid of Figure 3.28. Initially the focal length is at infinity and gradually moves in as the initial layer is formed. The focus spot eventually moves into the glass and the exposure profile is not flat as a consequence. The second possibility is that there is diffusion of the photoproducts. This supported by the fact that it is primarily with the use $(CH_3)_3SnI$ that a continuous profile is produced whereas with such compounds as titanocene and manganese carbonyl the flat-top profile is evident. An alternative

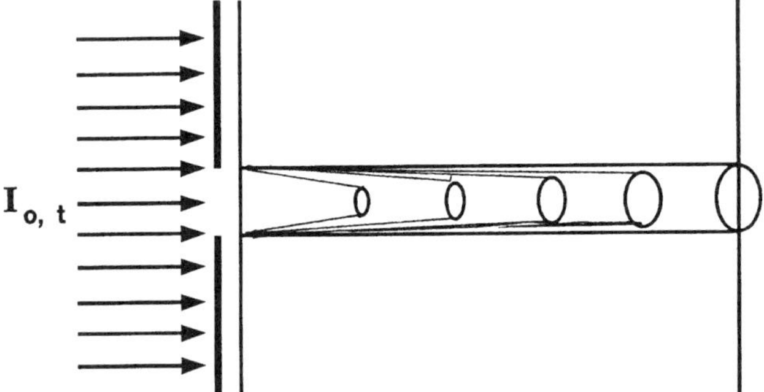

Figure 3.28 Possible explanation of how parabolic profile is produced from exposure through a circular opening mask shown at the left of the drawing. Initially, a weak lensing effect is produced, maybe only at the edges near the front. This section bends the light more which produces stronger lensing, and so on.

process is somehow to use an exposure intensity pattern that yields a parabolic profile.

Because of this unknown mechanism by which the gradient evolves, it is difficult, if not impossible, to estimate the lens diameter and thickness that can be effectively achieved. All of the reported data refers to lenses with diameters of the order of 150 μm [16].

There is another approach to utilizing the porous glass as a medium to support ways to produce refractive index changes. Although it has not been applied to the fabrication of lenses, but rather to the making of holograms, nonetheless it could be adapted to do so by the approach described above. The technique pioneered by Sukhanov [30,31] was to utilize the porous glass and the material that fills it as a composite media. It should be appreciated that the difference in this approach from that discussed above is that here the porous glass is to be infiltrated with another phase or phases which are the photosensitive species. In the loading process discussed above, the doping is via a physisorbed molecular species on the pore walls.

The refractive index of the composite is then some function of the refractive index of the glass, the filler material, and the volume fraction. For example, one might express the composite index as

$$n_c = \left\{ \frac{[(1 + 2F)n^2 + 2(1 - F)n_m^2]}{[(1 - F)n^2 + (2 + F)n_m^2]} \right\}^{1/2} \tag{3.50}$$

where n is the index of the impregnated phase with volume fraction F, and n_m is the index of the glass. What this allows you to do is to understand that if either, or both, n and F can be changed by the exposure to light, then an overall change in the composite index will occur. In the case of polymeric materials, photoinduced changes in volume can be a means to a "photorefractive" effect since it influences F. We will deal more with this material in Chapter 7.

Photosensitive Glass

There is a class of inorganic glasses containing silver, sodium, and fluorine that can be made to undergo what amounts to a phase separation of a crystalline microphase under the action of uv light and a subsequent thermal treatment. These glasses are often termed "photosensitive" glasses [32]. The mechanism involves the photoexcitation of an electron from a donor such as Ce^{+3} or Cu^{+1}, which is contained in the glass, and which is ultimately trapped by the Ag^{+1}. After a heat treatment, the silver agglomerates into a colloid that serves as a nucleus for the crystallization of a NaF phase. The schematic of the process is shown in Figure 3.29. The crystal size can be maintained small enough to render the glass transparent. The average crystal size is in the order of 5–6 nm, and the total amount of crystal phase is in the order of 1%.

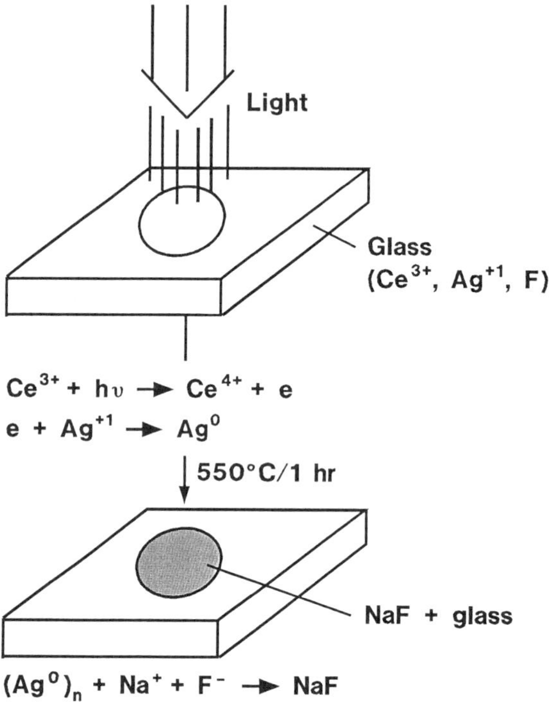

Figure 3.29 Schematic representation of the the way a refractive index change is produced in a photosensitive glass. After exposure and thermal treatment, a NaF microcrystalline region is produced which has a higher refractive index than the surrounding unexposed, undeveloped region.

The refractive index difference between the exposed and unexposed regions arises from the differing contribution of the fluoride ion between when it is incorporated in the glass, or when it is separated into the crystalline phase. The incorporation of fluoride ions into oxide glasses decreases the refractive index [32]. In the exposed and crystallized region, the fluorine is tied up in the crystalline phase where its depressing effect on the refractive index is less.

The development of a parabolic profile is difficult to achieve with any simple

Figure 3.30 Photomicrographs of images formed by gradient index lenses produced in photosensitive glass by the method shown in Figure 3.29. For this particular glass the exposure was 20 min under a 1000-W Hg-Xe lamp, and heated to 550°C for 1 h.

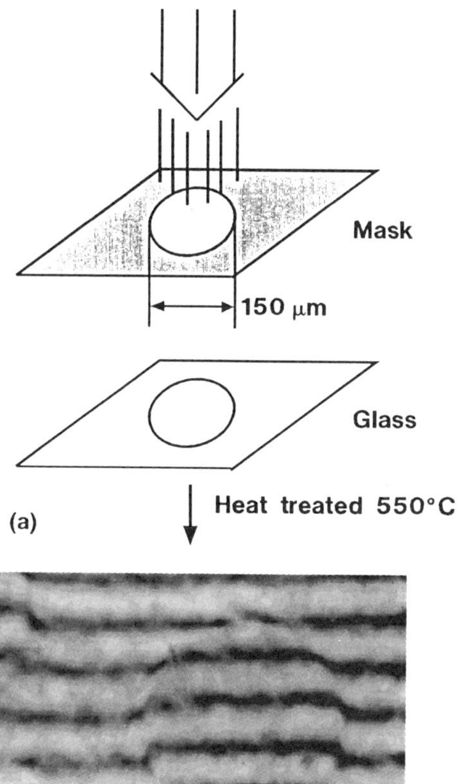

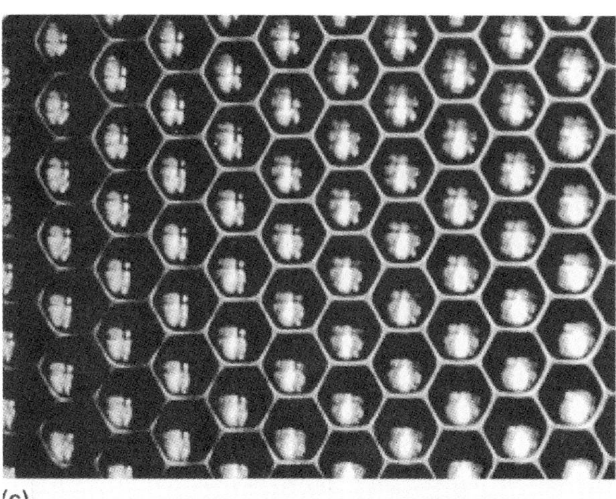

(c)

masking procedure because of the need to provide an intensity profile. With a simple clear opening mask one obtains an index profile, as shown in Figure 3.30. The Δn is 3×10^{-4} across the 150-μm diameter.

Miscellaneous

Marchand in his 1978 book [2] refers very briefly to a number of methods, some of which are not included in the above. Since these methods have not ever moved forward, we merely mention the reference here and offer no further comment.

3.5 ANALYSIS AND COMPARISON

As we did in Chapter 2 on refractive lenses, we would like to evaluate the performance of GRIN lenses made by the various methods discussed above in the same way, and to indicate how certain features of the GRIN lens performance might be related to a particular fabrication method. The methods used to evaluate and characterize performance of conventional refractive lenses discussed in Chapter 2, such as interferometric Zernike polynomial analysis, or the intensity profile in the image plane, are every bit as useful and important for the characterization of GRIN lenses as they are for refractive lenses. We will present this data as it is found in the literature. The SELFOC product line manufactured by Nippon Sheet Glass has been the source of the bulk of the published quantitative data.

As we had to be concerned with the true surface figure of the microptic refractive lenses and how it depended on the method of fabrication, here we have to be concerned with the actual refractive index profile and its dependence on how it is made. One cannot decouple in all cases the effects of profile from those arising from pure aberrations. In what follows, to the extent that it can be done, we will first deal with the consequence of the profile and then to the limitations set by aberrations.

3.5.1 Profile

We have seen from the analysis given above in Section 3.2.2 that there are certain limitations in imaging performance of a GRIN lens based on the deviation from the ideal index profile. In the case of the parabolic profile of Eq. (3.19) we see that there the existence of a common focus is based on the assumption of the constancy of the l_0 ray parameter defined in Eq. (3.15). As pointed out above, this parameter is essentially constant when Δn is small, and when the square of the input ray slope is small compared to unity (equivalent to a paraxial approximation). To show this graphically, we show a case of a GRIN lens imaging in a one-to-one erect mode in Figure 3.31. Here, to effect the righting of the image, a small relay image is formed at the midplane of the lens. (A two-dimensional array of lenses functioning in this mode is of significant practical utility as bar

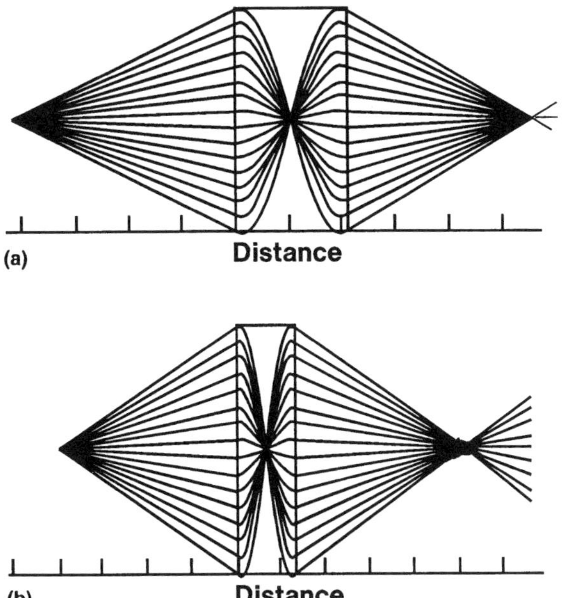

Figure 3.31 Computed ray trace for lens with a parabolic profile, upper curve for the case of $\sqrt{A} = 0.44$ mm^{-1}, and the lower $\sqrt{A} = 0.66$ mm^{-1}. The image formed by the stronger gradient lens shows evidence of spherical aberration in that the image is somewhat blurred.

lenses for a number of paper reading devices. These applications will be discussed in Chapter 5.) The computed ray trace shown in Figure 3.31a is for a lens with a $\Delta n/n$ of 0.001, whereas in Figure 3.31b the value is 0.0044. The working distance is the same in both cases (distance from lens surface to object), so that the external NA is constant. The ray trace was obtained from a numerical solution to Eq. (3.4) assuming an index profile of the form given in Eq. (3.19). The values of $\sqrt{A}$, which is a measure of the gradient, is obtained from the radius of the lens and the value of Dn/N_0.

$$\sqrt{A} = \left(\frac{\sqrt{2Dn}}{N_0} \right) \left(\frac{1}{r_0} \right) \tag{3.51}$$

It is clear from the ray trace that there is evidence of aberration for the lens with the steeper gradient, $\sqrt{A} = 0.44$ mm^{-1}, curve b, compared to curve a with a value of $\sqrt{A} = 0.66$ mm^{-1}. One has a situation entirely equivalent to that of spherical lenses. We can also use this case to look at the consequence of a deviation of the profile from a parabola. This is done by assuming the equation for the refractive index profile as

$$n = N_0\left[1 - \left(\frac{A}{2}\right)r^2 + \mathrm{B}r^4\right] \tag{3.52}$$

for the profile and letting the value of B be such that the contribution of the quartic term represents 1% of the maximum value of Dn/N_0. The ray trace of this case is shown in Figure 3.32, where the upper curve represents the pure parabola, $B = 0$. The blurring of the image spot is clearly evident. It should be pointed out that the example case used is quite a bit more sensitive to the profile distortion than others might be, but it is, nonetheless, a case of a major application of GRIN lens arrays.

Experimentally, the situation is somewhat more complicated because the measurement of the spot size, by whatever technique, would include diffraction as well as the contribution from the profile distortion. In the example given above, one is provided with some advantage in that the diffraction-limited spot size is

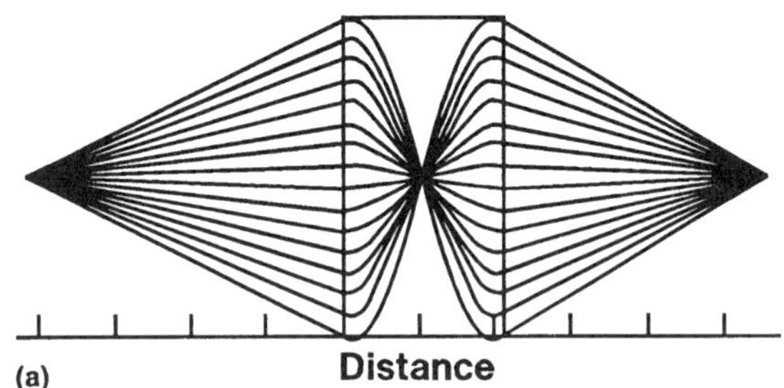

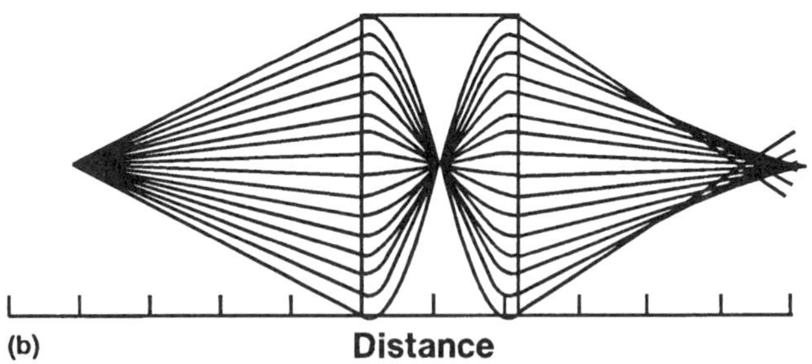

Figure 3.32 Computed ray trace for imaging for a lens with a pure parabolic profile compared to one with a nonzero quartic term. See Eq. (3.46).

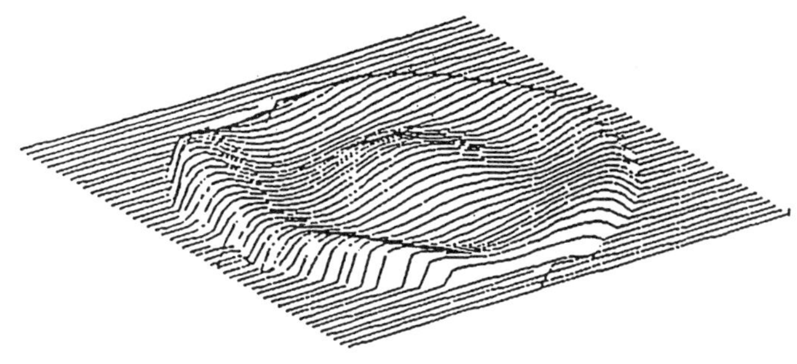

Figure 3.33 Zernike profile analysis of SELFOC GRIN lens as reported by NSG.

the same for the two cases, so any difference in the resolution could be attributed to the profile effect.

In Section 3.4.1, we briefly discussed how the process parameters are used to optimize the index profile. A good example of the nature of the empirical optimization method was shown in Figure 3.13 where the time in the ion-exchange bath was the parameter of interest. In addition, as discussed more extensively in Iga et al. [3], the process is further refined by a heat treatment that follows the ion exchange. This adds an additional step in the diffusion process where one can now solve the diffusion equation using essentially the adiabatic boundary conditions (no material flowing out, so derivative of concentration at boundary is zero), with the initial distribution that of the rod after the ion exchange. The performance parameters are optimized with respect to the heat treatment which is characterized by the same parameter as used in Figure 3.13, DT/r_0^2 but now T refers to the heat treatment time rather than the ion exchange time.

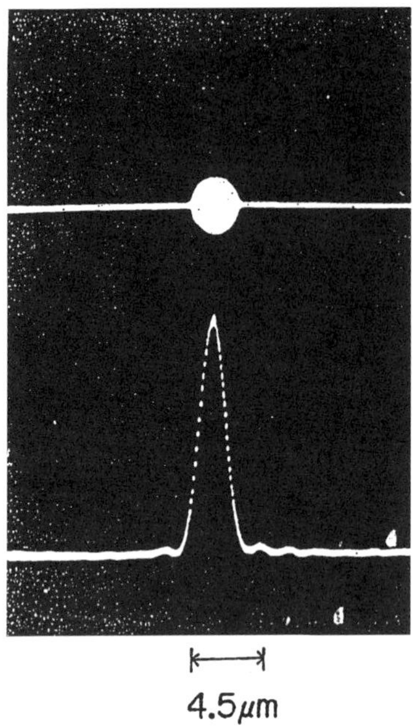

$$\lambda = 0.63 \ \mu m$$

Diffraction Limited Spot Size

$$2a = \frac{1.22\lambda}{N.A.} = 4.29 \ (\mu m)$$

Figure 3.34 Irradiance profile in the focal plane of a SELFOC GRIN lens as reported by NSG.

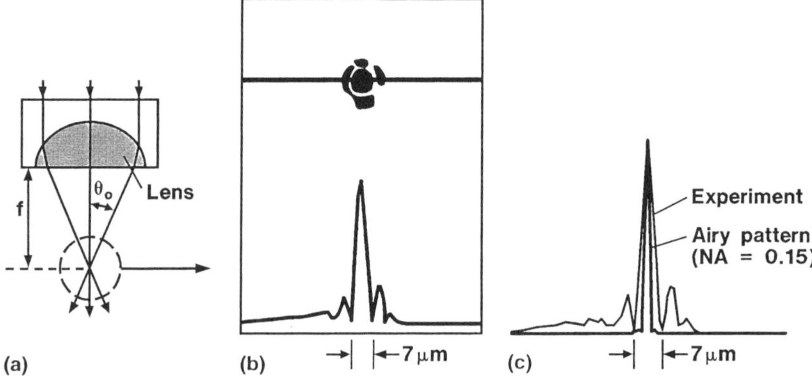

Figure 3.35 Irradiance profile in the focal plane of a stacked planar GRIN lens (as reported in Ref. 32).

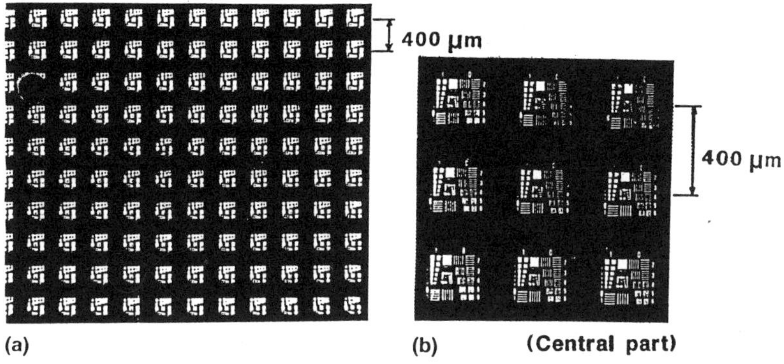

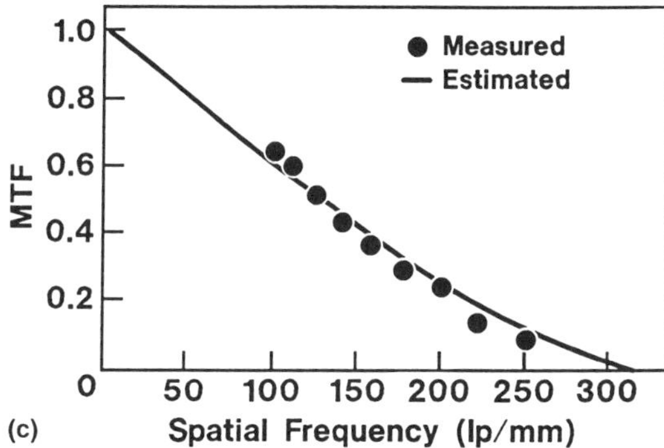

Figure 3.36 Multiple images in coherent light from a planar array of SELFOC GRIN lenses, along with the contrast vs. spatial frequency for an individual lens as reported by NSG.

3.5.2 Aberrations

As mentioned above, the standard Zernike interferometric analysis for the GRIN lenses has only been reported for the SELFOC. Their reported typical rod-type lens result is shown in Figure 3.33. A reported irradiance profile is shown in Figure 3.34. The same mistake in the conclusion is made in this case as is generally made. As we pointed out in Chapter 2, the dimension of the first Airy ring is not particularly sensitive to the amount of aberration.

For the planar lenses, the performance is indicated in two ways. The image plane irradiance function is shown in Figure 3.35. Here one clearly sees the serious aberration/profile problem from the large wing contributions compared to

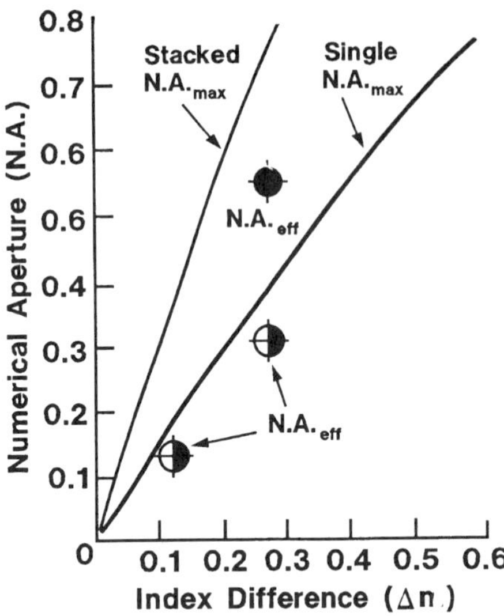

Figure 3.37 Relation between the NA and the index difference of the GRIN lens for an individual planar GRIN lens compared to the stacked configuration. (See Ref. 31.)

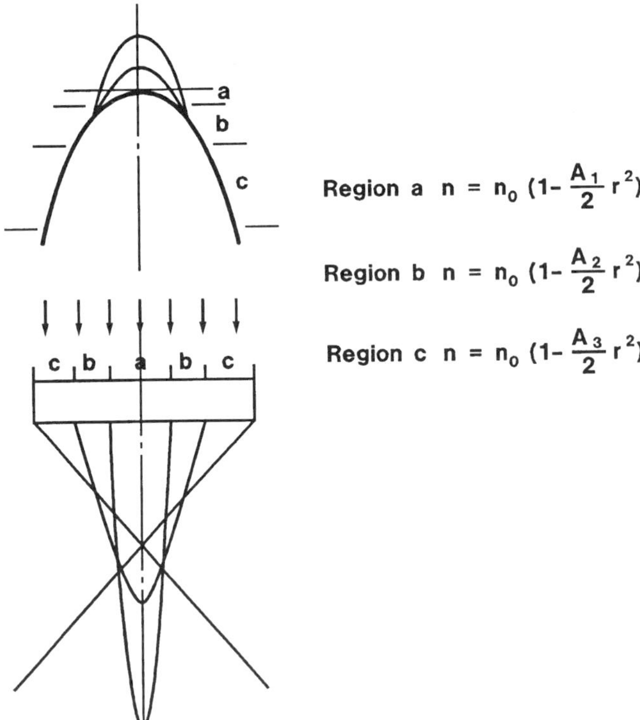

Region a $n = n_0 \left(1 - \dfrac{A_1}{2} r^2\right)$

Region b $n = n_0 \left(1 - \dfrac{A_2}{2} r^2\right)$

Region c $n = n_0 \left(1 - \dfrac{A_3}{2} r^2\right)$

Figure 3.38 Imaging from a lens with a refractive index profile that deviates from a parabola. Each portion of the profile provides imaging over its parabolic range.

the theoretical Airy profile. Another performance criterion is the resolution as measured by the standard MTF method of a multiple image array which is shown in Figure 3.36. Another important aspect is the achievable NA. Misawa et al. [33] report data on the NA as a function of the value of Δn for both the single and stacked lens configuration, which we show in Figure 3.37.

The reader is referred to Iga et al. [3] for a fuller description of the aberration testing of the SELFOC lenses, both rodlike and planar. As far as the results on lenses made by the other methods, there is little quantitative data reported in the literature. Resolution results that have been presented for the photosensitive glass method are plagued with what seems to be severe profile problems. One obtains images from different portions of the lens profile, as that portion approximates a parabola. The is schematically shown in Figure 3.38. As a consequence the object stays in focus for a long distance, as each region of the lens contributes. However, the image is always of poor contrast because of the light coming from the ill-focused regions.

References

1. R. W. Wood, *Physical Optics*, Macmillan, New York, 1905.
2. E. W. Marchand, *Gradient Index Optics*, Academic Press, New York, 1978.
3. K. Iga, Y. Kokubin, and M. Oikawa, *Fundamentals of Microoptics*, Academic Press, Tokyo, 1984.
4. S. Kawakami and J-I Nishizawa, An optical waveguide with the optimum distribution of the refractive index, *IEEE Trans. Microwave Techniques*, MTT-16(10), 814 (1968).
5. For the reader unfamiliar with the calculus of variations, it is sufficient to know that the solution of the Euler-Lagrange differential equations yield the path, or function, which corresponds to the minimum condition expressed in the least action integral. For example, see G. A. Bliss, Calculus of variations, *Math. Soc. Am*, Open Court Publ. Co., LaSalle, IL, 1923.
6. K. Iga, Theory for gradient-index imaging, *Appl. Opt. 19*(7), 1039 (1980).
7. E. T. Kornhayser, and A. D. Yaghjian, Modal solution of a point source in a strongly focusing medium, *Radio Science 2*(3), 299 (1967).
8. E. G. Rawson, D. R. Herriot, and J. McKenna, Analysis of refractive index distributions in cylindrical GRIN rods used in Image Relays,'' *Appl. Opt. 9*(3), 753 (1970).
9. F. P. Kapron, Geometric optics of parabolic index-gradient cylindrical lenses, *J. Op. Soc. Am. 60*(11), 1433 (1970).
10. Nippon Sheet Glass (NSG) product sheets for SELFOC rod lenses, Clark, NJ.
11. D. Gregoris and K. Iizuka, Ray tracing method for refractive index profiling, *Appl. Opt. 22*(12), 1820 (1983).
12. M Oikawa, K. Tanaka, and T. Yamasaki, Highly integrated distributed-index planar microlens and its characteristics, *SPIE*, vol. 554, International Design Conf. 314 (1985).

13. T. Sakamoto, Ray tracing in a planar lens with spherically symmetric quadratic index profile, *Appl. Opt. 22*(10), 1598 (1983).

14. K. Sono, "SELFOC technology, *IFOC Handbook* 1980–1981 edition.

15. LightPath Technologies, Inc. Albuquerque, NM, 87109.

16. N. F. Borrelli and D. L. Morse, Planar gradient optics Fourth Topical meeting on Gradient-Index Optical Imaging Systems, paper D1, Kobe, Japan (1983).

17. N. F. Borrelli and D. L. Morse, Photosensitive impregnated porous glass, *Appl. Phys. Lett. 43*(11), 992 (1983).

18. Y. Koike and Y. Ohtsuka, Studies on the light-focusing plastic rod prepared by photocopolymerization of a ternary monomer system, *Appl. Opt. 22*(3), 418 (1983).

19. A. K. Varshneya Fundamentals of Inorganic Glasses, pp. 339–344, Academic Press, San Diego, CA, 1994.

20. H. S. Carslaw and J. C. Jaeger, *Conduction of Heat in Solids*, Oxford Press, 1959.

21. B. Messerschmidt, B. L. McIntyre, and S. N. Houde-Walter, Desired concentration-dependent ion exchange for microoptic lenses, *Appl. Opt. 35*(28), 5670 (1996).

22. W. Jost, *Diffusion*, Academic Press, 1952.

23. R. J. Araujo, Colorless glasses containing ion-exchange silver, *Appl. Opt. 31* 5221 (1992).

24. J. Crank, *The Mathematics of Diffusion*, Clarendon Press, 1975, p. 231.

25. S. N. Houde-Walter, J. M. Inman, A. J. Dent, and G. N. Greaves, Sodium and silver exchange environments and ion-exchange process in silicate and aluminosilicate glasses, *J. Phys. Chem. 97*, 9330 (1993).

26. K. Nishizawa, Planar optical elements on glass substrate, Nippon Sheet Glass (1994).

26a. S. Misawa, M. Oikawa, and K. Iga, Maximum and effective numerical apertures of a planar microlens, *Appl. Opt 23*(11), 1784 (1984).

27. Optical properties of glass, D. R. Uhlmann and N. J. Kreidl, eds. Optical fibers, M. A. Newhouse, p. 185, *Am. Cer. Soc.*, 1991.

28. J. A. Bello and D. P. Hamblen, U.S. Patent 5,122,314, 1992.

29. N. F. Borrelli, M. D. Cotter, and J. C. Luong, Photochemical method to produce waveguiding in glass, *IEEE J. Quantum Electronics*, QE-22(6) 896 (1986).

30. V. I. Sukhanov, Porous glass as a storage medium, *Optica Applicata*, XXIV(1–2), 13–26 (1994).

31. S. A. Kuchinskii, V. I. Sukanov, and M. V. Khazova, "Principles of hologram formation in capillary composites," *Optics Spectroscopy 72*, 383–391 (1992).

32. N. F. Borrelli and T. P. Seward, Photosensitive glasses and glass-ceramics, p. 439, *Engineered Materials Handbook*, vol. 4, Ceramics and Glasses in ASM, 1991.

33. S. Misawa, M. Oikawa, and K Iga, "*App. Opt. 23*, 1784 (1984).

34. K. Iga, M Oikawa, J. Banno, and Y. Kokubun, *Appl. Opt. 21*(19), 1127 (1992).

4

Diffractive Element Lenses

4.1 DIFFRACTION THEORY REVIEW

The development of methods to fabricate diffractive elements to function as microoptic lenses and lens arrays, as well as other optical elements, has become an active and productive area of research in recent years. For an excellent introductory review the reader is referred to a *Scientific American* article by Veldkamp and McHugh [1]. The reason this has occurred is the availability of high-resolution photolithography techniques and equipment developed by the IC industry. There are other diffraction-based elements used in fiber optic devices and computer back plane applications [2,3]. Holographic elements to direct light or separate wavelengths are examples of such devices. We will cover these diffraction gratings in Chapter 7. In this chapter we will be concerned only with those elements that act as lenses.

This is not intended to be an optics textbook, by any stretch of the imagination. However, in order to understand the design and performance characteristics of diffractive element lenses, we must devote some space to review the basic principles of diffraction theory as it pertains to the lens function.

4.1.1 Fresnel-Kirchhoff Diffraction Integral

Consider Figure 4.1 where light in the vicinity of the point $P(x, y)$ in the x-y plane reaches point $P'(x', y')$ after passing through an obstacle located in the x_0-y_0 plane a distance r from P and s from P'. The expression for the complex

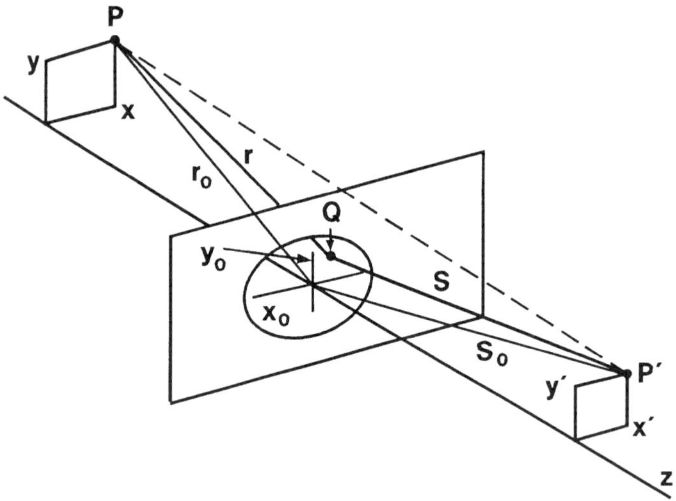

Figure 4.1 Reference diagram for the development in the text. *x-y*, the plane of the object; x_0-y_0, the plane of the obstacle; *x′-y′*, the image plane.

amplitude of the wave from *P* to a point *Q* in the plane of the screen is the spherical wave [4].

$$E(Q) = \left(\frac{E_0}{r}\right) \exp\,(-ikr) \tag{4.1}$$

where we have suppressed the time dependence term, $i\omega t$, and $k = 2\pi/\lambda$, λ being the wavelength of the light; the refractive index in which the wave is traveling is taken to be unity. Utilizing Huygens' principle, every point in the x_0-y_0 plane is a source of a new spherical wave propagating toward the point *P′*, which in differential form can be expressed as

$$d\{E(P')\} = (i/\lambda)E(Q)\,\frac{e^{-iks}}{s}\,d\sigma \tag{4.2}$$

Here, $d\sigma$ is a differential element in the x_0-y_0 plane near the point *Q*. For a derivation of this equation the reader is referred to Born and Wolfe [3]. After the substitution of (4.1) into (4.2), the integration of Eq. (4.2) over that part of the wavefront passing through the obstacle, designated as *S*, gives us the desired diffraction result.

$$E(P') = \left(\frac{iE_0}{\lambda}\right)\int_s \frac{\exp\left[-ik(r + s)\right]}{(rs)}\, d\sigma \tag{4.3}$$

This is a general expression except for the simplification of taking the angle between the normal to the wavefront and the direction s to be zero. In the more rigorous development this need not be so, but little generality is lost in taking the special case. The intensity at the point P' is the product of E with its complex conjugate,

$$I(P') = E(P')\, E^*(P') \tag{4.4}$$

4.1.2 Fraunhofer Diffraction

The availability of solutions to Eq. (4.3) depends on making certain assumptions. For example, consider the expansion of the distances r and s in a power series of the variables x_0 and y_0.

$$r = r_0\left[1 + \frac{x_0^2 + y_0^2}{2r_0^2} - \frac{x_0x + y_0y}{r_0^2} + \cdots\right] \tag{4.5a}$$

$$s = s_0\left[1 + \frac{x_0^2 + y_0^2}{2s_0^2} - \frac{x_0x' + y_0y'}{s_0^2} + \cdots\right] \tag{4.5b}$$

It is easy to see that if the spatial extent of the obstacle (x_0, y_0) is much smaller than the distances r_0 and s_0, one need keep only the first two terms in Eqs. (4.5). Similarly, one can assume that r and s in the denominator of Eq. (4.3) can be taken outside the integral since they do not vary significantly over the region S in the x_0-y_0 plane, and can be taken as both equal to z_0. Further, if the source is small and essentially on the z axis, that is, $P(x, y)$ is confined to small values, then one can further ignore the second terms of Eqs. (4.5). Hence, we can now utilize all the above approximations and write Eq. (4.3) as follows:

$$E(P') = C\int_s \exp\left[-i(k_xx_0 + k_yy_0)\right]\, d\sigma \tag{4.6}$$

Here, C is a constant phase term, and we have adopted the more useful notation, as we shall see shortly, $k_x = (2\pi/\lambda)(x'/z_0)$ and $k_y = (2\pi/\lambda)(y'/z_0)$. One can specify the region S in terms of a mathematical function which is called an "aperture" or "transmittance" function. We shall express this function in the following way:

$$t(x_0, y_0) = T(x_0, y_0)\, \exp\, i[\phi(x_0, y_0)] \tag{4.7}$$

where $T(x_0, y_0)$ represents the amplitude variation, and $\phi(x_0, y_0)$ the phase variation over the region S. Incorporation of Eq. (4.7) into Eq. (4.6) appropriately allows extension of the limits of integration to plus and minus infinity. Equation (4.6) is recognized as the 2D spatial Fourier Transform of the transmittance function, $t(x_0, y_0)$:

$$E(P') = c \int\limits_{-\infty}^{\infty}\!\!\int T(x_0, y_0) \exp{(i\phi)} \exp{[-i(k_k x_0 + k_y y_0)]} dx_0 \, dy_0 \qquad (4.8)$$

This transform representation will prove to be of considerable use when we extend this treatment to important diffraction geometries where the transmittance function can be easily written.

Just to be complete, when the approximations leading to the form of Eq. (4.6) are made, this is referred to as the far-field diffraction regime. More commonly, it is referred to as Fraunhofer diffraction. With less restrictive assumptions about the relative magnitude of the distance from the object relative to the size of the disturbance, one refers to this regime as near-field or Fresnel diffraction. For our subsequent development of diffractive microlenses, the far-field approximation is appropriate. We will be considering diffractive elements with a numerical aperture (approximately the lens radius divided by the focal length) of the order of 0.1. This justifies the assumptions made in ignoring the higher order terms in Eqs. (4.5).

4.1.3 Diffraction Gratings

As an example of the utility of Eq. (4.8), the diffraction from a narrow slit in an otherwise opaque screen is easily obtained by writing the aperture function $t(x_0, y_0)$ by a function that is termed the "rectangular" function [5]. It is defined as follows:

$$\begin{aligned} R(x_0, y_0) &= 1 \qquad - a < x < a, \, - b < y < b \\ &= 0 \qquad x > |a|, \, y > |b| \end{aligned} \qquad (4.9)$$

where a, b are the dimensions of the slit. The utility of the rectangular function should be appreciated in that it allows us to mathematically represent a sharply defined region, which constitutes the essence of the diffraction phenomenon. The Fourier transform of R is

$$F\{R(x_0, y_0)\} = 4ab[\text{sinc}(k_x a)][\text{sinc}(k_y b)] \qquad (4.10)$$

This expression, aside from being a constant, represents the diffracted amplitude obtained by integrating Eq. (4.8). We have used the common notation that $\text{sinc}\, x = \sin x / x$. The diffracted intensity is just the square of Eq. (4.10).

A transmission grating can be considered as an N-fold periodic extension of a slit with a separation distance of T. Mathematically, one can express this by use of the Dirac δ function for the case of a slit in the x direction:

$$t(x_0) = \Sigma\delta\left[\left(\frac{x_0}{T}\right) - n\right]^{*} R\left[\frac{x_0}{T}\right] \qquad (4.11)$$

where n takes on the values, $0, \pm1, \pm2, \ldots$.

The asterisk is written to signify a convolution. [Strictly speaking, the convolution of two functions, $f(x)$ and $g(x)$ is defined as the integral of $f(\tau)g(x - \tau)d\tau$.] The action of the δ function is to shift the center of the rectangular function by $T, 2T, \ldots, nT, \ldots, NT$. This is a useful way to write the transmission function since one can use the fact that the Fourier transform of a convolution is just the product of the individual transforms [6]; that is to say,

$$F\{t(x_0)\} = F\left\{\Sigma\delta\left[\left(\frac{x_0}{T}\right) - n\right]\right\} F\left[R\left(\frac{x_0}{T}\right)\right] \qquad (4.12)$$

It can be shown [5] that the Fourier transform of an N-sum of δ functions of the form $\delta(x_0/n\text{-}T)$ is

$$F\left\{\sum_N \delta\left(\frac{x_0}{n - T}\right)\right\} = \frac{\sin\,[2N + 1)(\pi T/\lambda z_0)x']}{\sin\,(\pi T/\lambda z_0)x']} \qquad (4.13)$$

We have the Fourier transform of $R(x_0/T)$ from above, so the diffracted amplitude from a transmission grating is the product of Eq. (4.13) and the one-dimensional form of Eq. (4.10), viz., sinc $k_x a$. For an example of an intensity pattern that would be obtained from Eq. (4.13) for an array of slits of width a, arrayed a distance T apart, see Figure 4.2.

Of more relevance to the understanding of how a diffractive element can act as a lens is the behavior of a phase grating. In the above we talked about an array of slits where the amplitude varies periodically, and now we want to investigate a similar structure where the phase is varied. The interest in the spatial variation in phase is because a refractive element acts as a lens because it produces a phase difference that varies parabolically with the radial distance. We will come back to this point in more detail in the next section. The phase grating represents a special case of the general spatial variation of the optical path length in the x_0-y_0 plane, as represented by $\phi(x_0, y_0)$ in Eq. (4.7). The aperture or transmittance function for a phase grating with a blaze (see Figure 4.3) can be written as

$$t(x_0) = \sum \delta\left(\frac{x_0}{T} - N\right) R\left(\frac{x_0}{T}\right) \exp\,(i\alpha x_0) \qquad (4.14)$$

Here, $R(x_0/T)$ is a rectangular function over the interval T. The tangent of the blaze angle is d/T, d being the depth of the grating, and the corresponding phase shift produced in the groove is $2\pi(n - 1)(d/T\lambda)$, where n is the refractive index

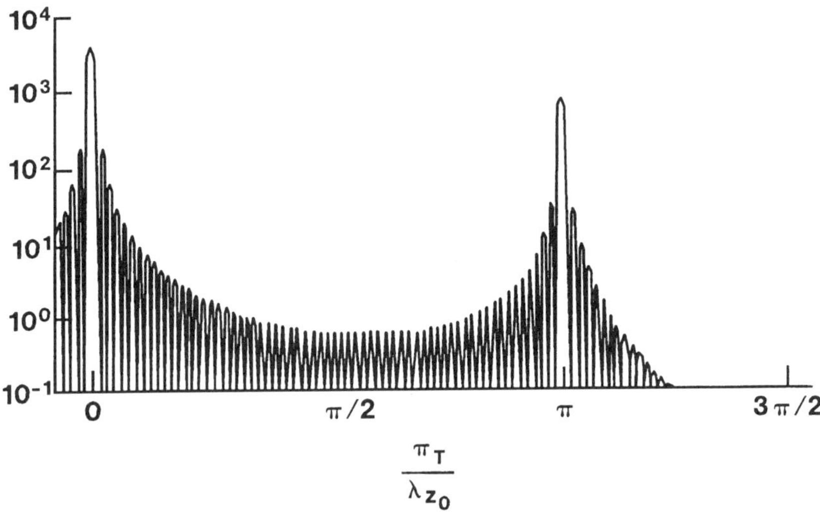

Figure 4.2 Diffracted intensity from a transmission grating according to Eq. (4.13) in the text. The individual aperture opening is x_0 and the separation distance is T. The intensity is measured at a distance z_0 from the grating, and x' measures the position in the image plane.

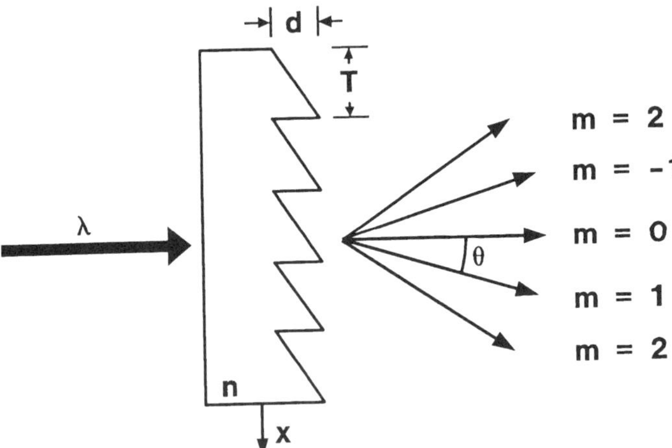

Figure 4.3 Blazed diffraction grating with an indication of the direction of the various diffracted orders.

of the material. The Fourier transform of this transmittance function represented by Eq. (4.14) is

$$\sum \delta\left\{\frac{2\pi N}{T} - k_x\right\} \operatorname{sinc}\left\{\left(\frac{T}{2}\right)(\alpha - k_x)\right\} \tag{4.15}$$

Each term in N represents the successive diffracted order. In general, the nth-order efficiency would be given by the expression

$$\left[\operatorname{sinc}\left\{\left(\frac{T}{2}\right)\left(\alpha - \frac{2\pi n}{T}\right)\right\}\right]^2 \tag{4.16}$$

From Eq. (4.15) one sees that the first-order efficiency will be 100% when the value of $\alpha = 2\pi/T$, which corresponds to a groove depth d of $\lambda/(n-1)$.

4.1.4 Diffractive Elements in Relation to Refractive Elements

As we mentioned above in introducing the phase term $\phi(x_0, y_0)$ into the transmittance function of Eq. (4.7), this could be any arbitrary function. We looked at a special case of a linear function for the blazed grating. (In Chapter 7 we will analyze the case for gratings with different forms, rectangular, sinusoidal, etc.) One should realize for this case that the refractive analogue to the blazed grating is a prism. Moreover, the diffractive phase is periodic in the refractive phase. This correspondence suggests that if we take the phase function that describes a refractive lens, then a diffractive element that acts like a lens should ensue if certain conditions are met [7].

What we are trying to do is find a way to write the diffractive transmittance,

$$t_d(r) = \exp\left\{2\pi i \phi_d(r)\right\} \tag{4.17}$$

in terms of the refractive phase term, $\phi(r)$. One proceeds by comparing an arbitrary refractive phase function to the diffractive phase modulated with a limiter function [7] as shown in Figure 4.4. The limiter function maintains the phase between the values of $-a/2$ and $a/2$ as shown. This function was done in the grating case by the depth of the groove. Again, one sees that the diffracted phase is periodic in the refractive phase which means that $t_d(r)$ is also periodic in $\phi(r)$. This suggests that one can write $t_d(r)$ in terms of a generalized Fourier series in $\phi(r)$ as follows [6]:

$$\exp\left\{2\pi i \phi_d(r) = \Sigma A_n \exp\left\{2\pi i n \phi(r)\right\} \tag{4.18}$$

The coefficient A_n is determined in the usual way for a Fourier expansion:

$$A_n = \int \exp\left\{2\pi i (\phi_d - n\phi)\right\} d\phi \tag{4.19}$$

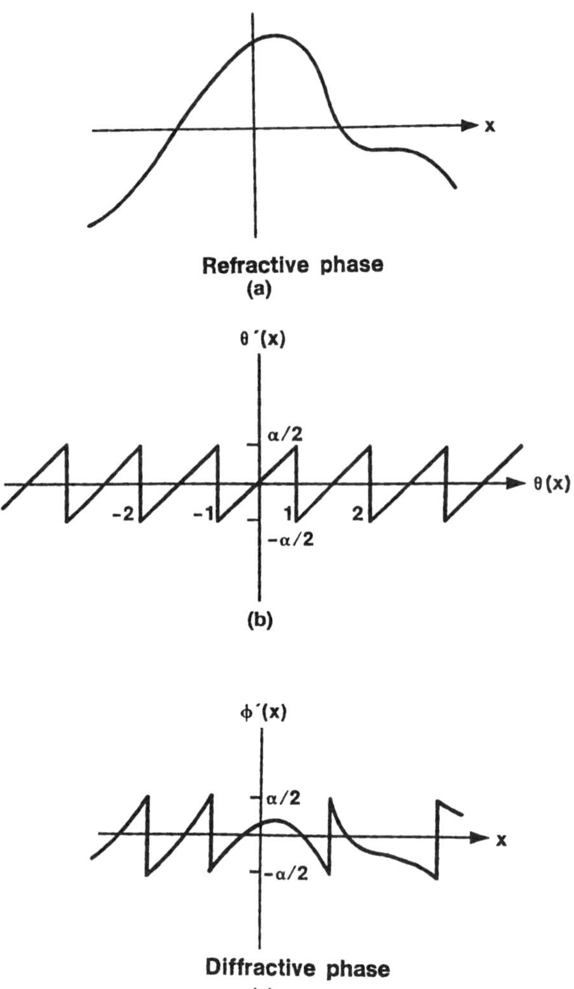

Figure 4.4 Upper curve, arbitrary refractive phase function. Middle curve represents the limiter function which limits the phase between $-a/2$ and $a/2$. Lower curve is the resulting diffractive phase function. (From Ref. 7.)

Further, from Figure 4.4 we see that $\phi_d = a\phi$ within the periodic interval, so that we can integrate Eq. (4.18) to obtain the expression

$$A_n = \text{sinc}\ \{\pi(a - n)\} \tag{4.20}$$

It should be clear that when $a = 1$, A_1 is equal to unity, and all other A_n's are zero. That is to say, we have found a phase function $\phi_d(r)$ that produces the same transmittance function as the corresponding refractive element. Whatever phase front is produced by the refractive element characterized by $\phi(r)$ will be exactly produced by the corresponding diffractive element characterized by $\phi_d(r)$. Putting it another way, we know that the Fourier transform of $\exp\ \{-2\pi i t(r)\}$ will produce a given far-field intensity pattern. Since we have shown that the same transmittance function is produced by a given diffracting structure, then the same far-field pattern will ensue.

4.1.5 Diffractive Lenses

For a spherical lens, the axial dependence of the phase follows the lens thickness; the thickness is given by the equation

$$h_r(r) = \left(\frac{1}{2R_c}\right)(r_0^2 - r^2) \tag{4.21}$$

Here, r_0 is the lens radius, and R_c is the radius of curvature. One can equally express this as referenced to zero at the center.

$$h(r) = \frac{-r^2}{2R_c} \tag{4.22}$$

The phase, expressed as a transmittance function, is written as follows:

$$\phi(r) = \frac{-2\pi r^2}{2R_c\lambda} \tag{4.23}$$

We have shown in the previous section that this also represents the required phase function for the diffractive lens if we limit the phase difference in each interval, as shown in Figure 4.5. We will go into considerable detail on how one produces such structures in the following sections.

Corrected Lenses

It is worthwhile to point out at this stage, before we get into describing the fabrication techniques, another potential advantage of the diffractive approach. This is

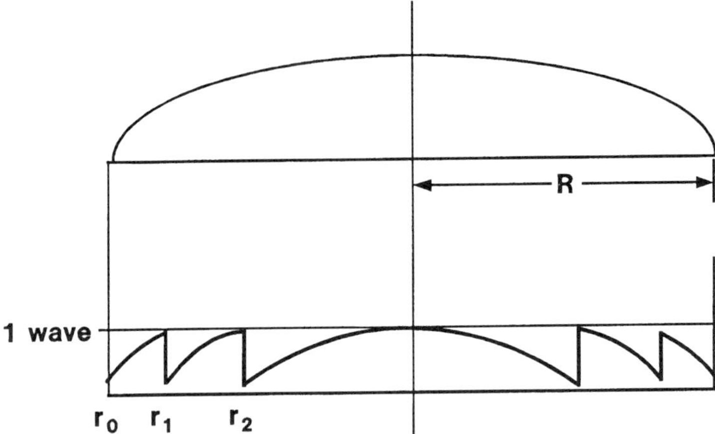

Figure 4.5 Upper curve, phase function of a refractive lens. Lower curve, corresponding phase function of equivalent diffractive lens.

that one can make corrected lenses with no additional difficulty of fabrication [7]. In the above we talked about making a spherical lens, that is, writing the appropriate phase term that would make a diffractive element produce the same phase front as that of the corresponding spherical lens. We pointed out in Chapter 2 the various aberrations that accompany the performance of a spherical lens. If one considers, in a simple example, the surface that would produce a perfect spherical wavefront,

$$\frac{-2\pi}{\lambda\sqrt{r^2 - R^2}} \tag{4.24}$$

and one expands Eq. (4.24) in a power series:

$$\left(\frac{\pi}{\lambda_0}\right)\left[\frac{-r^2}{R} + \frac{r^4}{4R^3} + \cdots\right] \tag{4.25}$$

One can see the spherical approximation (r^2 term) and the higher order terms that would amount to aberrations terms. In this case we specifically show a third-order spherical aberration term. However in principle, one can incorporate this correction into the diffractive element design. For example, the grating period would now come from solving Eq. (4.25) with as many higher order terms as one desires, or Eq. (4.24) directly. We shall see that this can be incorporated into the fabrication without difficulty.

4.2 METHODS OF FABRICATION

There are essentially two approaches to the fabrication of diffractive lenses. The first involves methods of making Fresnel-like structures, that is, trying to reproduce the refractive surface feature as a periodic grating, while the second is a relatively new approach spawned from the IC microfabrication technology, which is called binary optics. The word stems from the use of discrete levels to approximate the desired surface as shown in Figure 4.6. There are some trade-offs, for example, the efficiency vs. the ease and/or cost which we will try to point out as we cover each method.

4.2.1 Micro-Fresnel Lenses

The methods for making micro-Fresnel lenses and lens arrays are all based on being able to provide a gray scale of exposure. This coupled with a resist that has a linear response over a reasonable exposure range forms the basis for producing a prescribed surface curve. There are essentially two ways this can be done, which we cover below.

Direct Write

In this method a beam, laser or *e*-beam, writes directly onto the resist. There is no mask used. The resist-coated substrate is scanned, usually on air bearing stages, and exposed using an intensity modulated beam. The developed resist is then subjected to an etching, usually RIE [8], or is used to form a master for

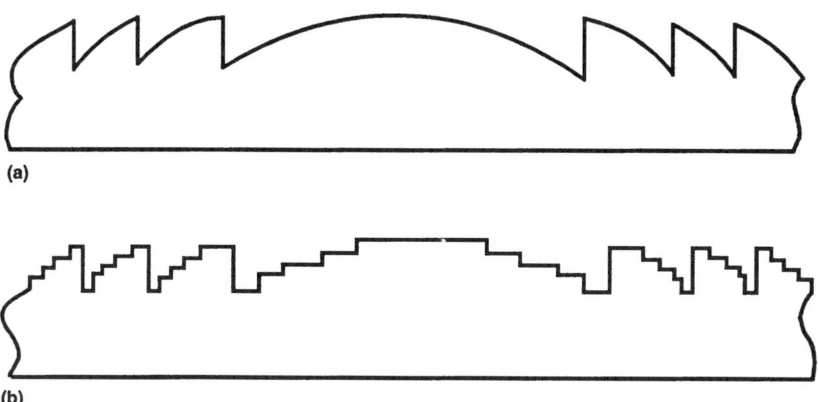

Figure 4.6 Upper curve, conventional diffractive lens with continuously varying phase. Lower curve, the binary approximation where the continuous phase is approximated in a stepwise fashion.

replication [9,10]. A focused HeCd laser is a typical source used as the exposure beam. A schematic of the process is shown in Figure 4.7. The resolution or pixel size is reported as 0.5×0.5 µm [8].

This method is an attractive way of making a master from which one can replicate the desired lens array. For those materials that would be difficult to mold or emboss, each array would have to be individually produced. This will be true for any process where direct writing is used.

Gray Scale Mask

In this case the desired graded exposure function is built into the exposure mask itself. As an example, the radial transmittance on the mask would look like that shown in Figure 4.8a. After exposure to this mask, the pattern in the developed positive resist would be as shown in Figure 4.8b. One way to do this is to utilize the gray scale capability of current *e*-beam writing systems. For one example consider the use of a process which patterns clear dots of varying size on a chrome mask. Because the dot pitch is sufficiently small, it cannot be resolved by the projection photolithographic system. Thus the amount of light that reaches the resist-coated substrate is a slowly varying intensity pattern proportional to the average dot size in any portion of the mask. An example of such a mask pattern is shown in Figure 4.9 where the resist used was EBR-9. Recall that this method is intended for projection systems. The optical pattern of the regions of Figure 4.9 is thus $5\times$ smaller than shown when exposed on the resist. In this pattern the dot pitch on the wafer was the order 0.3 µm and hence not resolvable. At the highest optical magnification, one can actually see the dots, and the eight regions. This is not the only way one can manipulate the *e*-beam exposure to produce the desired result. One could equally use the same dot size and let the exposure vary to enlarge or reduce the dot size, or any method which produces

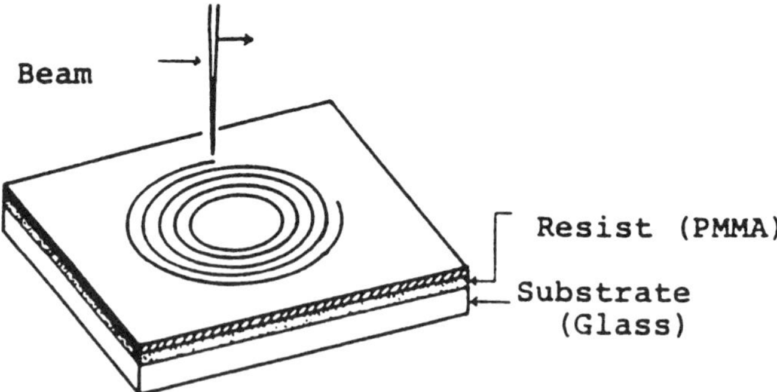

Figure 4.7 Schematic representation of direct laser beam writing onto resist to produce continuous relief pattern.

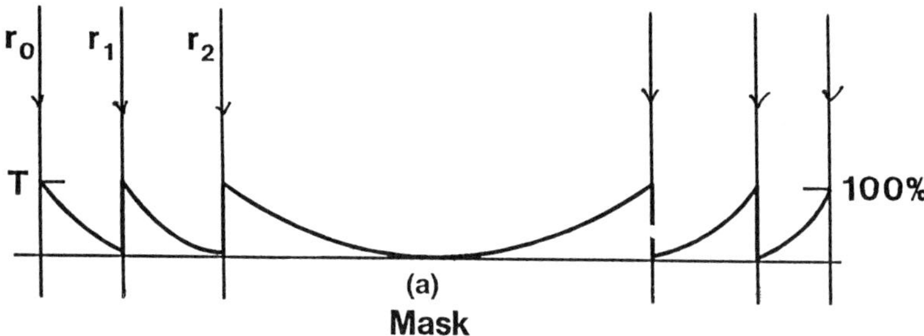

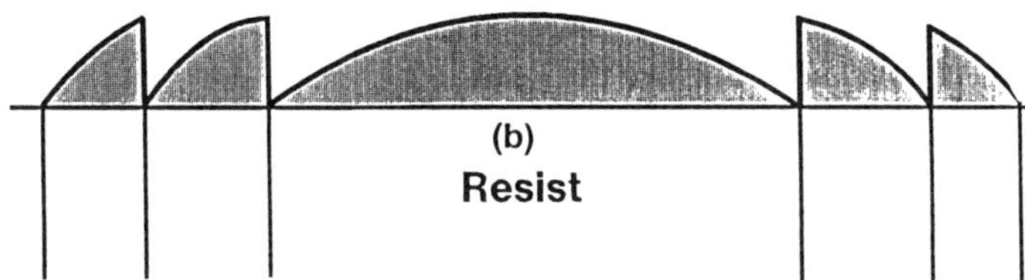

Figure 4.8 Schematic representation of an exposure through a variable transmittance mask. The upper drawing represents the mask where the transmittance is plotted in the vertical direction as a function of the horizontal position. The lower figure represents how a positive resist would develop in response to such an exposure.

a quasi-continuous exposure. These methods tend to be iterative in nature because one cannot measure the actual transmission pattern on the mask. One can develop, however, over a number of trials, methods that essentially calibrate the process.

Assuming one has a positive resist that has a linear response over a reasonable exposure level, one then has to fabricate the transmittance function of the mask to correspond to the desired surface figure. A positive resist is one that is subsequently removed in development where exposed. For a spherical surface the transmittance function should follow the lens thickness

$$h(r) = R_c \left\{ \sqrt{1 - \left(\frac{r}{R_c}\right)^2} - \sqrt{1 - \left(\frac{r_0}{R_c}\right)^2} \right\} \tag{4.26}$$

Figure 4.9 Photomicrograph of a section of a variable transmittance mask made by controlling the areal density in the radial direction of an *e*-beam writing pattern, 800×.

where R_c is the radius of curvature of the surface and r_0 is the radius of the lens. For most cases, and to simplify the example calculation, we can assume that $R_c \gg r_0$, so we can use the familiar approximation to Eq. (4.26):

$$h(r) = \frac{r_0^2 - r^2}{2R_c} \tag{4.27}$$

One wants more light where the lens is the thinnest, so the mask transmittance function should follow the following expression:

$$T_m(r) = K\{1 - (r_0^2 - r_2)\} \tag{4.28}$$

The value of K scales the actual transmittance, so we can take it to be equal to unity for this example. From Eq. (4.22) we calculate the radii of the rings for the desired radius of curvature. We will number the rings from the radius r_0 inward. As we start in from the lens radius, we want the transmittance to vary in the outermost ring as

$$T = ar^2 + b \tag{4.29}$$

and be unity at $r = r_0$ and zero at $r = r_1$. This means the function in the first interval is

$$T(r) = \frac{r^2 - r_1^2}{r_0 - r_1^2} \qquad r_0 < r < r_1 \tag{4.30}$$

In the next ring, one wants the transmittance to continue to follow Eq. (4.27), but swing from unity at r_1 to zero at r_2. In general, the form for the transmittance of the mask in region k is given in the equation below.

$$T(r) = \frac{r^2 - r_k^2}{(r_{k-1}^2 - r_k^2)} \qquad r_{k-1} < r < r_k \tag{4.31}$$

An atomic force microscope picture of the result of the exposure of a resist to a mask fabricated in the way just described is shown in Figure 4.10 [13]. The lower curve shows a quantitative measure of the profile.

Another interesting and unique method of making a gray scale mask involves the utilization of a special glass, termed HEBS [14]. This glass develops optical absorption in a thin layer near the surface upon exposure to an *e*-beam. A schematic of the process is shown in Figure 4.11. This patented material is used by Canyon Materials to produce masks. The reported induced optical density at 436 nm as a function of the indicated dose rates is shown in Figure 4.12a. An optical micrograph ($100\times$) of an *e*-beam written test pattern is shown in Figure 4.13. Canyon further reports the required exposure level that was used to write on the glass in order to expose two representative photoresists as reproduced here in Figure 4.12b. The beam-writing method is ideally suited for gray scale patterning since most *e*-beam writing packages allow the control of the current over different areas. The fact that the sensitivity of this glass is $10-20\times$ less than typical *e*-beam resists is not too important since one only has to write the photomask once, whereas in the resist case one writes for each copy.

4.2.2 Binary Optics

This approach of digitizing the desired surface figure, or binary optics as it has been called, has produced the biggest breakthrough in the fabrication of diffractive element devices, and in particular lens arrays [15–17]; see Figure 4.6b. The reason for this is because it follows the same method that is used in the multilevel fabrication of integrated circuits. There is a price to be paid, however, in that the diffraction efficiency depends on the number of levels used to approximate the continuous surface.

To the first order, one can estimate the diffraction efficiency of an N-level system by realizing that the discrete level approximation is really two structures superimposed. We will use a linear grating diagram, as shown in Figure 4.14 for simplicity's sake. If we had a blazed grating, the efficiency was shown above to be given by the expression

$$\left| \mathrm{sinc}\left\{ \pi\left(\alpha - \frac{N}{T} \right) \right\} \right|^2 \tag{4.32}$$

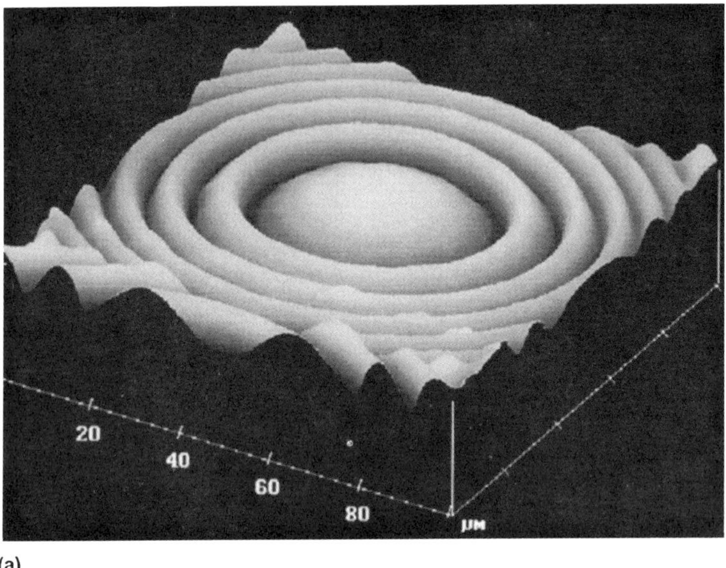

(a)

(b)

Figure 4.10 AFM image of the resist pattern formed after exposure to variable transmittance mask. The lower figure shows the trace of the pattern.

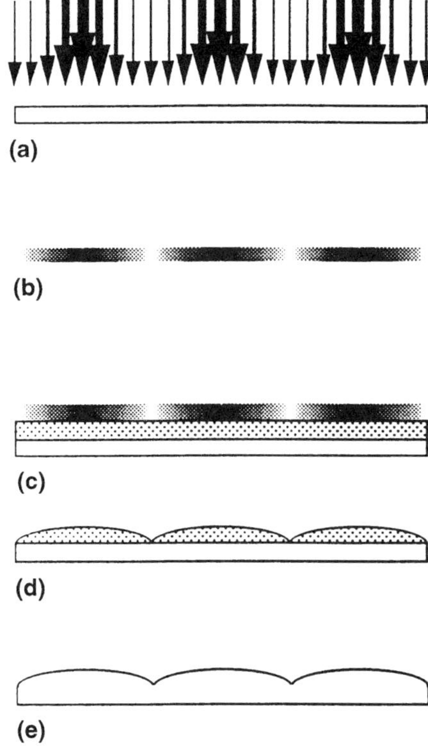

Figure 4.11 Canyon Materials, Inc. representation of the exposure process using their high-energy *e*-beam sensitive glass. The glass photodarkens when exposed to *e*-beam radiation as shown.

where $\alpha = (n - 1)d/\lambda T$. Superimposed, we have another grating of pitch T/N and a depth d/N. The diffraction efficiency of such a grating would be

$$\left| \text{sinc} \left\{ \frac{\pi a}{N} \right\} \right|^2 \tag{4.33}$$

The product of the two expressions represents the overall efficiency of the N-order binary approximation. For the lens case we can apply the same idea and assume that the efficiency of the Fresnel-like structure is 100%. Thus the efficiency of the N-fold binary approximation is simply

$$\text{Eff}(N) = \left| \text{sinc} \left(\frac{\pi}{N} \right) \right|^2 \tag{4.34}$$

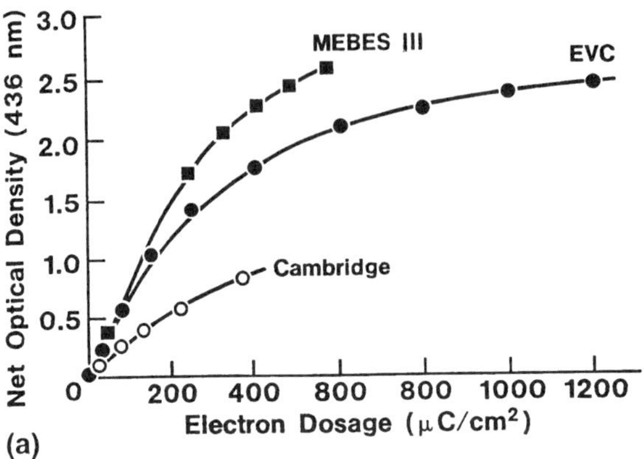

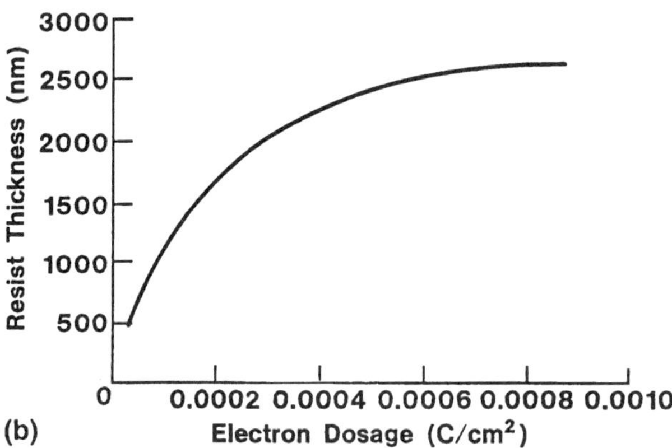

Figure 4.12 Information supplied by Canyon Materials, Inc. as to the optical density produced by given electron beam dosages, for three different machines. Also, the remaining resist thickness after the listed *e*-beam exposure is given in the lower figure. In this case the resist was Shiply S1650. (From Ref. 14.)

Here, we have assumed that the condition that $d = \lambda/(n - 1)N$ has been met.

How the efficiency expression relates to fabrication will become obvious in the next section.

Photolithography

For those readers unfamiliar with the photolithographic technique, we will try to incorporate into the following description some of the details. The general idea is to coat the substrate to be processed with a photosensitive organic film called

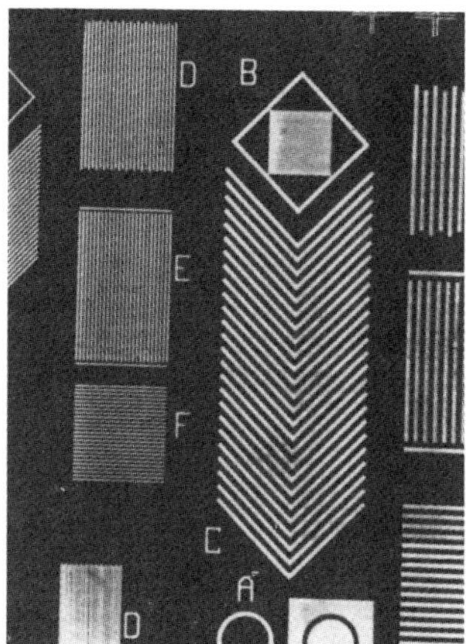

Figure 4.13   Photomicrograph of a portion of the *e*-beam produced pattern on the HEBS™ glass.

a photoresist, or resist for short. Typical thickness is in the tenths of micrometers range, although thicker layers are also possible when needed. The resist is usually spun on, that is, the resist is dispensed onto a spinning substrate to distribute the resist over the substrate. The resist can also be vacuum deposited. The resist when exposed to light either becomes soluble to a development solvent (positive resist), or insoluble (negative resist). Commercial resists exhibit a wide range of properties aimed at specific applications. For example, there are resists that are

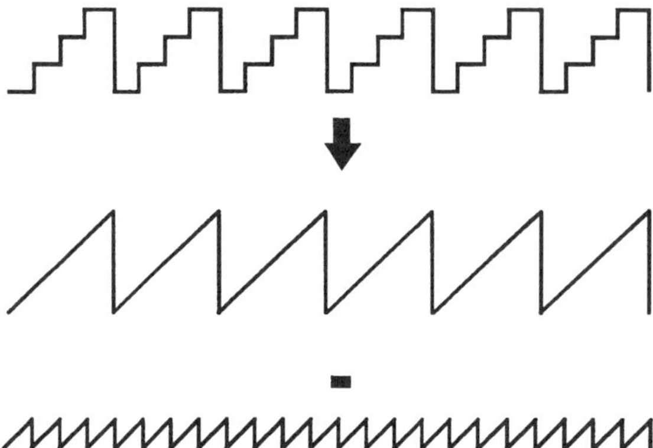

Figure 4.14   Diffraction grating representation of the behavior of a binary optic approximation to a blazed grating. There is the contribution of the higher-order grating coming from the period produced by the steps. (From Ref. 7.)

very sensitive: those with linear development behavior, those that are sensitive to shorter wavelengths, etc. The sensitivities are measured in terms of energy per unit area, typically in the range of mJ/cm^2 on up [18].

As an example we will use Figure 4.15 where the intention is to make the binary approximation to the Fresnel structure shown in Figure 4.15b. In the first case shown in Figure 4.15c, we will make what is called a two-level structure. A mask is made to conform to the desired geometric structure. An example is sketched in Figure 4.16. The pattern is usually made in a chrome film to provide the optically opaque regions. In most cases this pattern is made through the use of *e*-beam writing because of the required resolution. The *e*-beam writes onto an *e*-beam resist which responds in the same manner as described above. The exposed resist is developed and the Cr is removed where left unprotected by the resist. The mask is then used to expose the resist-coated substrate as shown in Figure 4.17a. The resist provides a protection for the subsequent etching step. The method of choice is called reactive ion etching, more commonly referred to as RIE [19]. This is a plasma-based process where the fluoride ion is produced from a precursor like CF_4 which then reacts with substrate to produce a volatile fluoride compound. The etch rates vary for different materials. As a typical number for fused silica, it is of the order of 15 min/μm. The required depth of $d = \lambda/2(n-1)$ is controlled by the etching time. The maximum efficiency that could be obtained from this two-level grating is calculated from Eq. (4.34) as 41%.

If one uses two masks, one can achieve a four-level approximation as shown Figure 4.15c. The sequence of exposures is shown in Figure 4.17b. The first mask exposure and etching produces the coarse structure. The etch depth here is $d/4$.

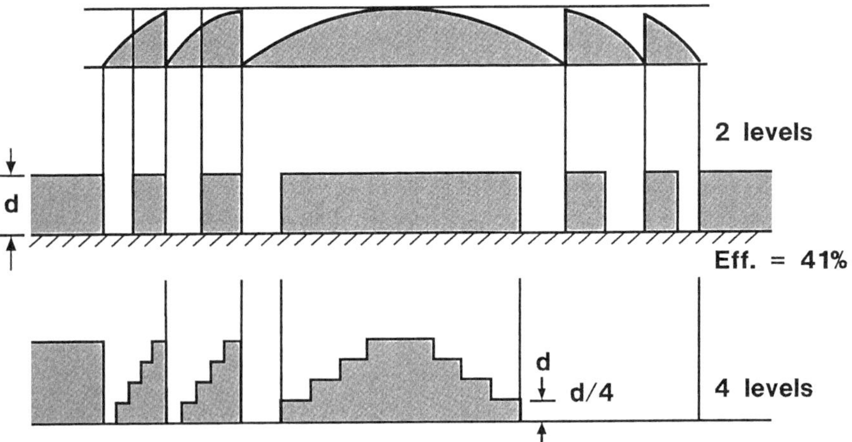

Figure 4.15 Schematic representation of a two-level and four-level approximation to a Fresnel structure.

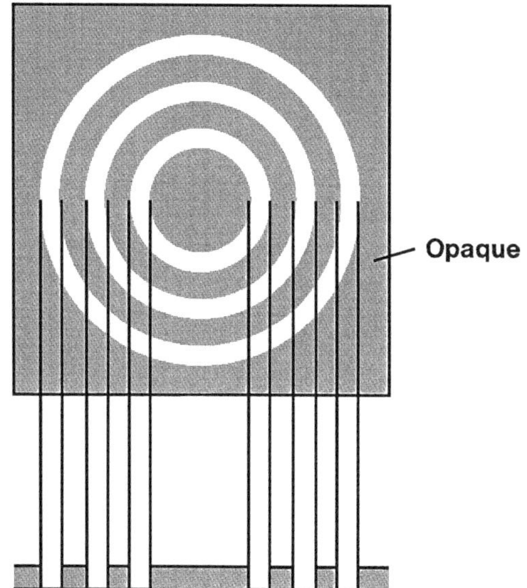

Figure 4.16 Example of a mask required for a two-level exposure.

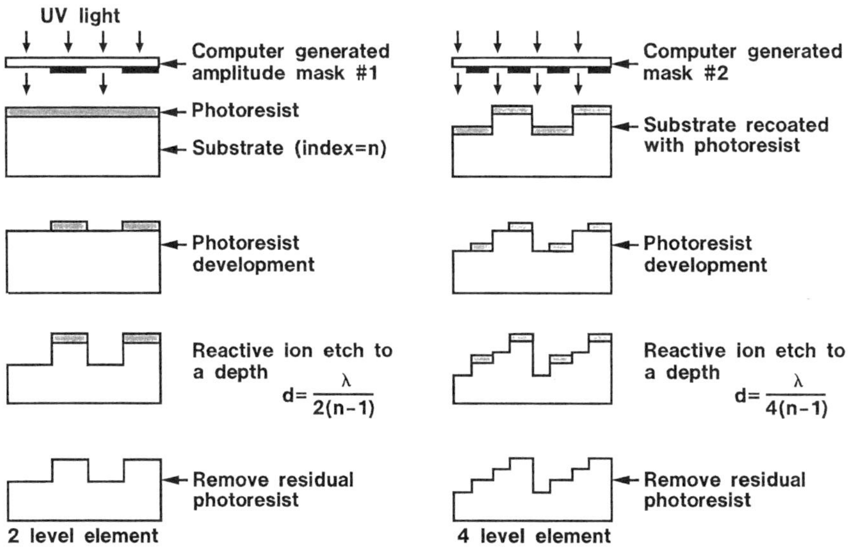

Figure 4.17 Schematic representation of what the exposure steps would be to produce a four-level binary approximation utilizing two masks as compared to a single exposure two-level.

This structure is coated with resist and exposed to the second mask which represents a higher resolution. Essentially every ring in the mask is split into two. The next $d/4$ step is etched as shown. The result is a four-step approximation with a maximum efficiency of 81%.

One can go to three masks, each one doubling the resolution of the preceding one which would produce eight levels with an efficiency of 95%. It is easy to see that there is a simple relationship between the number of masks used and the number of levels $L = 2^N$. One can begin to appreciate how the existing IC fabrication facility, which uses many separate masking steps to create the desired structures, lends itself to the multistep binary-optic approach. An SEM photomicrograph of a hexagonally packed array of eight-level lenses is shown in Figure 4.18.

In a related approach [20], one can produce a two-level lens by direct e-beam

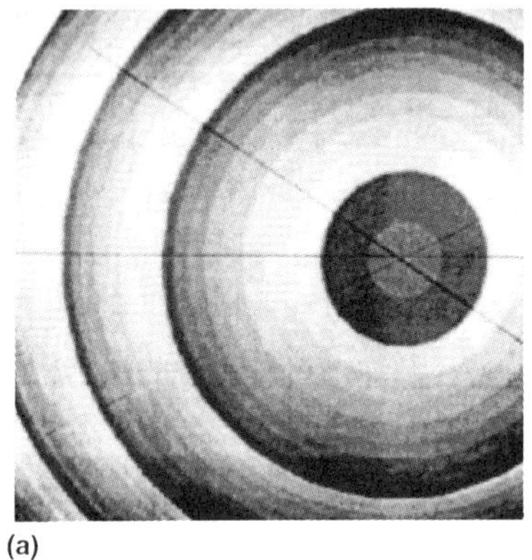

(a)

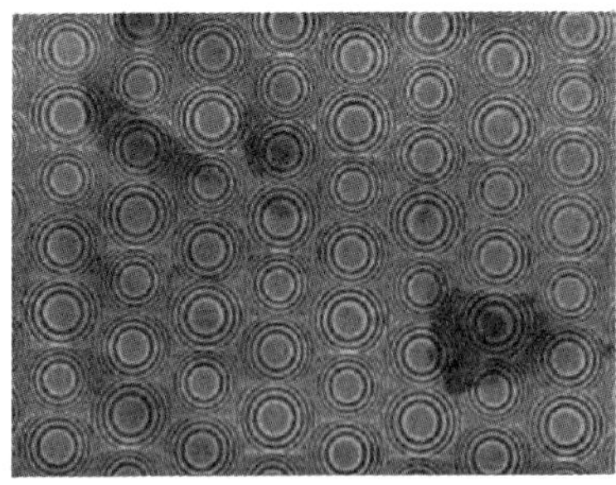

(b)

Figure 4.18 Photomicrograph of an eight-level exposure pattern.

writing onto an *e*-beam resist. In this reported case the 1.0-mm diameter lenses reside in the resist. The minimum feature size was 6.4 μm. The measured efficiency was 30%.

There have been a large number of publications reporting on lens arrays fabricated by the binary-optic approach [21]. Many have been fabricated in silicon for integrated optoelectronic devices. We will try to cover these in Chapter 6 which deals with applications.

Molding

Molding or replication into plastic from a master has been demonstrated for multilevel structures. A recent publication [3] described an eight-level 2 ×2 mm lenslet with a 6.5-mm focal length replicated onto a PMMA-based film pressed between the silica master and an optical flat. The master consisted of a diffractive lenslet array pattern which was fabricated with the opposite phase onto a 3-mm thick quartz glass substrate. The minimum feature size was 2 μm, and the step depth was the order of 1 μm. See Figure 4.19. Although this lens diameter was relatively large, it is the feature size resolution that is important for microoptic applications. In the next section we will estimate the required resolution for given applications. There was a brief discussion given of the thermal characteristics of PMMA, such as thermal expansion and shrinkage, and how these properties might affect the lens performance. The measured diffraction efficiency of the molded plastic lens was 91%, compared to the theoretical value for an eight-level structure of 95%. One major issue that was mentioned was the optical quality of the back side, and the wedge that is created in the relatively thick layer of PMMA. An excellent review of the status of replication in polymer-based materials is given in Ref. 22.

The molding of diffractive elements into inorganic glasses represents a more challenging problem. In addition to the mold life issues discussed in molded aspherics in Chapter 2, one has the issues associated with small submicrometer features. All this notwithstanding, there have been reports of molding diffractive elements such as gratings and holograms in glass, which will be covered in Chapter 7.

Laser Ablation

Another technique is to precisely remove the material by laser ablation [23,24]. In this case a focused excimer laser (both 248 and 193 nm have been used) is directed to a mask and then focused onto the substrate as shown in Figure 4.20. The reported smallest feature that can be written is <1 μm, and to a depth to 0.1 μm. The position accuracy is given as 3 μm. However, in a 10× system this is correspondingly reduced. The high-speed stage allows movement of centimeters in tenths of seconds. There is not much discussion of the materials that have been used although polyimide films are mentioned as an ideal material. The method, however, is not limited to this material.

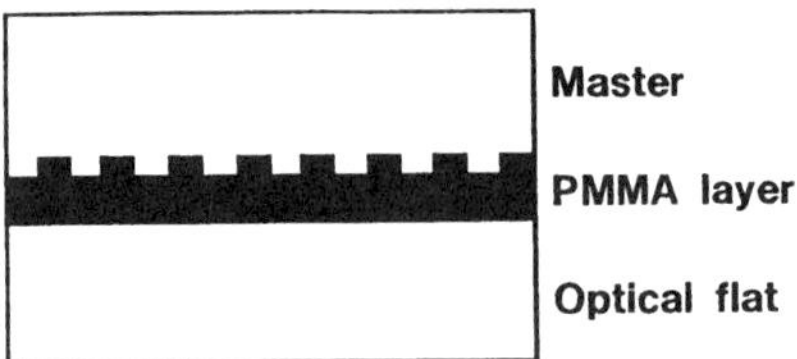

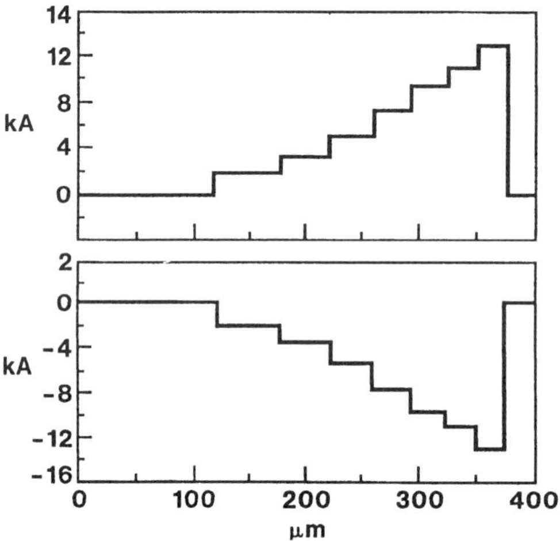

Figure 4.19 Representation of the molding of a binary optic lens. The upper figure schematically depicts the molding, whereas the bottom figures show the relationship of the mold pattern to the material to which it is pressed into. (From Ref. 22.)

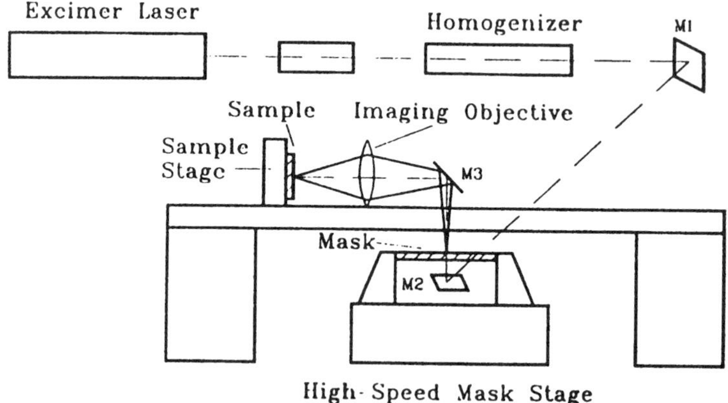

Figure 4.20 Drawing of laser ablation method of forming diffractive elements. (From Ref. 24.)

4.3 ANALYSIS

The analysis of the performance of microlenses made by any of the diffraction grating processes in terms of what would be equivalent to refractive lenses is not straightforward. Whereas in the case of refractive lenses the distortion produced in the transmitted phase front derives from the surface figure, it is no less critical than other geometric issues involved in the diffractive lenses. For the Fresnel type, the surface profile is still an issue, and the consequence of its deviation from the desired shape can be considered in the same way as one would for a refractive lens. Perhaps a simpler way to look at this effect is to go back to Eq. (4.18). Here we expressed the diffractive phase function ϕ_d in terms of the desired refractive phase function ϕ. To the extent that the condition ϕ_d is not proportional to ϕ, more than one term in Eq. (4.18) will have nonzero coefficients, as seen from Eq. (4.19). The practical result of this is that light will be diffracted to higher orders, reducing the efficiency in the desired first order. These deviations in the surface profile become serious when they correspond to magnitudes comparable to those of refractive lenses.

For the binary-optic lenses one must consider the effect on the performance, both resolution and efficiency, of small errors in the depth and shape of each step. This is true for the Fresnel-type lens as well, although to a lesser extent since it is a two-level structure. For examples, see Figure 4.21. This type of analysis is difficult to portray in any simple way, and is usually done by some sort of computer modeling code. This type of analysis is clearly outside the scope of this discussion [25].

What we will consider here is the issue of chromatic aberration which is severe enough to limit the use of diffractive lenses to relatively narrow band widths. Another consideration of consequence will be that of the limitation that the present fabrication techniques impose on the lens numerical aperture.

4.3.1 Chromatic Aberration

The fact that the diffraction grating spacing is derived from a single wavelength presents a problem for a diffractive lens in terms of what would ordinarily be called chromatic aberration. The focal length of the diffractive lens was given above in Eq. (4.23), where it is related to the grating spacings. One can easily see that the focal length times the product is invariant with respect to the grating spacing. Thus the focal length R_{c0} at wavelength λ_0, which corresponds to the desired wavelength, would change at any other wavelength λ by the expression

$$R_c = R_{c0}\left(\frac{\lambda_0}{\lambda}\right) \tag{4.35}$$

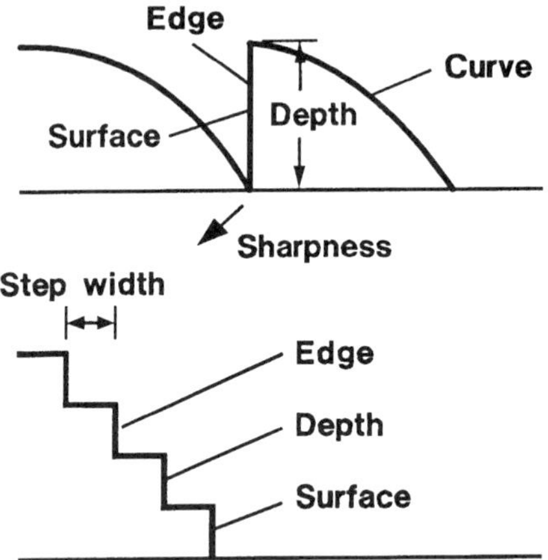

Figure 4.21 Schematic depiction of common geometric imperfections on a diffractive element that could impact the performance of a Fresnel structure, upper, and a binary optic, lower.

This represents considerably more change than a corresponding refractive element where the chromatic aberration would be solely determined by the wavelength dispersion of the refractive index. Consider the change that would occur from 400 to 700 nm. For a refractive element, a 10% change would be considered large, whereas for a diffractive element, the focal length would nearly halve. This problem limits the application of microoptic diffractive lenses to narrow wavelength sources such as lasers, although this does not seriously diminish the breadth of applications.

As far as the diffraction efficiency is concerned, the binary-optic lens is strongly dependent on wavelength whereas the Fresnel lens is not. For the binary-optic lens one can use the grating analog, the efficiency expressed in Eq. (4.32), at a wavelength other than λ_0 $[d = \lambda_0/(n - 1)]$, can be written as

$$\text{Eff} = \left| \text{sinc}\left[\pi\left\{ \frac{\lambda_0}{\lambda} - 1 \right\} \right] \right|^2 \tag{4.36}$$

which we show graphically in Figure 4.22. The efficiency of a Fresnel lens will have essentially the same dependence on wavelength as that of a refractive ele-

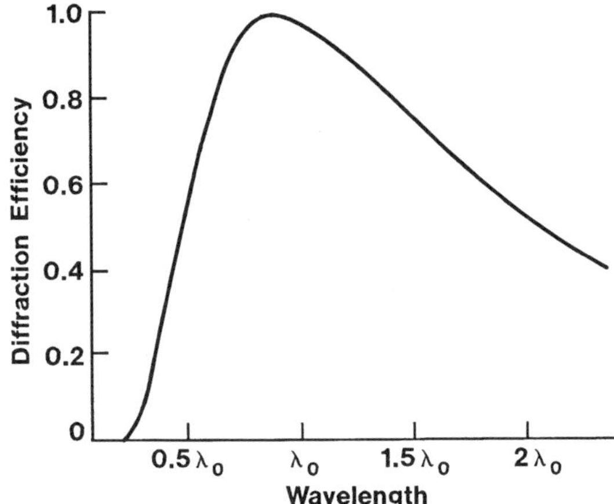

Figure 4.22 Computed diffraction efficiency of a binary-optic lens; see Eq. (4.35).

ment, so although its focal length is subject to the monochromatic condition, the efficiency is not.

4.3.2 Fabrication Limitations on Lens Design

In this section, we will try to relate the limitations imposed by the photolithographic fabrication techniques to the range of lens performance that can be achieved. What this means is that the photolithographic method has built-in resolution limitations, currently the order of 0.5 μm depending somewhat on how sophisticated the approach that is used. These resolution issues circumscribe the nature of lens one can make.

There are essentially three lens parameters of consequence, any two of which are independent: the lens diameter, the focal length, and the numerical aperture. Consider the numerical aperture, and what value might be achieved. For the moment, since we are concerned with microoptic applications, let us take the lens radius to be in the order of 100 μm. One must consider the resolution issue slightly differently for the cases of lens fabrication using a mask from that of the case where the direct-write e-beam or laser is used. This is because in the mask method a 5-1 or 10-1 reduction is commonly used. This means the features are 5 to 10 times larger on the mask than they will be in the resist plane. The ultimate resolution still is determined in the wafer plane, but there are aspects unique to each method.

Resolution and Numerical Aperture

The simplest way to consider this relationship is to think of how finely one can subdivide a given ring portion of the lens pattern to make the desired feature on the lens. Consider the graded transmittance mask method mentioned above and depicted in Figure 4.8, and actually shown in Figure 4.9. One has effectively taken the given grating regions Δr_i and divided them up into p subregions. Essentially the same approach is taken to produce the binary-optic structure; see Figure 4.17. Each region corresponding to a one-wavelength phase shift is to be broken up into subregions corresponding to smaller phase shifts. The system resolution will ultimately determine the smallest number this ratio $\Delta r/p$ can assume.

The values of Δr_i, and in particular in the outermost ring, will be determined by the desired extent of the lens, which is the diameter, and the desired focal length, or the numerical aperture. To take an example, consider the case of a 50-μm-radius lens. We show in Figure 4.23 the width of the outermost ring as a function of the numerical aperture, for each of two wavelengths, 550 and 1100 nm. One should now be able to see just how the lens design is limited by how far one can subdivide the rings considering the 0.5 μm resolution limit provided by the photolith method.

Let us consider the binary optic approach first. If one assumes that >90% efficiency is necessary for most applications then an eight-level structure is required. This means that the minimum width of the ring element must be larger than 4 μm. From Figure 4.23, this would put an upper limit on the NA of a lens

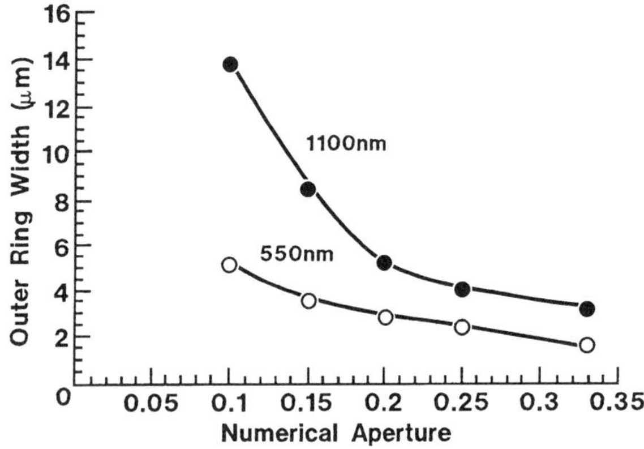

Figure 4.23 Outer ring width in micrometers vs. the numerical aperture for two wavelengths. The lens radius is 50 μm. The intention is to show the dimensional limitation that is imposed at high NA and short wavelength.

intended for visible wavelength use to about 0.15, and that for near-IR use at about 0.25. For many applications these correspond to slow lenses.

The direct-beam writing techniques suffer similar overall resolution limitation. The direct *e*-beam write could produce somewhat higher resolution, although this is the least practical method.

REFERENCES

1. W. B. Veldkamp and T. J. McHugh, Binary optics, *Scientific American*, pp. 92–97, May (1992).
2. M. R. Feldman, Diffractive optics move into the commercial arena, *Laser Focus World* (1994).
3. J. Jahns, System Design for Planar Optics, Optical Society of America, Topical Mtg. *Diffractive Optics: Design Fabrication, and Applications*, Series vol. 9, April 13–15, New Orleans, LA (1992).
4. M. Born and E. Wolf, *Principle of Optics*, Pergamon Press, London, 1956.
5. C. S. Williams and O. A. Becklund, *Optics*, Wiley-Interscience, New York, 1972.
6. I. N. Sneddon, *Fourier Transforms*, McGraw-Hill, New York, 1951.
7. G. J. Swanson, Binary optics technology: The theory and design of multilevel diffractive optical elements, Technical Report 854, MIT Lincoln Laboratory (1989).
8. RPC (Rochester Photonics Corp.), Rochester, NY.
9. R. E. Kunz and M. Rossi, Phase-matched Fresnel elements, *Opt. Comm. 97*, 6–10 (1993).
10. M. T. Gale, M. Rossi, and H. Schtz, Fabrication of continuous-relief microoptical elements by direct laser writing in photoresist, *Proc. SPIE*, vol. 2045 (1994).
11. W. Daschner, R. Stein, P. Long, C. Wu, and S. H. Lee, One-step lithography for mass production of multilevel diffractive optical elements using HEBS gray level mask, *SPIE*, vol. 2689, 153 (1994).
12. C. Wu, Method of making high energy beam sensitive glasses, U.S. Patent 5,078,771, 1992.
13. N. F. Borrelli, Corning Inc., unpublished data.
14. Canyon Materials Inc., Hebs-glass photomask blanks, CMI product information 96-18, San Diego, CA.
15. SPIE 1991 International Symposium, Miniature and micro-optics: Fabrication and system applications, Tech. Conf 1544, San Diego, CA, July 21–26, 1991.
16. SPIE 1992 International Symposium, Miniature and Micro-optics: Fabrication and system applications II, Tech Conf 1751, San Diego, CA, July 19–24, 1992.
17. Diffractive optics: Design, fabrication, and applications, 1994 Technical Digest, Series vol. 11, Opt. Soc. Am., June 6–9, Rochester, NY, 1994.
18. S. M. Sze, *VLSI Technology*, Chap. 4, McGraw Hill, New York, 1983.
19. S. M. Sze, *VLSI Technology*, Chap. 5, McGraw Hill, New York, 1993.
20. T. Fujita, H. Nishihara, and J. Koyama, Fabrication of micro lenses using electron beam lithography, *Opt. Lett. 6*(12), 613 (1981).

21. Diffractive and miniatrized optics, S. H. Lee, ed., SPIE Optical Engineering Press, CR49, Bellingham, WA, 1993, and references contained.
22. See Ref. 21, F. P. Shvartsman, Replication of diffractive optics, p. 165–186.
23. G. P. Behrmann, and M. T. Duignan, Excimer laser micromachining for rapid fabrication of diffractive optical elements, *Appl. Opt.* 36(20), 4666–4674 (1997).
24. X. Wang, J. R. Leger, and R. H. Rediker, Rapid fabrication of diffractive optical elements by use of image-based excimer laser ablation, *Appl. Opt.* 36(20), 4660–4665 (1997).
25. See Ref. 21, D. W. Ricks, Scattering from diffractive optics, p. 187.

Part II
APPLICATIONS

5

Erect One-to-One Imaging

5.1 INTRODUCTION

The applications of microoptic lenses and arrays can be conveniently classified into three categories. The classification is somewhat arbitrary, but it provides a useful way to compare and contrast the advantages of lenses fabricated by one method over that of another. We could discuss the applications for each of the fabrication methods separately, but in doing so we would lose some of the comparative aspects, as well as have the same device discussed in more than one place. For example, for some of the application categories GRIN lenses dominate the other available methods. This means that some applications may be better suited to a particular fabrication method. Often, such factors as achievable lens diameter or ease of producing a given geometrical pattern or overall cost are as important in the choice as the optical performance. In the way we have chosen to deal with the applications, it will be easier to point out these advantages/ disadvantages as they appear.

In this chapter we deal with the application category which involves erect one-to-one imaging. These are arrays of lenses designed to produce collectively a single image, as shown in Figure 5.1. The upper diagram shows the overlapping fields and the lower two show the ray diagram of erect one-to-one imaging, (b) for the GRIN lens and (c) for the refractive lens. The overlap of all of the individual image fields constitutes a single image. As we will see, they are used primarily for document scanning where the printed information is imaged onto a photosensor of some kind, as shown in Figure 5.2a.

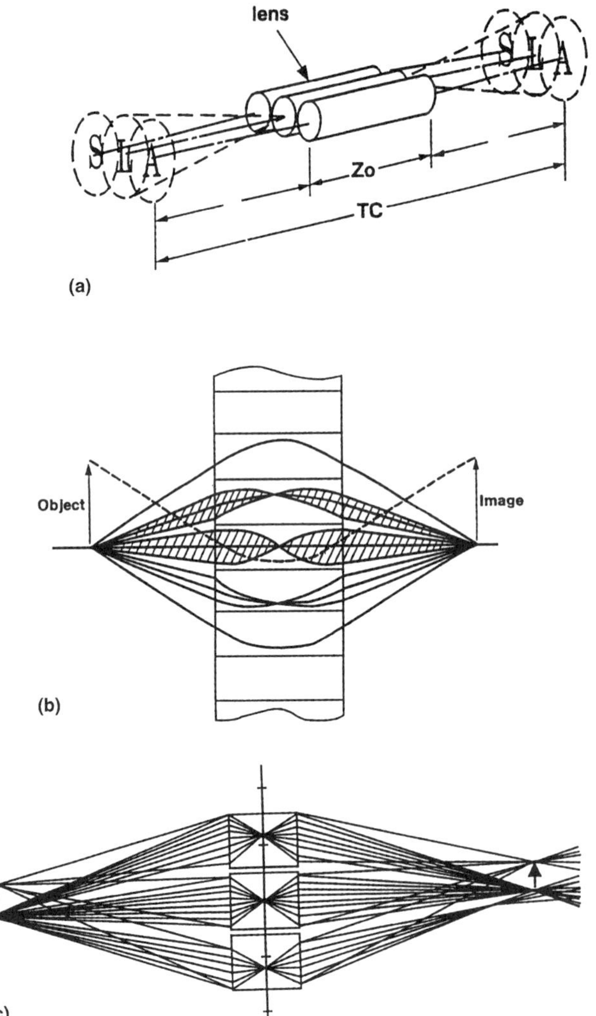

Figure 5.1 (a) Schematic of linear array of lenses operating in the erect one-to-one mode. Note overlapping of fields in the object and image planes. (b) Ray trace showing how erect one-to-one imaging occurs in a GRIN rod lens. (c) Ray trace showing how erect one-to-one imaging occurs in a thick biconvex lens.

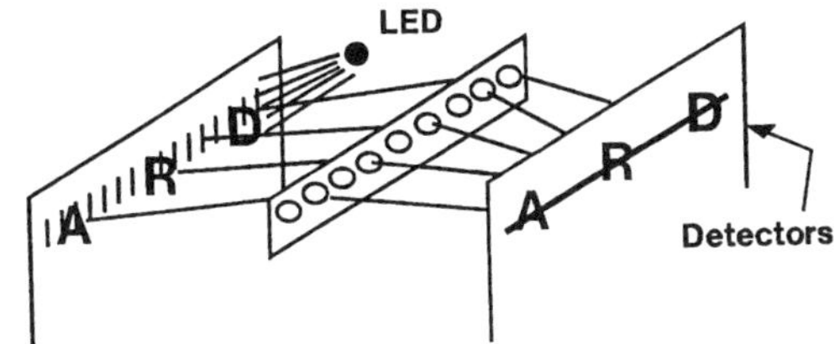

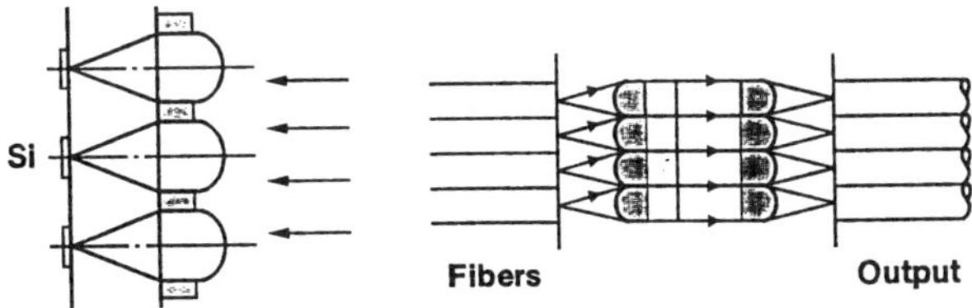

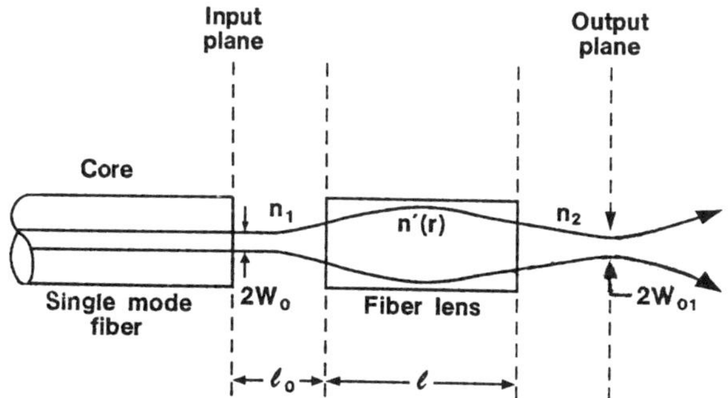

Figure 5.2 Representations of the microlens application categories: (a) One-to-one reading or scanning area; (b) 2D array for coupling light from laser arrays or fiber arrays; (c) single-element lenses used with single-fiber applications.

In Chapter 6 we cover the second broader category where precisely positioned two-dimensional arrays of individual lenses provide functions for individual detectors, light sources, or apertures. These applications essentially provide an optical interface for other microelectronic fabricated array devices, such as those based on CCDs (charge coupled devices), laser diodes, optical fiber arrays as indicated in Figure 5.2b, and optical fiber-based devices as in Figure 5.2c.

5.2 OPTICS OF ONE-TO-ONE ERECT IMAGING

As mentioned above, the one-to-one erect imaging lens array application derives primarily from the need for a spatially compact way to electronically read or scan a conventional 8.5×11 in. print document. From the simplest thin lens standpoint, it is straightforward to produce an erect one-to-one image. One places the object at a distance twice that of the focal length which would form a one-to-one inverted image at a distance $2f$ on the opposite side. By placing another identical lens $2f$ beyond this image, it would form the erect image as shown in Figure 5.3. There are more elegant ways to do this with compound lenses, but the problem is still essentially the same, viz., the total optical path length is long: greater than four times the focal length of a typical imaging lens whose focal lengths are in the range 150–300 mm. This becomes even more of a problem when a faster lens is required. Consider again our simple case with the total conjugate distance of $8f$; for an $f/5$ lens the total distance from the object to the image is 40 times the lens diameter. This distance would be impractical for a compact document reader. What is normally done in most photocopiers is to use a folded path as shown in Figure 5.4a. What the lens bar approach accomplishes is to allow the use of an array of small diameter lenses, which reduces the value of the total conjugate distance for the same f number. The imaging is accomplished through the overlapping of field of each lens in the array, as indicated in Figure 5.1, so that the area covered is comparable to that of a larger single lens.

In the next section we will derive the appropriate expressions for the one-to-one imaging situation for both the GRIN and conventional thick refractive lens cases. Diffractive element lenses are not particularly suited to this application

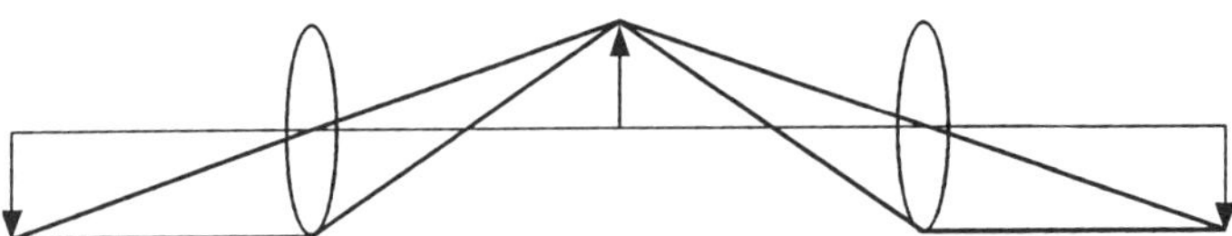

Figure 5.3 Conventional lens approach to forming one-to-one erect image.

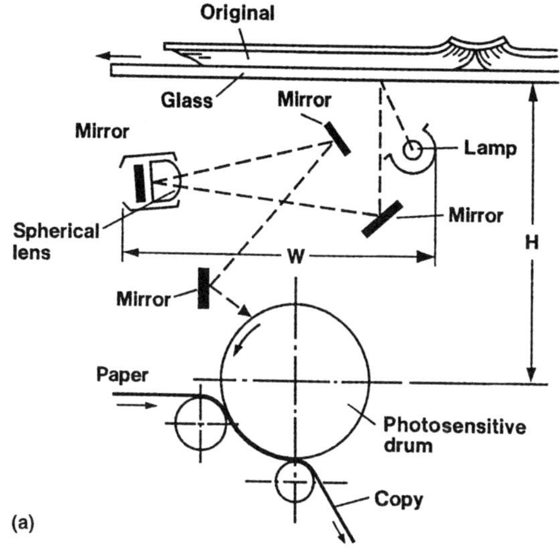

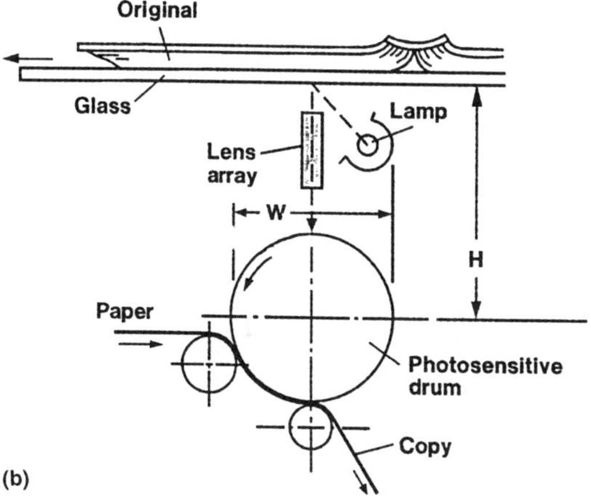

Figure 5.4 (a) Folded path arrangement for lens in photocopier application; (b) one-to-one lens array arrangement. (From Ref. 13.)

primarily because of their poorer imaging capability and speed limitation in the visible wavelength region.

5.2.1 Refracting Lens Array

To derive the expression for the erect one-to-one imaging condition for a conventional thick lens, it is convenient to use the matrix method suggested in Section 2.1.1 along with the diagram shown in Figure 5.5a [1,2]. We will use two rays emanating from the same point on the object a distance t_1 away from the lens: the *a* ray with initial position y_1 and slope of 0, represented by the vector $(y_1, 0)$, and the *b* ray which passes through zero at the lens with a slope of y_1/t, represented by the vector $(0, -y_1/t_1)$ [1]. Using each of these as input rays, one obtains the exit ray vectors $\mathbf{r}_a$ and $\mathbf{r}_b$. One translates rays to the position t_2 behind the lens by applying the translation matrix

$$\mathbf{T} = \begin{pmatrix} 1 & t_2 \\ 0 & 1 \end{pmatrix}$$

(5.1)

The total operation on the two input vectors is then

$$\begin{pmatrix} y_0 \\ m_0 \end{pmatrix} = \mathbf{TM} \begin{pmatrix} y_{in} \\ m_{in} \end{pmatrix}$$

(5.2)

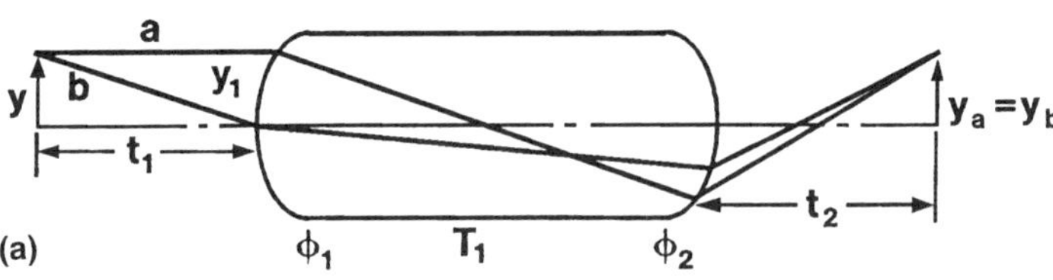

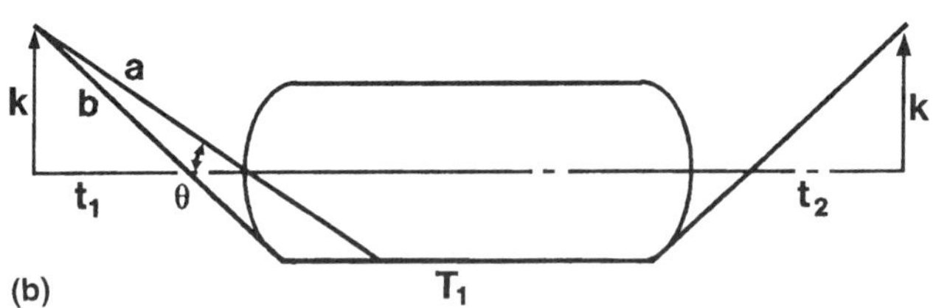

Figure 5.5 Diagrams showing the notation for the ray-trace method: (a) a and b rays defined for use with the martix method; (b) maximum field height defined.

where **M** is defined from Eq. (2.4) and reproduced here for convenience:

$$\mathbf{M} = \begin{pmatrix} 1 - T/nf_1 & T/n \\ t/nf_1f_2 - 1/f_1 - 1/f_2 & 1 - T/nf_2 \end{pmatrix} \tag{5.3}$$

The parameters contained in **M** are the front and back curvatures defined as $1/f_1$ and $1/f_2$ and the lens thickness T. By equating the y value obtained from Eq. (5.2) for the a and b rays, one obtains the expression for the working distance t_1:

$$t_1 = \frac{T/nf_2}{T/nf_1f_2 - 1/f_1 - 1/f_2} \tag{5.4}$$

$$\frac{t_1}{t_2} = \frac{f_1}{f_2} \tag{5.5}$$

For the symmetric case, $f_1 = f_2 = f$, Eq. (5.3) reduces to the simple form

$$t = \frac{Tf}{T - 2nf} \tag{5.6}$$

Another expression that will be used later when the radiometric performance is discussed is the so-called maximum object field height for image-forming rays. At the erect one-to-one distance t, it corresponds to the maximum height above the axis from which light will be captured. A corresponding condition that occurs is when the position of the relay image, the small inverted image formed inside the lens, appears at the lens radius as shown; that is, $y = -R$. This is shown in Figure 5.5b. To calculate the position of the relay image, one uses a portion of the matrix represented in Eq. (2.5). The matrix $\mathbf{TR}_1$ traces the ray after the first refraction, $\mathbf{R}_1$, and then translates the ray to a distance t within the lens. Using the a and b rays as shown, and setting $y_a = y_b = -R$, one can show that

$$k = \frac{2nRt}{T} \tag{5.7}$$

and the corresponding field angle of

$$\tan \Theta = \frac{2nR}{T} \tag{5.8}$$

One can also use two lens arrays stacked together as shown in Figure 5.6. The one-to-one distance formula of (5.6) is still true with the condition that T is twice the thickness of the individual elements labeled T_2 in the figure [1]. The internal lenses act as field lenses and have the effect of increasing the light throughput. A field lens [3] is a lens that is placed at the field stop, which in this

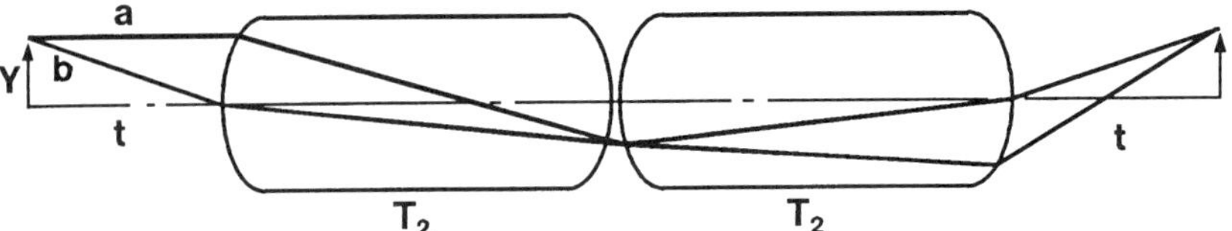

Figure 5.6 Stacked array to form erect one-to-one image. Intermediate lenses act as field lenses.

case is at the position of the relay image. One can see from the ray diagram how the field lenses at the midpoint of the lens redirect the rays that otherwise would be lost to the back lens. It has no other effect on the optical system performance.

5.2.2 GRIN Lens Arrays

We can obtain the one-to-one imaging condition for a GRIN lens by using the paraxial matrix approach outlined in Chapter 3 (Section 3.2.3) (4–6). We want to know what the distance l_0 should be to produce a one-to-one image for a GRIN

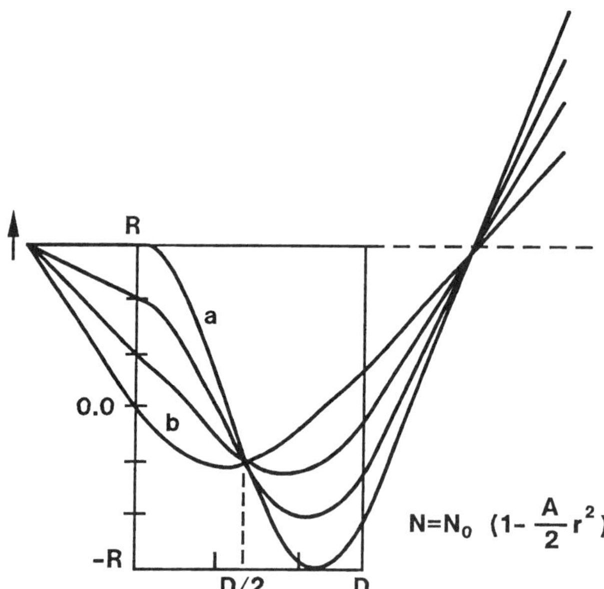

Figure 5.7 Ray trace showing the erect one-to-one imaging of a GRIN lens with parabolic profile. Also shown is the definition of a and b ray for use with the matrix method.

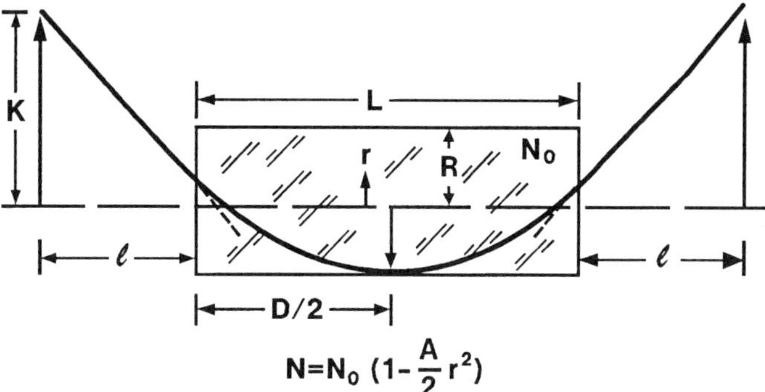

Figure 5.8 Diagram showing the ray that determines the maximum field height for a GRIN lens with a parabolic index profile.

lens of thickness D with a given parabolic index profile characterized by the parameter A in the form of Eq. (3.19). We will use two rays eminating from a point R above the axis at a distance l_0 from the front of the lens. The transformation of the a ray $(R, 0)$, and the b ray $(0, -R/l_0)$ through the lens is obtained through the matrix equation (3.20), reproduced here for convenience:

$$\begin{pmatrix} y' \\ m' \end{pmatrix} = \mathbf{M} \begin{pmatrix} y \\ m \end{pmatrix} \tag{5.9}$$

Here, the primes are the exit values and $\mathbf{M}$ is written in terms of the distance traveled, z, in the graded index medium,

$$\mathbf{M} = \begin{pmatrix} \cos(\sqrt{A}z) & (1/N_0\sqrt{A})\sin(\sqrt{A}x) \\ (-N_0\sqrt{A})\sin(\sqrt{A}z) & \cos(\sqrt{A}z) \end{pmatrix} \tag{5.10}$$

Equating the y values for the two rays yields the equation that the rays will come together at a distance z_0 inside the lens:

$$l_0 = -\frac{1}{N_0 \sqrt{A}} \tan \sqrt{A}z_0 \tag{5.11}$$

Realizing that the action of the lens must be symmetric, one can immediately equate z_0 to $D/2$. Further, one can see that the value of the angle of $\sqrt{A}D/2$ must fall in the second quadrant, that is, $\Pi/2 < (\sqrt{A}D/2) < \Pi$. Similar to the refracting case, the maximum field height occurs when the relay image forms at the lens boundary as shown in Figure 5.8. One can show this from either of the a or b ray equation by setting the position of the relay image equal to $-R$. One obtains the maximum expression

$$k = -R \sec \frac{(A)^{1/2} D}{2} \tag{5.12}$$

The field angle is essentially k/l_0, which is given by the following expression:

$$\tan \theta = \frac{N_0 R(A)^{1/2}}{\sin[(A)^{1/2}D/2]} \tag{5.13}$$

It is worthwhile noting that the quantity $N_0 R(A)^{1/2}$ is the normal definition of the maximum NA $= (A)^{1/2}(2N_0 \, \Delta n)$

5.3 DEVICE APPLICATIONS

One can catergorize the mode of operation of one-to-one erect imaging arrays into two classes, line scan and field scan [7,8]. Schematically, one can represent these two modes as shown in Figure 5.9. The essential difference of these two modes of operation is in the radiometric aspects. In the line scan, as the name implies, the amount of light captured by the array and delivered to the image plane from any point in the object plane is simply governed by the NA of the array. This is the common way the speed of a lens is measured, array or otherwise. Alternatively, one often uses the f number or radiometric efficiency which are related as follows: NA $= (\varepsilon)^{1/2}, f/N = \frac{1}{2\mathrm{NA}}$. Later on in this section we will derive the explicit expressions for the efficiencies.

For the field scan mode the image position is maintained on the photoreceptor over some fixed interval of time, since both the object and image are moving together over some distance. For a measure of the sensitivity one integrates the irradiance function, which is proportional to the NA as above, over the distance of travel corresponding to the time of exposure. In Figure 5.9b, this is the x direction. This can be made somewhat more clear from Figure 5.4a, where it is seen that both the object and the photosensitive drum move together. The maximum distance that the image remains stationary on the drum is limited by either the width of the array in the direction of movement, or the drum curvature. Field scan application requires high resolution, with speed a secondary attribute.

Common applications are listed in Table 5.1, along with important features and properties. We deal with each of these devices in the following sections.

5.3.1 Field Scan/Copiers

For the application of lens arrays to copiers one has to be concerned with both resolution and radiometric issues [2,5]. For copiers, the image quality must be sharp which is usually quantitatively measured by an MTF (modulation transfer function) measurement. The MTF is a measure of the contrast of the image as a function of the spatial frequency. One is also concerned about how sensitive

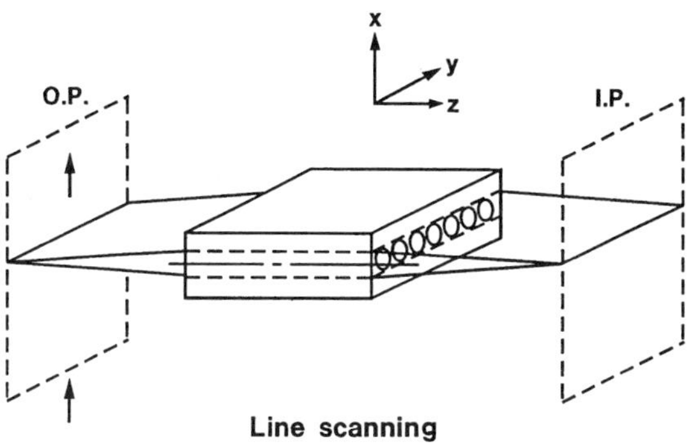

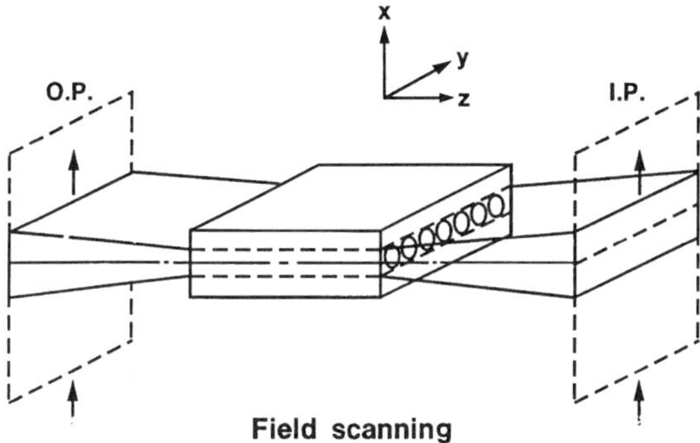

Figure 5.9 Two types of scanning applications. The top is the line scan mode where each line is directly imaged to detectors. The bottom is the field scan mode where the photoreceptor moves with the object for some distance.

Table 5.1 Applications of Erect One-to-One Imaging Lens Bars

Device	Type	Important properties
Copying machines, printers, scanners	Field scan	Compactness, resolution, depth of focus, illumination uniformity
Fax readers	Line scan	Compactness, speed, resolution, depth of focus

this function is when one moves away from the optimum object or image distances. Copiers have to be relatively forgiving to various copying conditions without serious loss of quality. For example, when one is copying a book with a thick binding, it is often difficult to maintain the printed page in intimate contact with the glass platten. One would still hope to make a decent copy under these circumstances.

From the radiometric aspect, what is equally important as the speed of the array is the uniformity of the irradiance. This is a function of the irradiance profile of the individual lens and how the lenses are spaced [2,9]. Copiers are relatively sensitive devices, and if the irradiance modulation is too great, then this is translated into unwanted lines or streaks on the copy. This is especially true when one wants to make a dark copy.

5.3.2 Line Scan/Facsimile Readers

A cross section of typical fax reader assembly, known as a CIS (contact image sensor), is shown in Figure 5.10 [9,10]. The speed of the unit is the most significant property, because it controls the data rate. The resolution requirement is not high since the printed word is the most common feature to be read. There are, of course, more expensive fax readers containing conventional lens systems that

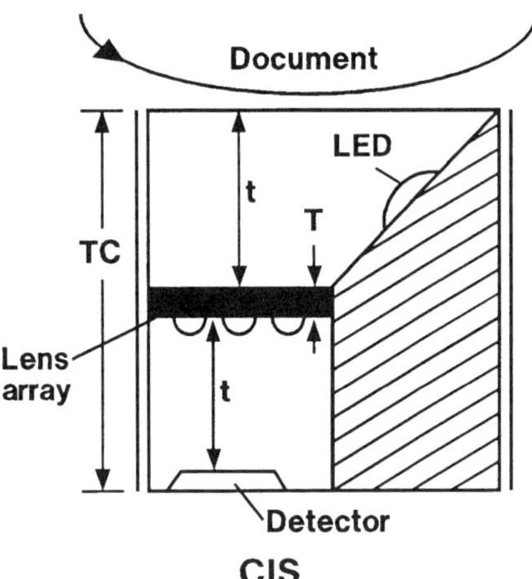

Figure 5.10 Contact image sensor (CIS) using fax reading lens for facsimile machines. The lens array tranfers the the information from the document to the detector array via the line scan mode. The typical distance for *t* is 14 mm.

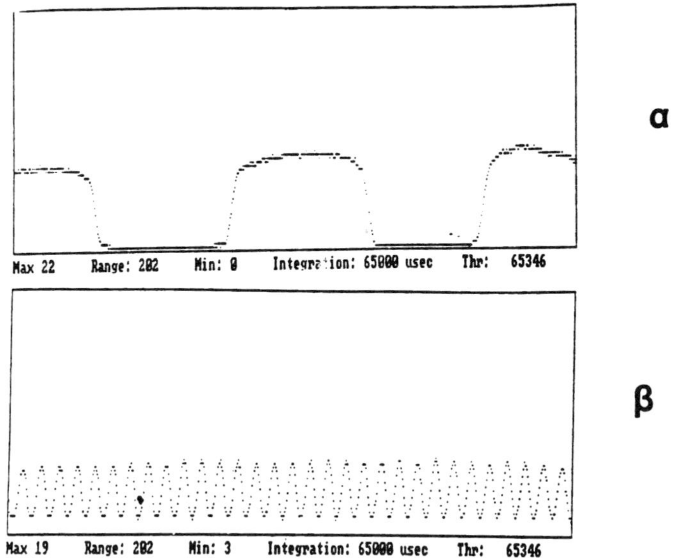

Figure 5.11   Representative output from a Corning SMILE fax lens array at two spatial frequencies. (See Ref. 10.)

can read documents with resolution equal to that of copiers. The device shown here is the version that appears in many inexpensive fax machines.

What is important for these devices is that there exists a sufficient signal differential between what corresponds to black and what corresponds to white on the document. The threshold trigger voltage is set midway between the two signal values and must accommodate light level variations that occur from document to document. The typical output from a CIS unit using a Corning SMILE lens array is shown in Figure 5.11 for two spatial frequencies. A comparison of the maximum and minimum signal levels achieved in the CIS unit using the Selfoc SLA-20 array and the Corning SMILE™ array is shown in Figure 5.12a. Another comparison that is made is that of the irradiance uniformity that is shown in Figure 5.12b. This more uniform irradiance is what would be expected because the SMILE array is made up of a larger number of closely spaced smaller lenses. We will see the quantitative explanation of this fact in the next section. Another property of interest is the change in contrast as a function of altering the object distance. This is because when the document is fed into the reader roll assembly, there is some "slop" in the object distance. A typical measurement is shown in Figure 5.13.

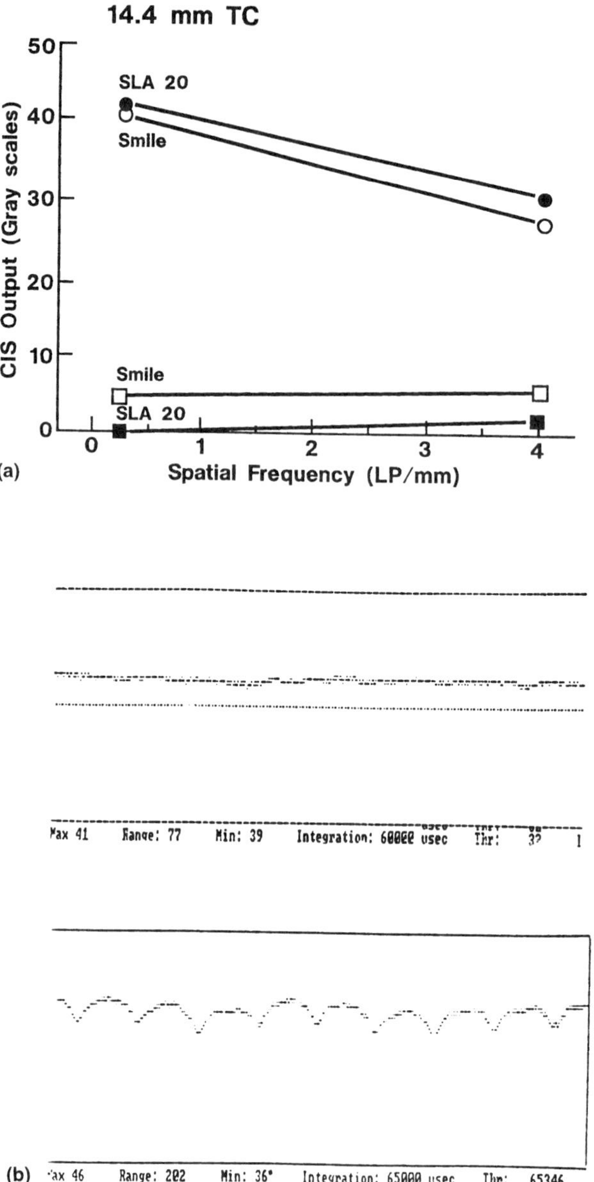

Figure 5.12 Comparison of fax reading contrast performance of SELFOC lens array and SMILE array. The upper line is read as white, and the lower is read as black. In the two curves below are shown the uniformity of the output to a uniform white source. The upper curve is the SMILE array and the lower in the SELFOC array. (See Ref. 10.)

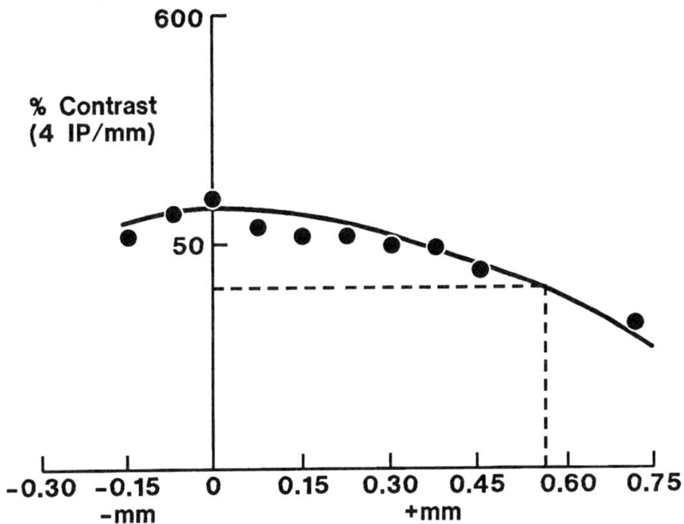

Figure 5.13 Variation in the % contrast as a function of the offset of the object distance for fax reader. (See Ref. 10.)

5.4 RESOLUTION/CONTRAST

5.4.1 Measurement Technique and Typical Results

As mentioned above, the key property of any copying lens is the quality of the copy produced, or in optical terms, what would be referred to as the resolution. The quantification of the resolution is usually an MTF measurement where test patterns of given spacings are imaged by the lens which is then scanned to produce an intensity modulation. This measurement technique is shown in Figure 5.14. A test target, usually chrome lines on clear glass, is illuminated with a lambertian source, here approximated by a diffuse surface integrating sphere. The target has a series of light-dark lines of progressively higher spatial frequency. The target is moved across the lens as shown. The image produced by the test lens illuminates a narrow slit in front of the detector and is displayed as a temporal modulation as a consequence of the target motion. The typical output is shown schematically in Figure 5.15. The inset shows how the contrast would diminish with increasing spatial frequency.

An example of the contrast and depth of focus performance of a two-row GRIN-based Selfoc array, as specified by the inset to the graph shown in Figure

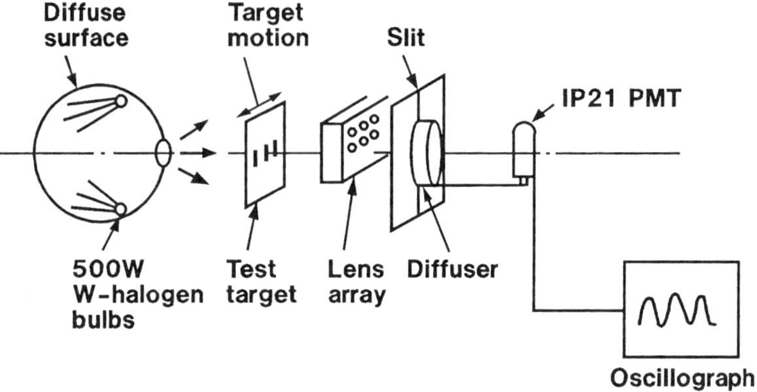

Figure 5.14 Diagram of setup to measure MTF for lens array.

5.16 [13]. The NSG-type lens is an SLA-12 signifying the field angle of 12°. The contrast results for three conditions of depth of focus are shown in Figure 5.17. In the first case of, Figure 5.17a, the object and image plane are symmetrically displaced; in the second, figure 5.17b, the image plane alone is moved; and in the third case, figure 5.17c, the total conjugate distance is maintained and the lens is moved. One can see that the third situation is the most demanding in that both object and image distances are altered. The result shown in Figure 5.17b is often called the ''through focus measurement'' and is of particular significance to copier applications because of the variation in the image distance due to the wobble in the photoreceptor drum as it rotates.

An example of the contrast and depth of focus performance of a refractive element-based Corning SMILE array is shown in Figures 5.18 and 5.19 [14]. In Figure 5.18, the contrast vs. symmetric working distance is plotted for four spatial frequencies. The array geometry is shown at the bottom. In Figure 5.19 the through focus behavior taken at 4 lp/mm is shown. Another direct, yet somewhat qualitative test, is to observe a copy of a standard test pattern such as shown in Figure 5.20. The lines in the pattern are marked with respect to the spatial frequency from 2 to 10 lp/mm.

5.4.2 Factors That Determine Contrast

The resolution of a one-to-one erect imaging array depends not only on the resolving capability of the individual lenses which makeup the array but also on the factors that influence the coincidence of their collective images, that is, array factors. As we have seen, depending on the maximum field height, a relatively large number of lenses are contributing to a given point of the image. Any varia-

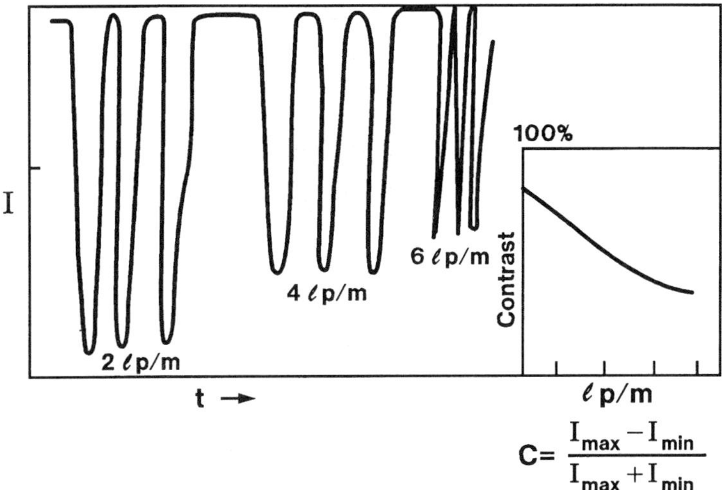

$$C = \frac{I_{max} - I_{min}}{I_{max} + I_{min}}$$

Figure 5.15 Typical MTF results at three spatial frequencies and resulting % contrast plotted against spatial frequency.

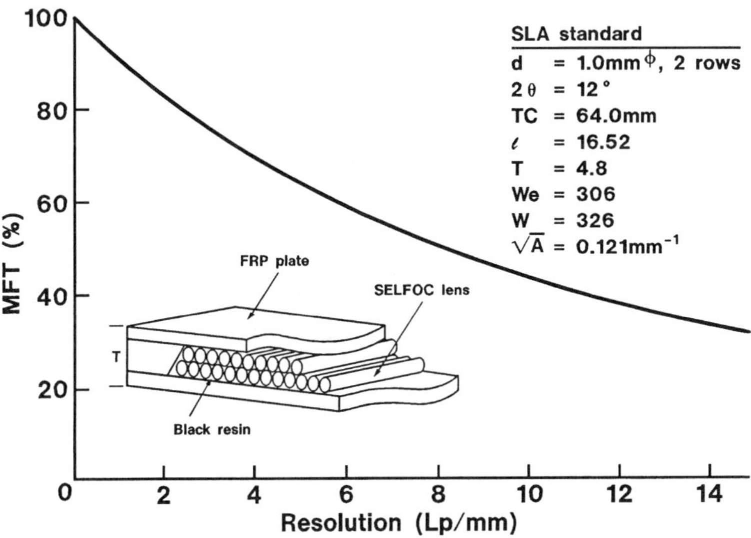

Figure 5.16 Experimental contrast vs. spatial frequency data for a SELFOC SLA lens array. Insert shows the construction of the two-row array. (From Ref. 13.)

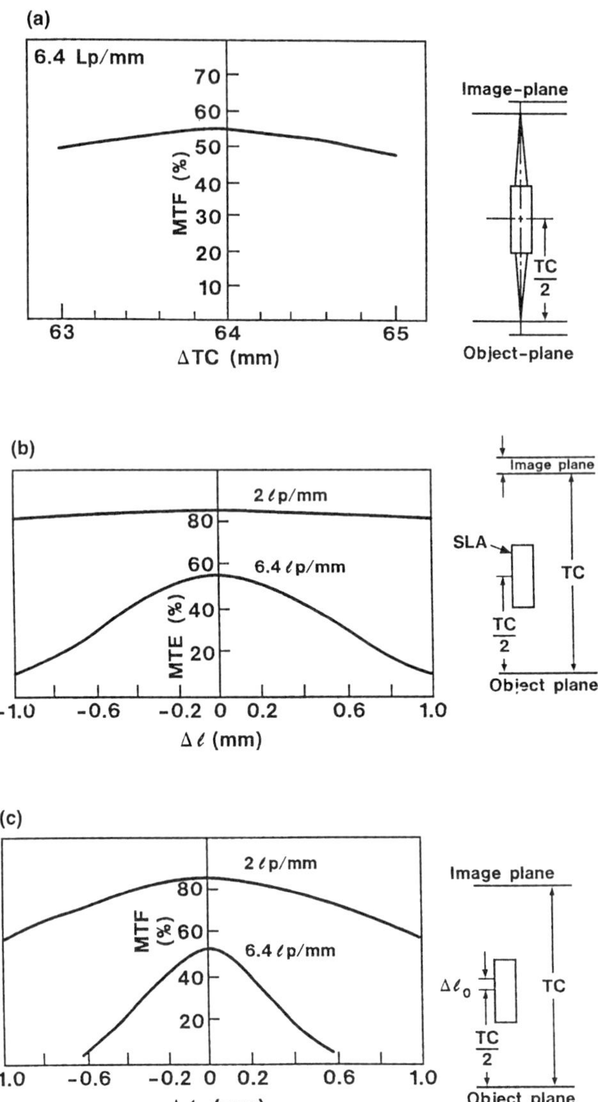

Figure 5.17 Effect of offset from optimum position on the contrast for SELFOC SLA lens array: (a) symmetric offset of total conjugate distance; (b) offset of image plane holding object distance constant; (c) unsymmetric variation of object distance and image distance holding TC constant. (From Ref. 13.)

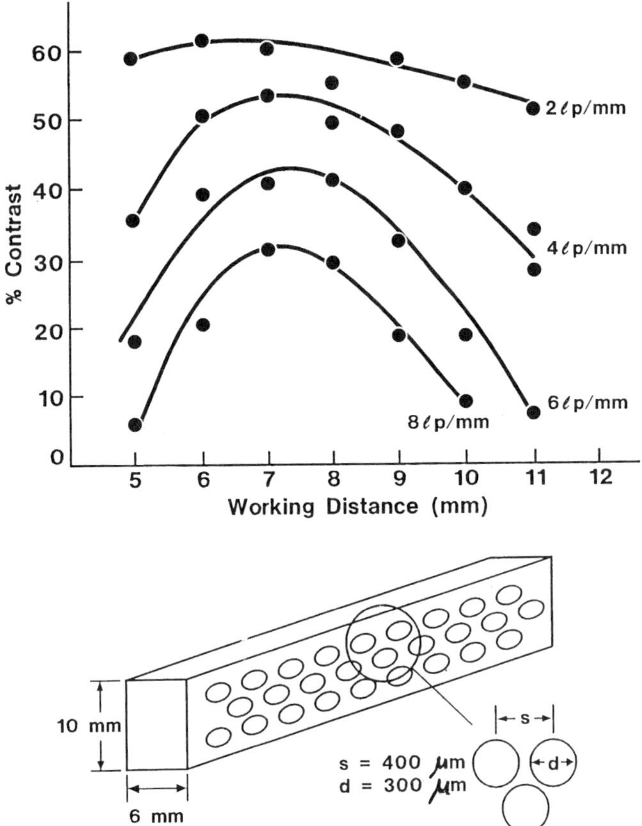

Figure 5.18 Dependence of contrast on working distance measured at four spatial frequencies for SMILE lens array. The structure of the array is shown in the inset.

tion in the optical parameters of these contributing lenses lead to a blurring of the collective image. To make this point we show some data in Figure 5.21, where the MTF was measured on a single microlens from a Corning SMILE array and then compared to the contrast measured with the array. One can see the large difference in contrast performance. One can estimate that about 12–15 lenses contribute to any given image point [2]. This estimate is obtained from the drawing in the inset where the maximum field-height radius is superimposed on the lens arrangement. Now consider the relationship between the image working distance and the lens parameters given above for the refractive lens case, viz., Eq. (5.6). If one takes the derivitive of the expression with respect to the lens thickness T, or the lens focal length f, one obtains an estimate of the changes

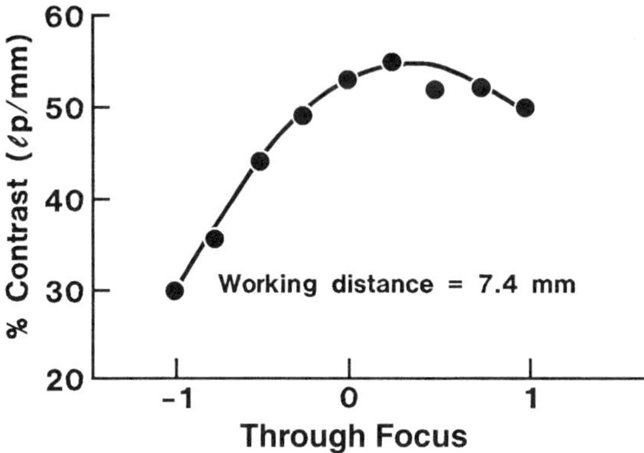

Figure 5.19 Effect of offset of object distance on the contrast for SMILE lens array.

that would occur in the image plane distance for given variations in the these two parameters.

$$\delta t = \{nf^2(T - 2nf)^2\}\, \delta T = \left(\frac{t^2}{T}\right)\left(\frac{\delta T}{T}\right) \tag{5.14a}$$

$$\delta t = \left\{\frac{T^2}{(T - 2nf)^2}\right\}\delta f = \left(\frac{t^2}{f}\right)\left(\frac{\delta f}{f}\right) \tag{5.14b}$$

Figure 5.20 Copy of test pattern formed by SMILE lens array used in a photocopier with total conjugate distance of 34 mm.

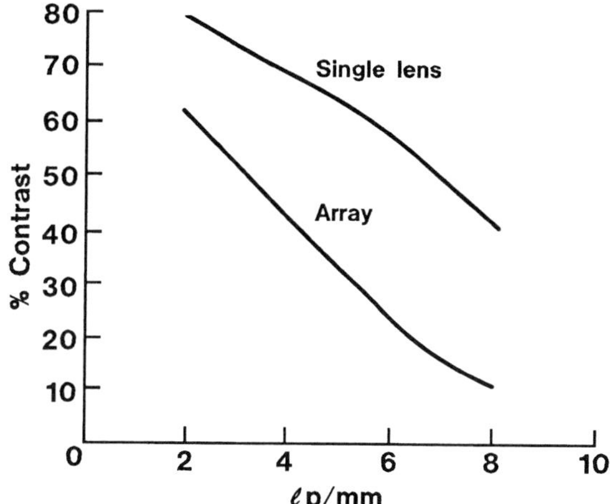

Figure 5.21 Comparison of the contrast vs. spatial frequency of a single lens from the SMILE lens array and the array as whole.

One can appreciate how variations among the contributing lenses can cause degradation of the image, especially for designs where the total conjugate, hence the working distance *t*, is long. This is experimentally shown in Figure 5.22 where the 4 lp/mm contrast is shown for arrays with lengthening optimum working distances.

For the Selfoc GRIN arrays the problem is not as bad because the ratio of the field height to the lens radius is smaller and there are only two rows. Only about 3 or 4 lenses contribute to a given point in the image plane. One can go through the same anaylsis with Eq. (5.11) as we did to obtain Eqs. (5.14), with the independent variables being the lens thickness *D* and the gradient parameter $\sqrt{A}$; e.g.,

$$\delta l_0 = -\left(\frac{1}{2N_0}\right) \sec^2 \left[\frac{(A)^{1/2}D}{2}\right](\delta D) \tag{5.15}$$

One can see that this term will be large when $\sqrt{A}D/2$ is close to $\Pi/2$, which is equivalent to the refracting case when *T* is close to *nf*. It turns out that for the working distances used in devices, the latter is more likely the case than the former. The explanation for this is not mysterious; it is because the GRIN arrays were used first in scanning devices and were incorporated in the optimal way.

This discussion of what affects the contrast provides an excellent example of how the fabrication method influences the ultimate performance for a given

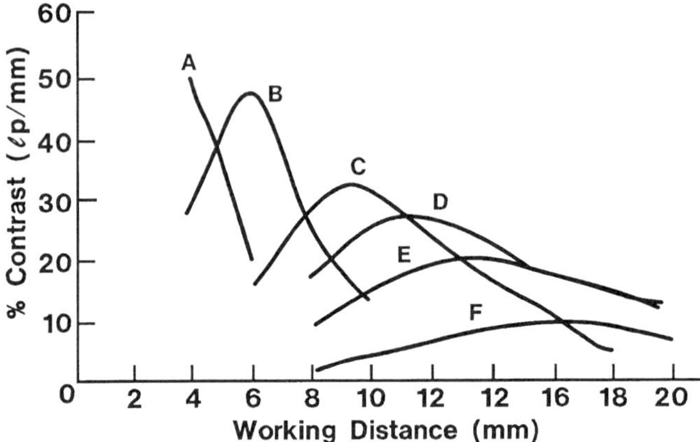

Figure 5.22 Contrast vs. spatial frequency for a number of SMILE lens arrays with different working distance. Longer working distance has poorer contrast because of greater sensitivity to image overlap.

application. The subtle point is that although the overall function is equivalent, the limitations provided by the specific method, like lens diameter, thickness, and strength, are significantly different. The GRIN method of assembling cut ion–exchanged rods of glass into two rows to form an array has certain limitations. For example, handling rods as small as 200–400 μm is impractical, or assembling more than a two-row stack is difficult. On the other hand, for the refractive array made by the photosensitive process, lens diameters above 400 μm begin to deviate from perfect sphericity, and having the thickness more than 3 mm is impractical because of the optical exposure step. Each application of one-to-one arrays has its own set of desirable and necessary properties which may not be met by any one fabrication method.

5.5 RADIOMETRY

The initial radiometric challenge is to determine the irradiance profile of a single lens. The overall profile will come from the addition of the individual profiles according to their two-dimensional layout. The method used to calculate the irradiance profile can be found in the literature. We will briefly review the method and results following that of Borrelli et al. [2]. The basic idea is to treat the exit pupil as a Lambertian source with radiance N. For a given object point, at the one-to-one image distance, the exit pupil is found by tracing a fan of rays to the edges of the entrance pupil, through the lens to the rear surface. This is shown in Figure 5.23 for three cases: the simple refracting lens, the same lens with field lenses at the midplane, and finally the GRIN lens. The irradiance can be expressed in the paraxial approximation as

$$h(r) = h_0 \pi R^2 A(y) \tag{5.15}$$

where R is the radius of the lens, h_0 is the axial irradiance $N\Pi R_2/t_2$, and $A(y)$ is the function determined by the overlap of the projection of the front aperture to the back face, as shown in Figure 5.23c. The expression for the shaded area is given by

$$A(y) = (2/\pi) \left\{ \cos^{-1}\left(\frac{\Delta}{2R}\right) - \left(\frac{\Delta}{2R}\right)\left[1 - \left(\frac{\Delta}{2R}\right)^2\right]^{1/2} \right\} \tag{5.16}$$

One can relate the displacement $\Delta/2R$ to the lens parameters by the carrying through the method described in Figure 5.23 by the matrix ray trace outlined above for the three cases. For example, for the refractive case it is easy to show that the expression

$$\frac{D}{2R} = \frac{y}{k} \tag{5.17}$$

where k is the maximum field height. For the GRIN case the derivation appears in Kawazu and Ogura [5]. Using Eq. (5.15), one can obtain the irradiance profiles for the two lens types.

Refractive lens:

$$h(r) = \left(\frac{2}{\pi}\right)\left(\frac{N\pi R^2}{t^2}\right)\left\{ \cos^{-1}\left(\frac{r}{k}\right) - \frac{r}{k}\left[1 - \left(\frac{r}{k}\right)^2\right]^{1/2} \right\} \tag{5.18}$$

where $k = 2\pi R\tau/T$.

$$\text{GRIN lens: } h(r) = h_0\left[1 - \left(\frac{r}{k}\right)^2\right]^{1/2} \tag{5.19}$$

$$h_0 = \pi N n_0^2 A R^2 \cos\left[\frac{(A)^{1/2}D}{2}\right]$$

where $k = -R \sec (\sqrt{A}D/2)$. The profiles are shown graphically in Figure 5.24 along with a third case of the refractive lens with field lenses. We have chosen to show the case where all the lens curvatures are equal; this is not the optimum field lens curvature case. The optimum case is when $f_{\text{int}} = T/n$, which results in all the light that enters the lens, up to the maximum field height, exits the lens. In this case the irradiance profile is flat [2].

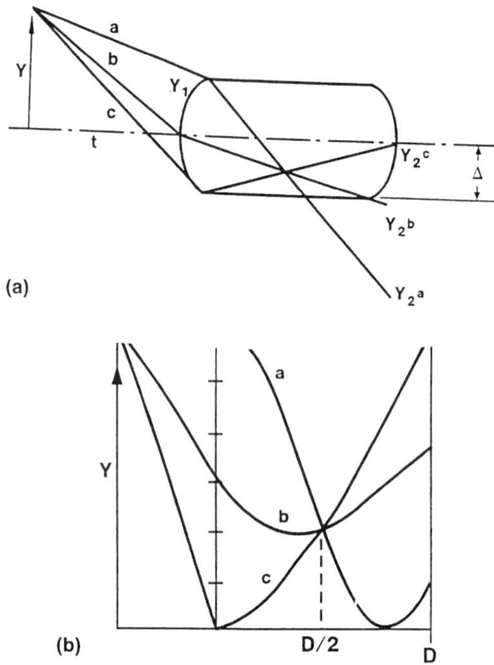

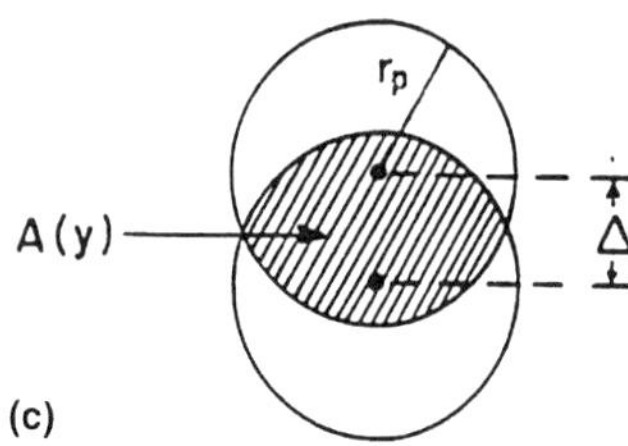

Figure 5.23 Rays showing how the radiometric efficiency is calculated. In all cases the procedure is to ray-trace the two boundary rays and the central ray from a given point in the one-to-one object plane to the back lens face. The overlap is defined as shown. The refracting case is shown in (a), the refracting case with field lenses is shown in (b), and the GRIN case is shown in (c).

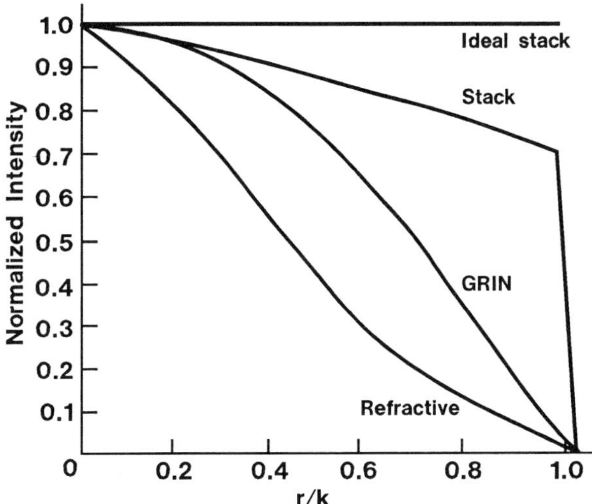

Figure 5.24 The computed irradiance profile for the three cases shown in Figure 5.23. Actually, two cases are presented for the stacked case, one where all the four lens curvatures are the same, and the other, labeled ''ideal,'' where the field lens curvature is optimum.

The overall power delivered by a single lens is then the integral of the irradiance profile:

$$P_s = \int h(r)r \, dr \, d\theta \tag{5.20}$$

For the refractive lens this is given by the following expression:

$$P_s = \left(\frac{\pi}{4}\right) h_0 k^2 = NR^2 \left(\frac{\pi n_0 R}{T}\right)^2 \tag{5.21}$$

and for the case with perfect field lenses it is 4 times this value.

For the GRIN case, the power is given by the following:

$$P_s = h_0(\pi k^2) \tag{5.22}$$

To determine the radiometric speed and uniformity of an array of such lenses, we must consider the overlap of the irradiance profiles according to the specific geometric arrangement. For example, consider the continuous hexagonal close-packed (hcp) arrangement shown in Figure 5.21a. The total power delivered in the area A containing n_x lenses in the x direction and n_y lenses in the y direction is simply

$$P = n_x n_y P_s \tag{5.23}$$

where the area is given by the following expression as indicated from Figure 5.25a.

$$A = (2bRn_x)((3)^{1/2} \, bRn_y) \tag{5.24}$$

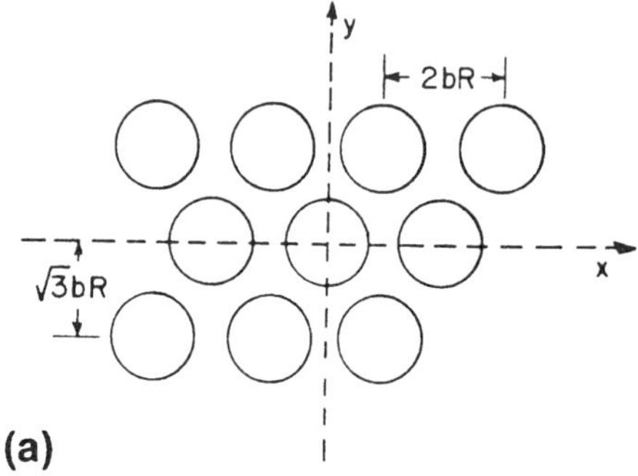

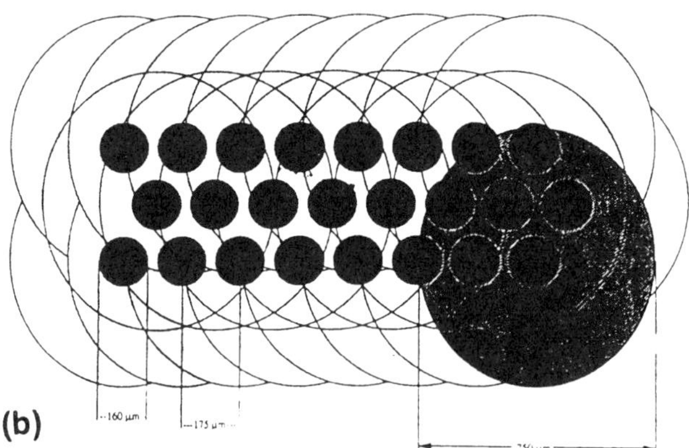

Figure 5.25 Geometric arrangement of lensets in SMILE array and the representation of the overlap of their fields. (From Ref. 2.)

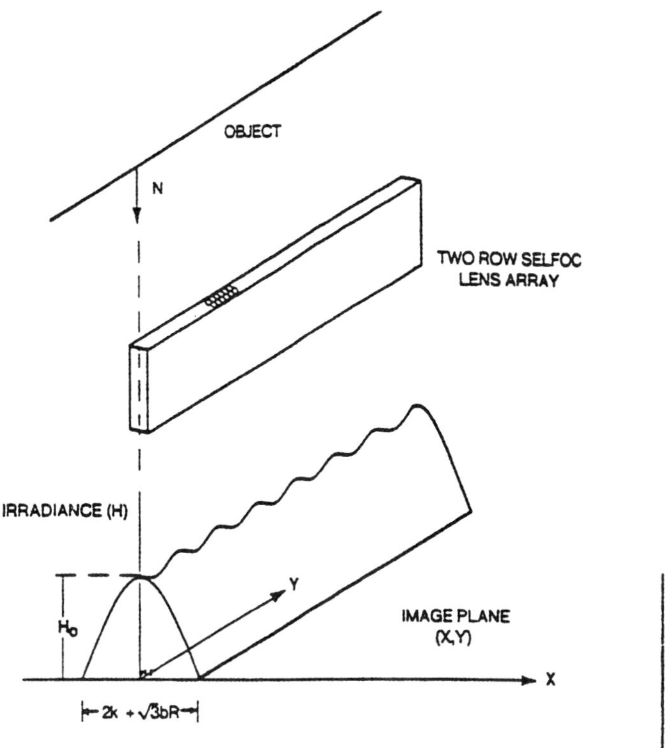

Figure 5.26 Irradiance uniformity of a two-row SELFOC lens bar. (From Ref. 9.)

The average irradiance of the array $\langle h \rangle$ would be P/A and the radiometric efficiency would be this number normalized to the radiance ΠN. Thus, the average radiometric efficiency of an array of lenses arranged in a hcp structure would be

$$\varepsilon = \left(\frac{1}{\pi N}\right)\left(\frac{1}{2(3)^{1/2}b^2R^2}\right)P_s \tag{5.25}$$

where P_s is the power delivered by a single lens.

For Selfoc lenses made up of p rows, Lama [15] has shown that the average irradiance is given by the following expression:

$$ph_0\left(\frac{\pi R}{4bk}\right)\left[1 - (p-1)^2\left(\frac{b^2}{4a^2}\right)\right] \tag{5.26}$$

Using the same method, but not performing the averaging, one can estimate the uniformity of the irradiance. As an example, the irradiance profile is shown in Figure 5.26 for a two-row Selfoc array.

5.5.1 Experimental Measurement

The radiometric efficiency of an array can be measured by the relatively simple experimental arrangement shown in Figure 5.27. By defining the Lambertian source and detector areas, and setting the reference distance at the correct total conjugate distance, one can obtain reasonably accurate estimates of the efficiency. Some results obtained for SMILE arrays by using this method are shown in Table 5.2, where they are compared to the theoretical values obtained from Eq. (5.25).

The radiometric efficiencies of the GRIN lens arrays can be obtained from Table 5.3, where the efficiency is equal to the square root of the numerical aperture.

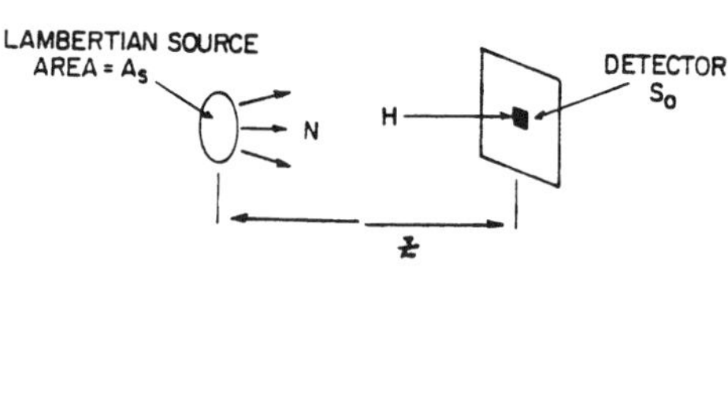

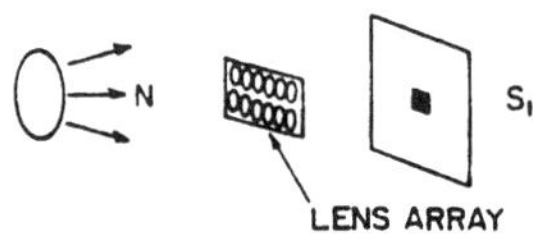

$$\epsilon = (A_s/z^2\pi)\,(S_1/S_0),$$

Figure 5.27 Representation of experimental method to measure radiometric efficiency.

Table 5.2 Listing of Properties of SMILE Lens Arrays

Lens diameter/spacing (μm/μm)	Thickness (mm)	TC (mm)	Measurement[a] (%)	Calculation[b] (%)
Thick single element				
400/480	6	25	0.13	0.16
350/420	6	25	0.09	0.12
350/420	6	25	0.10	0.12
450/540	6	30	0.10	0.20
450/540	6	25	0.12	0.20
Thick double element				
310/350	5.5	21	0.27	0.45
310/350	7	24	0.20	0.28
310/350	7	20	0.21	0.28
310/350	6	25	0.23	0.38
400/480	8	21	0.28	0.35
Thin single element				
160/195	0.83	5	2.7	1.4
160/195	1.85	11	0.5	0.27
200/240	2.75	13	0.40	0.21
200/240	2.75	12	0.45	0.21
Thin double element				
160/195	1.45	9	1.1	1.6
160/195	1.45	8	1.5	1.6
160/195	1.45	16	1.5	1.6
160/195	1.45	11	1.2	1.6
200/240	1.40	9	2.2	2.9
200/240	1.30	11	1.9	3.3

[a] Radiometric efficiency measured by method shown in Figure 5.27.
[b] Calculated from Eq. (5.25).

5.5.2 Field Scan Exposure

We mentioned above that in the field scan mode the image is maintained on the photorecepter for some time. This means that the one has to integrate the irradiance function over this time interval. Formally, one would write that the field scan exposure E would be expressed as

$$E = \left(\frac{1}{v}\right) h(x, \mathbf{y}) dx \tag{5.27}$$

Table 5.3 Listing of Properties of Selfoc Lens Arrays

		Type		
	Symbol	SLA-06	SLA-09	SLA-20
Lens diameter	D (mm)	1.055	1.055	1.055
Refractive index	N_0	1.538	1.606	1.563
Numerical aperture	NA	0.1	0.15	0.37
Gradient constant	A (mm^{-2})	0.016	0.032	0.201
Fiber length	Z (mm)	27–31	18–22	7–9
f Number	f/N	5	3.5	1.0
Thickness	t	4.8	4.8	4.0
Total conjugate	TC (mm)	64–75	46–55	16–20
Irradiance uniformity[a]	%	7	7	15
Number of rows		2	2	2

[a] See Figure 5.27 for definition.
Source: From various product information sheets supplied by Nippon Sheet Glass Co.

where x is the direction of travel with velocity v [2]. In order to deal with the geometry of the lens layout the irradiance profiles like those of Eqs. (5.18) and (5.19) are initially averaged in the y direction, transverse to the motion direction. We will do this for the single-row GRIN case as an example following the method of Lama [15]. We have seen that the irradiance profile for a single lens is given by Eq. (5.19). To average over the y direction we have to evaluate the integral

$$h(x, \mathbf{y}_0) = \left[1 - \frac{x^2 + y^2}{k^2} \right]^{1/2} \tag{5.28}$$

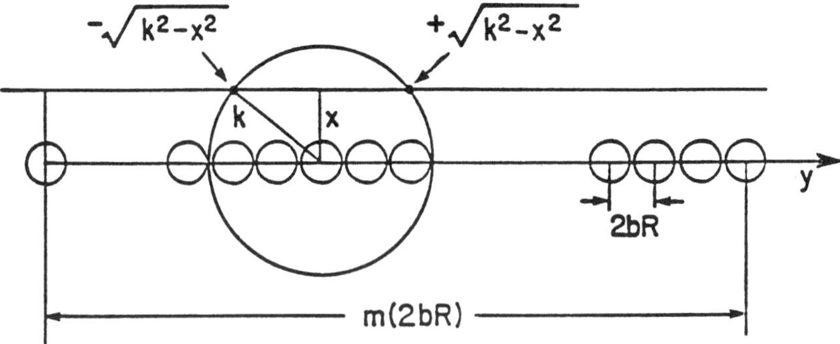

Figure 5.28 Method of determining the irradiace profile of a single row of lenses.

from $\pm (k^2 + x^2)^{1/2}$. (See Figure 5.28.) This yields the following expression:

$$h_0 \left(\frac{\pi}{2k} \right) (k^2 - x^2) \tag{5.29}$$

Now one adds the irradiance from total number of lenses in the row, say m, and divides by the total length $m(2bR) + 2k$ and gets

$$\frac{h(x, \mathbf{y})}{h_0} = \frac{\pi}{4kbR} (k^2 - x^2) \tag{5.30}$$

We can now substitute Eq. (5.30) into (5.27) to obtain the expression for the single-row exposure:

$$E_1 = \frac{\pi N T n_0 A R^3}{3bv} \tag{5.31}$$

For p rows, the answer is essentially p times this result.

For the refractive lens case, where the lenses are smaller and closer spaced so that p times the interrow spacing is essentially the width, the exposure is

$$E_r = \frac{\pi^2}{2(3)^{1/2}} W(nR/bT)^2 \tag{5.32}$$

Here W is the width of the array.

5.6 FURTHER ASPECTS OF FABRICATION

In addition to the general description given in Chapter 2 of how the Selfoc GRIN and the Corning SMILE arrays were made, it is worthwhile here to discuss how the fabrication technique is optimized for the specific application, in this case one-to-one arrays. From a historical perspective, the first lens array intended for use in the one-to-one mode is attributed to Moorhusen [16], who proposed a three-piece molded plastic structure.

For the Selfoc arrays, the inset in Figure 5.16 provides a good picture of the two-row array. A single row of lenses is laid up on a plate and one layer is placed over the other to form the two-row array in the fashion shown. The array is then potted in a black resin and ground and polished to the correct thickness. NSG makes a number of different versions of these two-row arrays. The major variable is the Δn which is controlled by the kind and amount of ion-exchanged dopant. The parameter that reflects this is the index profile constant labeled A. The range of SLA arrays made by NSG under the Selfoc trade name is shown in Table 5.3. One can see that the number following the SLA corresponds to the field angle which as we have seen above is related to the numerical aperture. In general, one can say that the mechanical fabrication of the Selfoc arrays does not significantly differ for the different applications, but the chemical part does.

For the Corning SMILE arrays the fabrication for the field scan applications, where radiometric efficiency is not a real issue, is done by double side exposing a final thickness bar of glass, followed by the thermal development shown in Figure 5.29a. Typically, the thickness is 6 mm, as shown in Table 5.2. One notes that radiometric efficiency is quite low, in the order of 0.2%. However for field scan devices it is the exposure that counts, not the radiometric efficiency. The relative exposure number, obtained from Eq. (5.32) for a 10-mm wide, 6-mm thick, 400-μm lens-diameter array with a separation parameter of $b = 1.2$ is 43. For the Selfoc SLS-6, with a radiometric efficiency of 1%, the relative exposure is 33, obtained from Eq. (5.31).

For the line scan applications the story is quite different. The speed of the array is paramount and the Corning SMILE arrays for this application are made from stacked thin arrays, as shown in Figure 5.29b. From Eq. (5.25), one sees that the efficiency is increased with the inverse square of the thickness, and that the power delivered by an array with field lenses is 4 times that of one without. From Table 5.2, one can see that using this geometry yielded efficiencies as high as 2% compared to the SLA-20 with a 7% efficiency.

One of the disadvantages of the lens bar approach is that it only can image one-to-one. Copiers with folded conventional lenses have magnification and reduction features. Rees et al. [17] proposed to produce the reduction and magnification

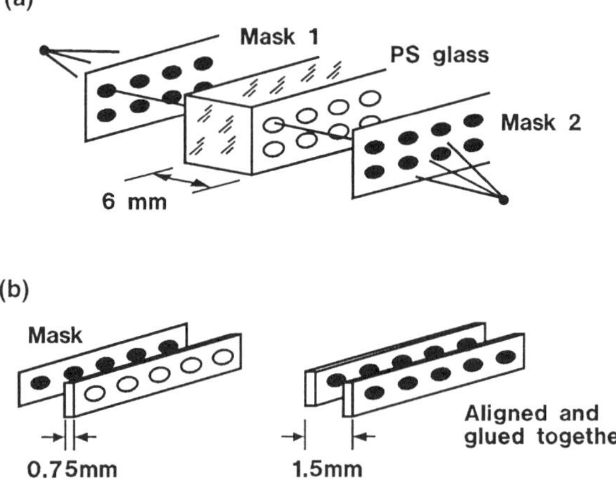

Figure 5.29 Schematic representation of the exposure method used to make the photosensitive SMILE™ one-to-one imaging lens arrays. (a) The double sided exposure method and (b) the structure of the stacked array.

features by having the individual GRIN rod lenses move away from the perpendicular direction as one moves away from the center of the array. This brings the images together before (or after) the normal image plane; thus the individual images are smaller (larger) at the point of coincidence. One problem with this idea is that the rod lengths must be gradually adjusted to maintain the same image distance.

REFERENCES

1. N. F. Borrelli, D. L. Morse, R. H. Bellman, and W. L. Morgan, Photolytic technique for producing microlenses in photosensitive glass, *J. Appl. Phys. 24*, 2520 (1985).
2. N. F. Borrelli, R. H. Bellman, J. A. Durbin, and W. Lama, Imaging and radiometric properties of microlens arrays, *Appl. Opt. 30*(25), 3683 (1991).
3. R. W. Ditchburn, *Light*, Blackie and Sons, Ltd., London, 1952.
4. F. P. Kapron, Geometric Optics of Parabolic Index-Gradient Cylindrical Lenses, *J. Opt. Soc. Am. 60*(11), 1433 (1970).
5. M. Kawazu and Y. Ogura, Application of gradient-index fiber arrays to copying machines, *Appl. Opt. 19*(7), 1105 (1980).
6. J. D. Rees, Non-Gaussian imaging prop. of GRIN fiber lens arrays, *Appl. Opt. 21*(6) (1982).
7. K. Sono, SELFOC™ Technology, *IFOC Buyers Guide Handbook*, 1980–1981 ed.
8. Nippon Sheet Glass (NSG), *SELFOC™ Rod Lens data sheet*, 1981 (NSG America, Clark, NJ 07066).
9. K. Matsushita and M. Tovama, Unevenness of illumination caused by GRIN arrays, *Appl. Opt. 19*(7) 1070 (1980).
10. R. H. Bellman, N. F. Borrelli, L. G. Mann, and J. M. Quintal, *Fabrication and Performance of a 1-to-1 Erect Imaging Microlens Array for FAX*, SPIE, vol. 1544, *Miniature and Microoptics* (1991).
11. K. Matsushita and K. Ikeda, *Newly Developed Glass Device for Image Transmission*, SPIE Conf., San Mateo, CA (1972).
12. Micro-optics has macro potential, *Laser Focus World*, p. 93, June 1991.
13. Nippon Sheet Glass Co. (NSG, Clark, NJ 07066) Selfoc™ linear lens array, *Data sheet*, September. 1979.
14. Corning, Inc. Private data.
15. W. L. Lama, Optical properties of GRIN fiber lens arrays, *Appl. Opt. 21*, 2739–2746 (1982).
16. R. W. Moorhusen, U.S. Patent 3,544,130, 1970.
17. J. D. Rees, D. B. Kay, and W. L. Lama, U.S. Patent 4,331,380, May 1982.

6

Two-Dimensional Arrays

6.1   INTRODUCTION

In this chapter we deal with microlens array applications where the major feature will be the ability to precisely fabricate and position microlenses in two dimensions. Linear arrays are a special important case, requiring no less precision. We will try to cover the important lens array applications that include the various lens fabrication techniques discussed in preceding chapters. In the description and discussion of these applications, we shall try to make clear why one fabrication method might be better suited over others for a particular application.

For any application, adequate lens performance will always be required, but it will be no more important than the ability to precisely pattern the lenses accordingly. In other words, one may not be able to form a lens with ideal performance, and at the same time produce thousands of them per square millimeter precisely positioned in a particular two-dimensional pattern. As we have seen in the previous chapters which dealt with the fabrication methods, some fabrication methods are better suited to make very small lenses, others better to be produced at precise locations, and still others with better imaging performance. It will be the trade-off of one or more of these factors that will determine the best way to accomplish the desired end. The applications that will be covered are listed in Table 6.1, along with the fabrication method.

6.2   APPLICATIONS

The applications that we cover in the next sections are intended to be representative of the wider field. In no way is this intended to be an exhaustive study. The intention is to give the reader a feeling for the areas of applications, the reasons

Table 6.1 Microlens Array Applications

Application	Size[1] (μm)	Array type[2]	Key feature[3]
Linear			
Laser diode array collimators	200	GRIN	NA
Expanded beam for SMF arrays	120	GRIN, refractive	Diffraction limited
Switching 4 × 4 crossbar	>120	Refractive	Alignment, cross talk
Computer backplane	>200	Refractive	NA
Camera autofocus	165	Refractive	Alignment, cross talk
LED print bar	85	Refractive	NA
2D			
Shack-Hartmann	>300	Refractive	Long focal length
Parallel processing (VCSELS)	>20	Diffractive, GRIN	Lens diameter
Projection LCD-TV	100	GRIN, refractive, diffractive	Alignment, large area
Laser protection	>300	Refractive	Viewing angle
3D photography	>1000	Refractive	Large area

[1] The lens diameter represents the range used, not necessarily the optimum
[2] Type of lens used; GRIN means gradient index
[3] The key feature is to indicate the dominant property that the array should possess for the given application.

that they are important, and the particular performance criteria required. To facilitate the process, we will use Table 6.1 as the guide to the discussion of the various applications of two-dimensional microlens arrays that we have chosen to study. The typical lens dimension is listed in the second column, although this really refers to the separation distance ultimately determined by the device that the lens is to be aligned to. For example, if it were a CCD array it would be the center-to-center distance of the active detector areas on the silicon wafer. In the third column, we list the type of microlens that has been reported for the particular application, and finally, in the last column we list the key, or limiting property for the given application, if there is one. As an example, for the laser diode application, the numerical aperture of the lens is crucial because the NA of the laser diode is usually >0.5. However, in a fiber-to-fiber application, the fiber NA is low, and the major issue is insertion loss. This often translates to alignment.

The first five applications that are listed are primarily aimed at optical fiber-based systems. Because of many common requirements, these will be discussed as subsections in the same overall section. The other applications are unique and will be discussed separately. The general approach will be to give a brief introduction to the particular application and the key requirements, and then to follow with the experimental results obtained from one or more of the fabrication techniques that have been reported.

6.2.1 Optical Fiber-Based Applications

Laser Diode Array Collimators

In this application it is desired to efficiently collect light from a high NA laser diode (LD) array and provide this as an input to an array of single mode fibers (SMF), as shown in Figure 6.1 [1–3]. The application requires two things. The first is to efficiently collect the light from the LD and the second to image it so that it is efficiently captured by the SMF. For the first, it is necessary to consider the numerical aperture of the microlens, which is a measure of the amount of light that is collected. For the second, it is important to consider those factors that control the coupling efficiency into the fiber, for example, the spot size in the focal plane of the fiber.

The problem for ordinary spherical lenses, or what would be equivalent to them in GRIN microlenses, is that these two properties are in opposition. To

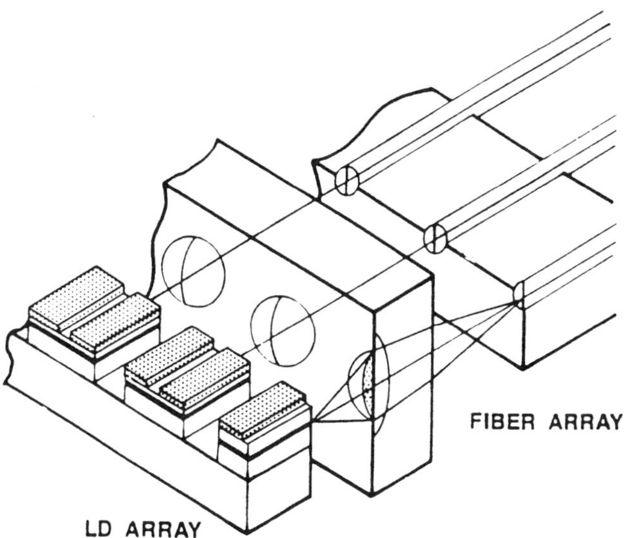

Figure 6.1 Schematic drawing of a linear lens array being used to couple the output from a laser diode array to a single-mode fiber array.

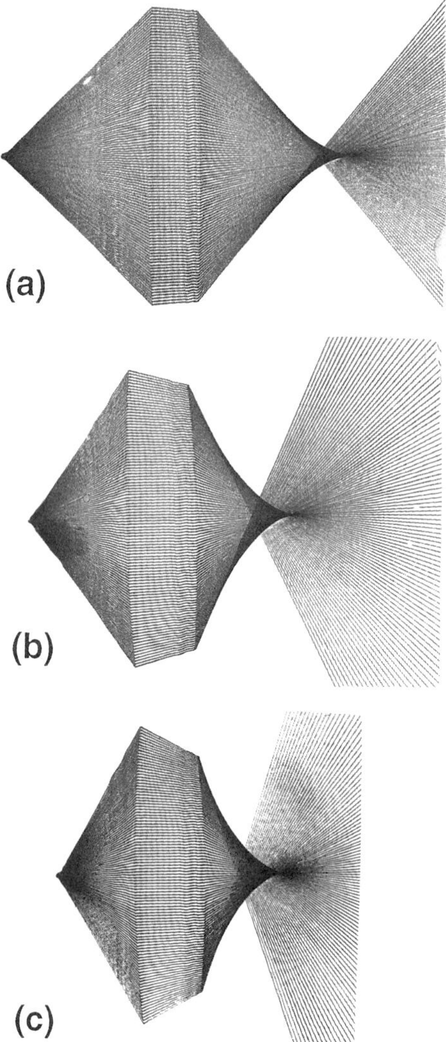

Figure 6.2 Computed ray-trace diagrams for microlenses with different numerical apertures to illustrate the effect of spherical aberration. The ray trace has a different scale for the vertical and horizontal directions. The lenses are 200 mm in diameter with a biconvex structure. The numerical aperture for the trace (a) is 0.14 with an EFL of 0.356; for (b), the numerical aperture is 0.18 and a EFL of 0.273; and for (c) the NA is 0.21 with an EFL of 0.21.

increase the efficiency of capturing the light from the high NA source, one naturally encounters a poorer focal spot definition as a consequence of added spherical aberration. The higher the numerical aperture, the greater the contribution of spherical aberration and thus the more blurred the focus spot. This is shown from the ray traces of Figure 6.2 for a biconvex (spherical curvature) structure having a 200-μm diameter. The extent of spherical aberration as indicated by the blurring of the focal spot is shown as a function of the increasing paraxial numerical aperture (radius of the lens divided by the effective focal length).

As we have seen in Chapter 4, diffractive element microlenses can be corrected for spherical aberration and thus would appear to be the better choice for this application. The problem is primarily in obtaining the high numerical aperture. As we have pointed out in Chapter 4, the limitation here stems from the resolution of photolithography.

Oikawa et al. [2] have shown the use of a GRIN microlens for this application. Their results are given in Figure 6.3. They plot the insertion loss in db against the NA of the microlens for five cases of the beam spread of the source. The inset defines the asymmetric angular beam spread of the source. Their data indicates that a 1-db loss would occur for a microlens with an NA of near 0.4 when used with a 0.5-NA source.

The emitting area of the LD is in the order of 2×2 μm, which yields a diffraction-limited beam spread of >0.5. This estimate comes from the diffraction-limited beam spread emanating from a 2-μm slit, yielding roughly a full angle of λ/D. This means that the microlens should have an NA of 0.5 or greater.

To deal with the situation, Oikawa et al. [2,3] used a special kind of GRIN lens. They combined the normal GRIN lens discussed in Chapter 3 with a re-

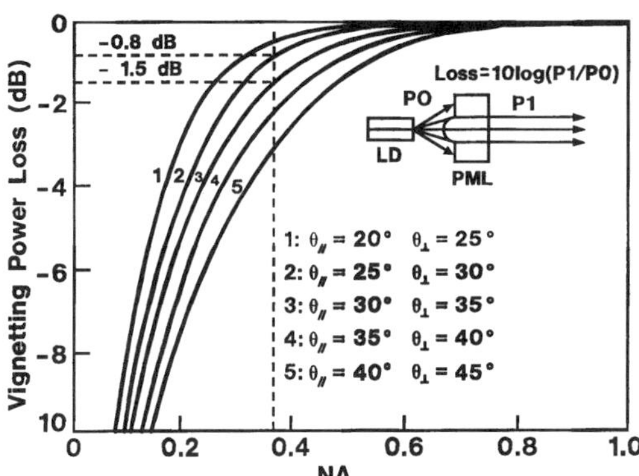

Figure 6.3 Power lost from a source with a beam spread angle indicated by the θ as a function of the numerical aperture of the microlens. (From Ref. 3.)

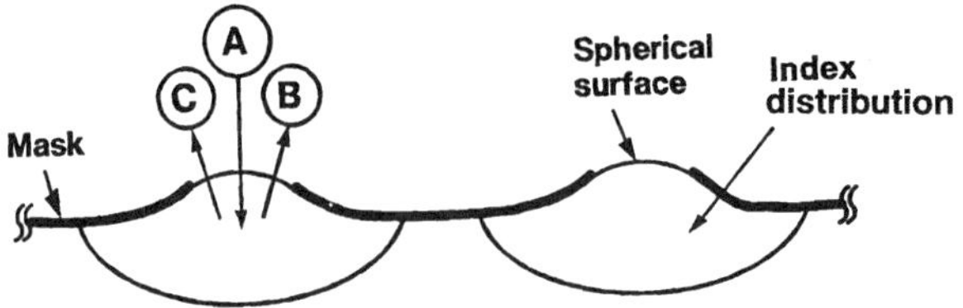

(Lens Diam 2a = 85 μm)

Diffusion Area (μm)	Diffusion Length (μm)	focal length (μm)	NA
250	85	110	0.37
400	156	78	0.48
422	168	55	0.61

Figure 6.4 Schematic drawings of the GRIN lens with "swell." The ion-exchange process produces a volume change which causes the lens to bulge. This can be used as a refracting component to the GRIN lens to significantly increase the NA. The insets indicate the effective performance numbers. (See Refs. 2 and 3.)

fracting effect brought about by retaining the swelling effect produced by the ion exchange. This is represented in Figure 6.4 by the sketch, together with the specific lens parameters that were used. What is most interesting and instructive is the insertion loss data. Radiometric lens performance is one thing; aligning and registration of the microoptical elements is an equally important issue. They show the loss with respect to variation of the object distance in Figure 6.5a, the lateral displacement of the object (LD) in Figure 6.5b compared to no lens, the effect of the variation of the image distance in Figure 6.5c, and the lateral displacement of the image (fiber) in Figure 6.5d. One can see the extreme sensitivity to any misalignment.

Now consider the output end where the light is to be focused onto the single mode fiber. The fiber has typically a 5–10 μm core diameter and an NA of the order of 0.1–0.2. For maximum coupling, one wants to match the NA of the exit focus beam with the fiber. This makes sense since any light ray making an angle greater than this would not get trapped by the fiber. One way to do this is to magnify the image, that is, working with an arrangement that is not symmetric with respect to object and image. We show an example with the aid of Figure 6.6. The lens properties are listed in Table 6.2. The schematic drawing of the

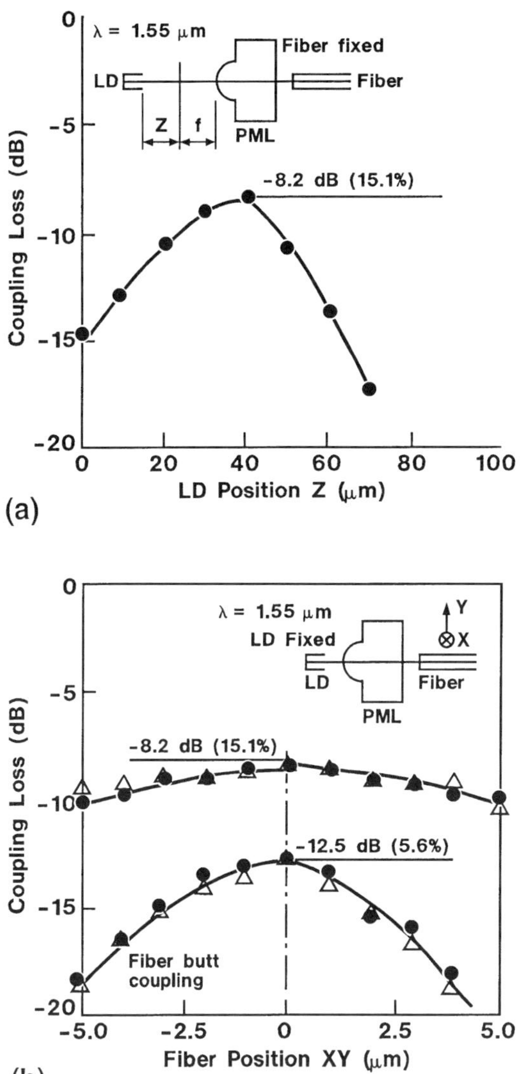

Figure 6.5 Effect of misalignment of the lens on the loss. The particular kind of offset is indicated by the drawings. (From Ref. 3.)

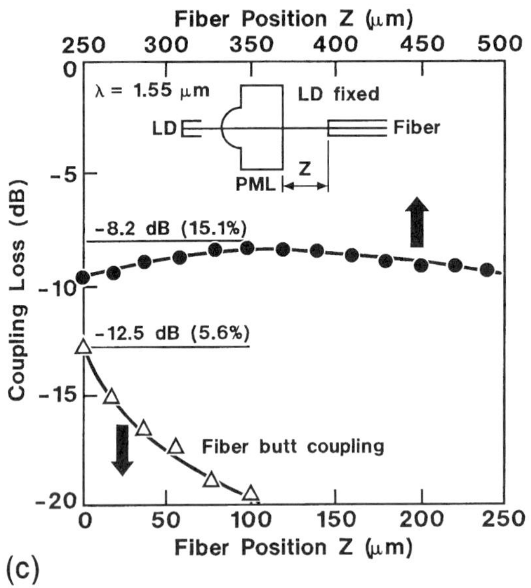

(c)

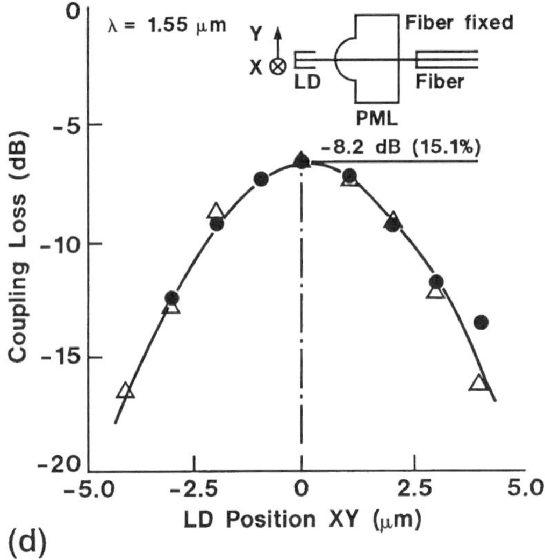

(d)

Figure 6.5 Continued

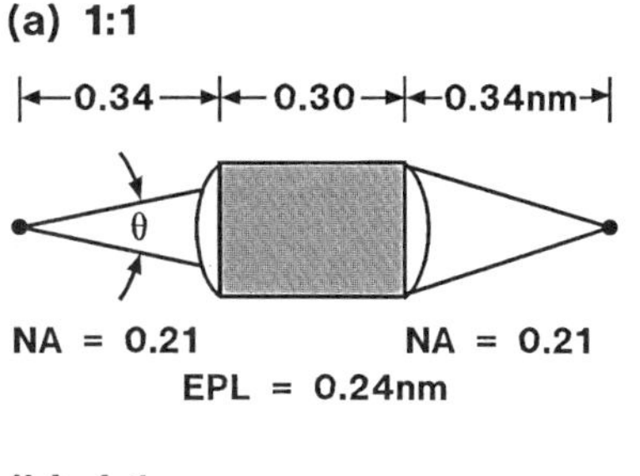

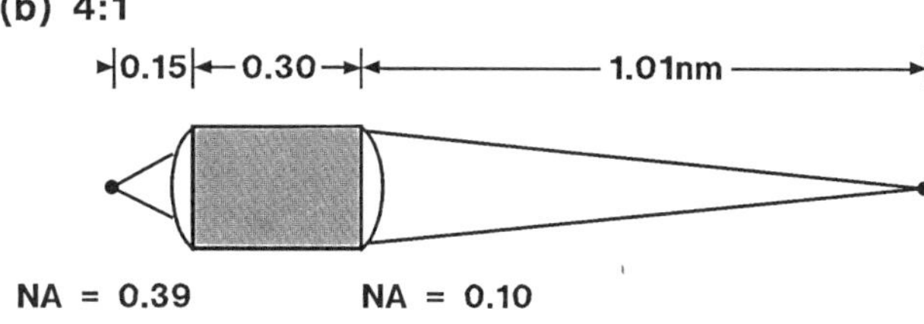

Figure 6.6   For the lens whose properties are listed in Table 6.2, A shows the schematic of the one-to-one imaging arrangment and B is the computed ray trace for this situation. In C is shown the four-to-one reduction arrangement where the NA of the fiber is matched.

Table 6.2   Lens Properties for Example Design

Lens type	Biconvex refraction
Diameter	0.2 mm
Lens radius of curvature	0.17 mm
Lens thickness	0.3 mm
EFL	0.24 mm
Principal plane	0.14
1-1 Configuration	
Object working distance	0.34 mm
Image working distance	0.34
Object and image side NA	0.21
4-1 Configuration	
Object working distance	0.15 mm
Image working distance	1.02 mm
Object side NA	0.39
Image side NA	0.10

one-to-one arrangement is shown in Figure 6.6a [4]. The 4-1 reduction schematic is shown in Figure 6.6b. The design was done by using the expression relating the object and image distances to the effective focal length.

$$\frac{1}{\text{EFL}} = \frac{1}{X_0} + \frac{1}{X_i} \tag{6.1}$$

and making $X_i = 4X_0$. The result is that $X_0 = 1.25$ EFL. Recall that the distances are measured from the principal planes; thus we have $X_0 = 1.25$ (0.24 mm) = 0.29 mm and the object working distance is then 0.29 mm -0.14 mm = 0.15 mm, while the image working distance is 4(0.24 mm) $-0.14 = 1.02$ mm. The object NA is 0.39 and the exit NA is 0.10. One can see the big improvement afforded by this design. The input NA is increased by moving the object closer, while the exit NA is reduced to match that of the waveguide. One can also view this from the spot size argument as well. By lowering the NA of the exit beam the spherical aberration is lessened; thus the blur of the image is less. All of this is from a purely geometric optic standpoint. One must also consider the diffraction aspect as well. With the lower exit NA, the diffraction limited spot size will be larger. Thus, although there is less aberration, the inherent spot size will be bigger. It does not take much misalignment to cause intolerable insertion loses.

Expanded Beam

The application is to collimate the output beams of an array of SMFs for a distance sufficient for the insertion of a functional optical element between them. The initial lens array collimates the light from the input array and a corresponding lens array to refocus the beams onto the output array. All this is to be done with the minimum of loss of signal. The application is represented schematically in Figure 6.7. The optical element to be inserted can be a passive element like an optical filter, or a more complex element like an optical isolator, or an active element like a liquid crystal array that could gate the passage of light from any of the fibers. A computed ray trace of such a lens function is shown in Figure 6.8a. The numerical aperture of single mode waveguides can range from values of 0.1 to 0.3. The ability to collimate the output as a function of the fiber NA is illustrated by way of a ray trace in Figure 6.9. The microlens here is a biconvex structure using a 200-μm-diameter SMILE array as a prototype. The upper trace (a) is for a numerical aperture of 0.16, while the lower (b) is designed for a NA of 0.25. One can see the poor collimation at the higher NA.

When one is dealing with single-mode waveguides, the more appropriate treatment is what is termed Gaussian optics [3a]. The single-mode intensity profile emanating from the 10-μm core is Gaussian and diffraction should be taken into account. The Gaussian development gives a picture as shown in Figure 6.8b. The consequence of diffraction is that a minimum in the beam waist occurs. The

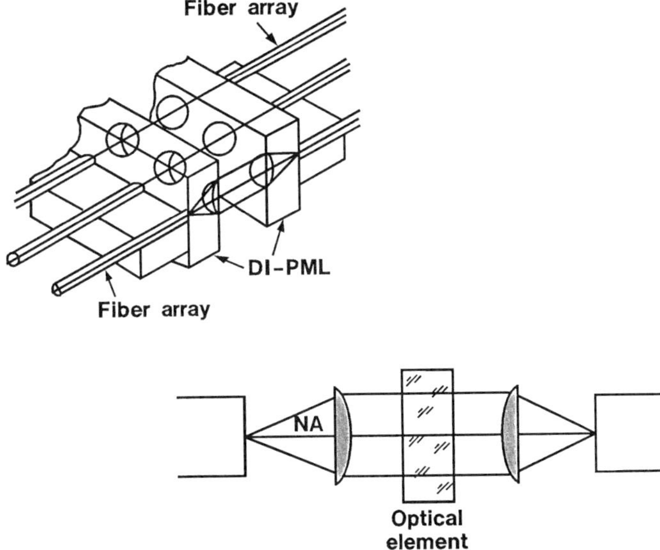

Figure 6.7 Schematic of the optical path for beam expanding application, and a proposed V-groove structure to implement the function. Single-mode fibers are placed in etched V grooves in Si to produce the precise spacing. The first lens array is aligned to collimate the light from each guide, and the second to refocus the light onto the output array. The intervening parallel beam opening is to be used to insert hybrid elements.

distance to this minimum limits the separation distance that can be opened up. This distance can be expressed as

$$L = \frac{\lambda f^2}{2\pi r_0^2} + f \tag{6.2}$$

where f is the focal length of the microlens, λ is the wavelength, and r_0 is the beam size at the fiber. One can easily see that to collimate light from single-mode fibers, there is a maximum distance over which this can be carried out and it is the order of millimeters.

Oikawa et al. [3] reports the insertion loss as a function of the separation distance L, as shown in Figure 6.10. Here they use a swelled GRIN lens as the collimator for the LD and a conventional planar GRIN to refocus the light onto the fiber array. This represents a more difficult situation than the fiber-to-fiber because of the higher NA requirement for the LD collimating lens.

As a matter of comparison, in the single-element case where GRIN rod lenses are used as the collimators, the insertion loss can be quite small, <0.5 db [2].

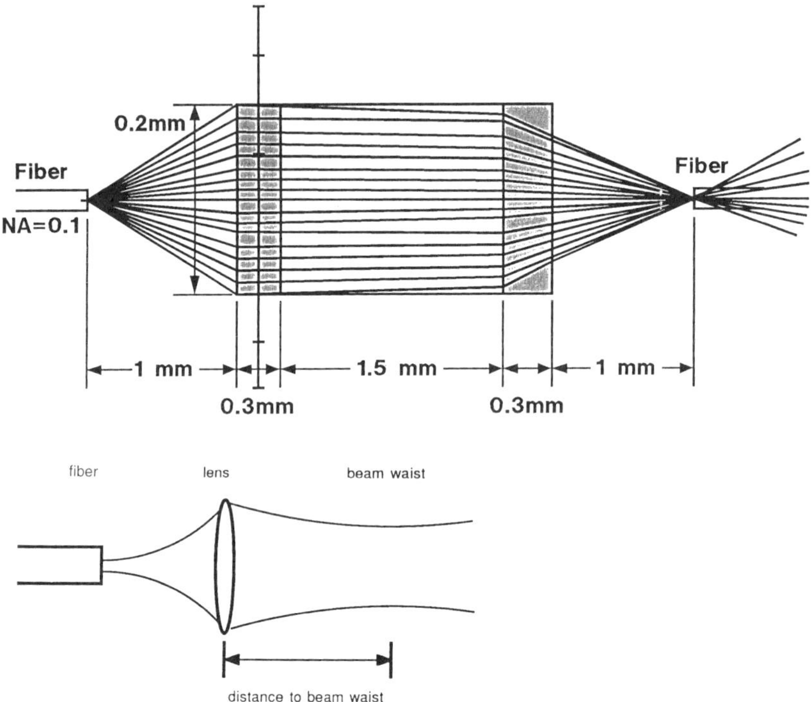

Figure 6.8   (a) Computed ray trace for lens array to perform expanded beam function. For this case the lens array was designed to accept light from a single-mode fiber with an NA of 0.1. The lens diameter was 0.2 mm, and the opening was 1.5 mm. (b) Actual Gaussian treatment of collimation where beam waist is shown.

As an example, we show in Figure 6.11 the utilization of this structure with a filter which directs light to the appropriate output fiber. The alignment is a critical parameter.

Crossbar Switch

This application is to take light of wavelength λ_i contained in one of N input fibers and direct it to any arbitrary one of N output fibers. This is shown schematically in Figure 6.12 for what would be called a 4×4 crossbar [5]. The input of 16 fibers is grouped into four sets of four, each fiber in the set carrying one of the wavelengths. The output fibers are also grouped into sets of four. Any combination of output wavelengths can be made to appear from each set.

In this case the switch is fabricated from a thin silicon film technology [6–

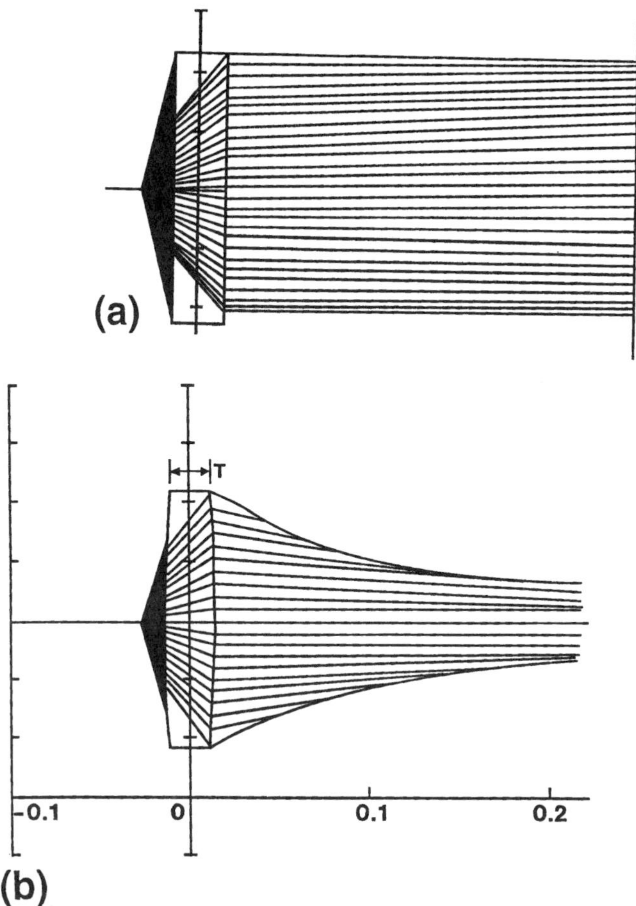

Figure 6.9 The effect of NA on the collimation. The upper ray trace is designed for an input NA of 0.16, the lower for an input NA of 0.25. One can see the effect of spherical aberration on the poorer collimation performance of the higher NA lens array.

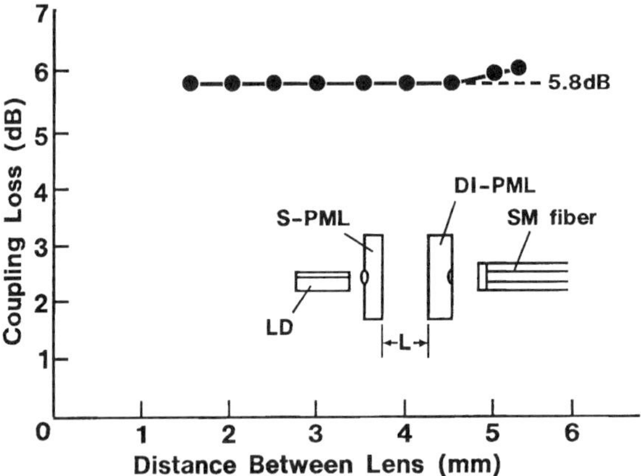

Figure 6.10 Coupling loss as a function of separation L. (From Ref. 3.)

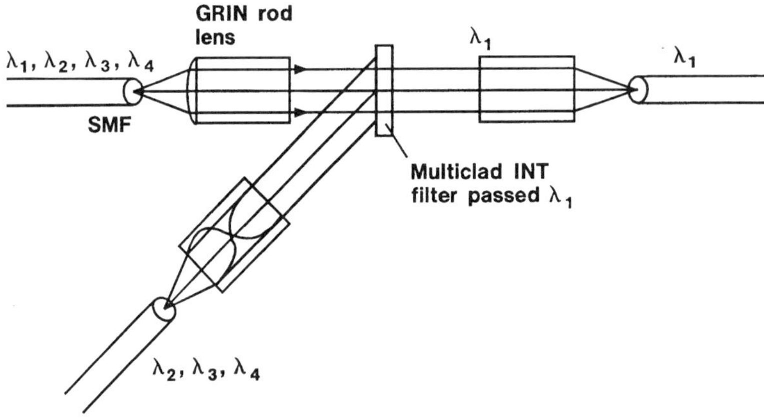

Figure 6.11 Example of the use of GRIN rod lenses in expanded beam dielectric filter-based wavelength demultiplexer. The lenses here are rod lenses, not the planar array type. Function here is to separate one wavelength and send it to another location.

9]. The photomicrograph of the actual silicon structure is shown in Figure 6.12c. The silicon element is electronically addressed, which has the effect of electrostatically deforming the silicon membrane, as shown in Figure 6.12b. The membrane has two positions, either allowing the light to get to the output fiber or not. By the appropriate switching sequence any wavelength can appear at any of the four outputs.

The role of the microlens array is to collimate the light from each of the 16 fibers and focus it onto the deformable mirror and then to reimage the focal spot to the output fiber as shown in the figure. The realization of this device by TI [5] was achieved using the 16-element SMILE™ lens array [10] shown in Figure 6.13. The lenses were planoconvex, 0.6 mm thick, with a focal length of 0.4 mm. The lens diameter/spacing was 160 μm/195 μm. The effective NA of this arrangement was only 0.1 which was, however, consistent with the multimode fiber.

The reported operation of the device produced greater than 30 db contrast in all channels. The overall insertion loss was reported as 24 db, but only 6 db was attributed to the optical system, the rest having to do with the splitting and recombining of the signals.

Computer Backplane

As computer designers seek ways to increase speed, it was suggested that the time it takes to communicate between chips, boards, and shelves could be reduced

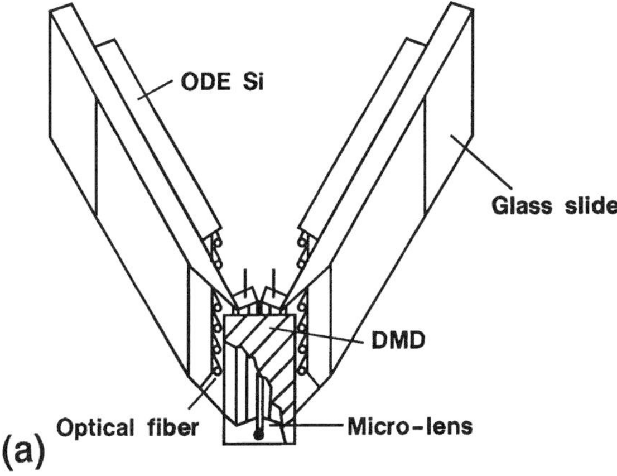

Figure 6.12 Example of crossbar switch application using a lens array. Signal brought in on the linear array of 16 fibers. The light is focused on to the micromirror as shown. Depending on the position of the mirror, it determines whether the light is transmitted or not. (See Ref. 9.)

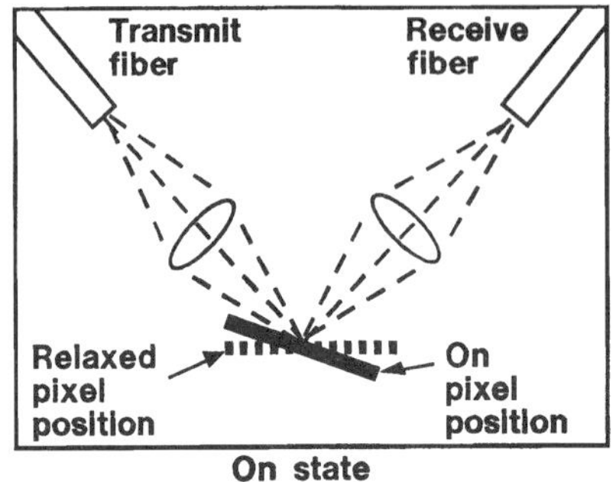

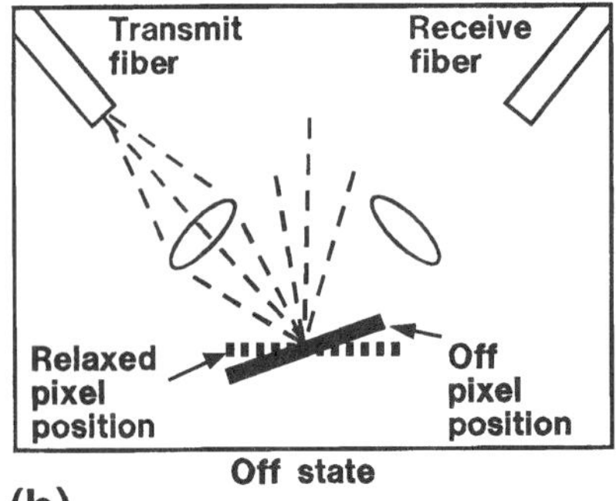

(b)

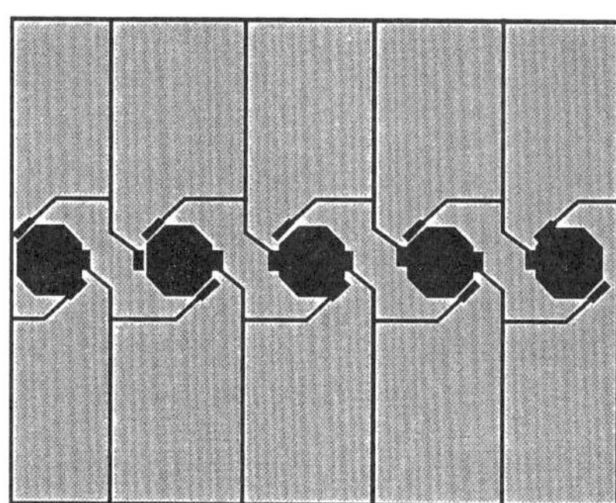

(c)

Figure 6.12 Continued

Figure 6.13 The SMILE lens array used for the crossbar switch application. The lens diameter is 160 μm on 195-μm centers with a focal length of 0.4 mm. The upper is the actual array, while the bottom is a SEM picture.

if it were done optically rather than electronically. This argument is essentially based on getting around the limitation posed by the RC time of connecting cables. The idea is to have the output of a chip or board within the computer converted to light and then have this transmitted to the detector array on the other chip or board. There are a number of ways that this transmission can occur, but we will only be concerned with the multimode fiber optic method. The other major method is called free space and involves diffractive elements to provide a directed beam pattern. (This will be covered in Chapter 7.) The fiber optic method does provide a more mechanically flexible approach, but otherwise there are advantages to both methods.

The optical application is similar to that of the collimator application discussed at the beginning of section 6.2.1, in that the source will have a large NA. But here the fiber is usually multimode; thus a higher NA is required, >0.3. Otherwise, the design method is exactly the same as that presented above. An example is shown

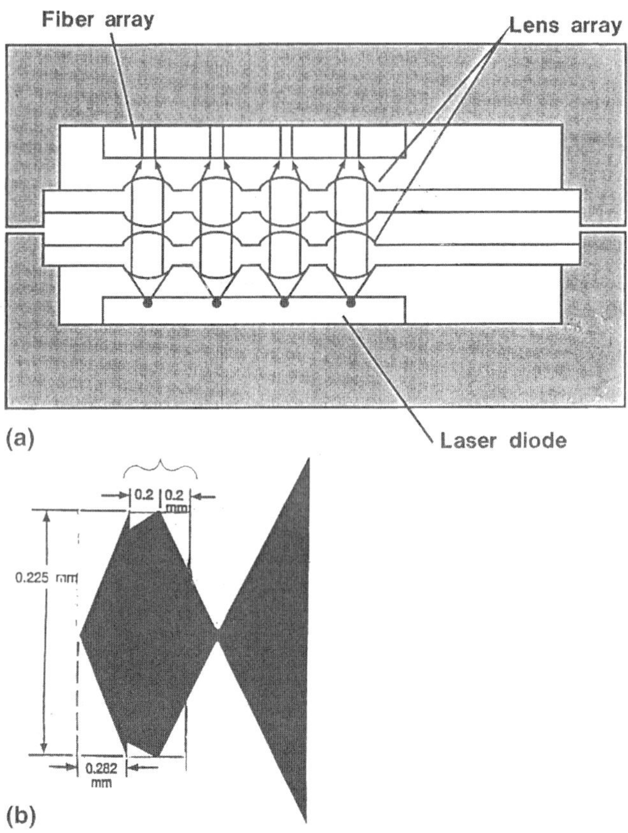

Figure 6.14   Schematic drawing of laser to fiber array. A computed ray trace is shown in the lower figure with the use of the stacked biconvex design. The lens diameter is 0.225 mm, and the radius of curvature of each surface is 0.326 mm. The effective focal length is 0.36 mm.

in Figure 6.14a. The light-emitting diode array is coupled to a fiber array through the use of a stacked pair of biconvex lens arrays. A simulated ray trace of the stacked biconvex lens array (SMILE) designed for this use is shown in Figure. 6.14b [11]. The input NA of this design is 0.34.

6.2.2   Other Related Applications of Linear Arrays

LED Print Bar

The application is to transmit the light from an array of red LEDs onto a detector array. This is a fast optical printing head with the resolution determined by the

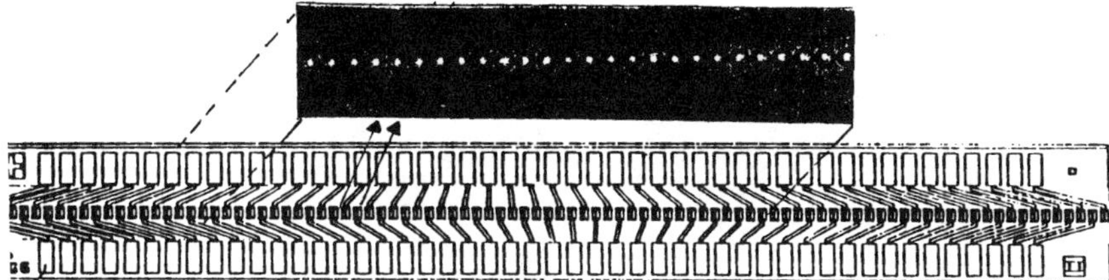

Figure 6.15 LED print bar application. The 300-DPI LED array is illustrated with the lens array shown in perspective. The lens array was made up of 75-μm lenses on 84.66-μm centers.

pitch of the LEDs which in this case is 85 μm, which translates into 300 dpi (dots per inch) for the example shown in Figure 6.15. An actual photograph of the lens array is shown above the LED array. The diameter of the microlenses were 75 μm on the 85-μm pitch [12]. A photograph of one of the individual images formed by the microlens is shown in Figure 6.16. The numerical aperture of the microlenses was 0.1, and the overall separation between the LED array and the detector array was 0.78 mm.

One of the problems with designing a lens for this type of application is that the lens diameter can only be as large as the pitch of the LED. The diameter represents the limitation of the SMILE lens array and is pushing the limit of the planar GRIN approach as well. For diffractive elements the problem is that, although this is not a small diameter from a fabrication point of view, it becomes increasingly difficult to fabricate a lens with a high NA.

Camera Autofocus

One microlens array that has reached production is that which was used by Honeywell [13] to produce an autofocus element for camcorders and 35-mm SLR cameras. It provided a through-the-lens autofocus technique that is a desirable feature because of its better accuracy. This is done with the use of a beam splitter that takes some of the light that enters the objective lens to the sensor, as shown in Figure 6.17. It should be noted that the distance from the lens to the film plane is equal to that to the sensor plane.

The autofocus concept is based on a radiometric rather than imaging principle. When an object is in focus the amount of light arriving at the image plane is equal from any arbitrary halves of the imaging lens. The Honeywell device uses this fact in the way that we will show with the use of the blowup of the sensor shown in Figure 6.18. There are two detectors behind each microlens. The method will be to compare the outputs from the string of A detectors to the outputs from the B string. When the object is in focus the image appears at the lens plane and

Figure 6.16   Actual photograph of the light-emitting area of one diode and the corresponding image produced by a lens of the array.

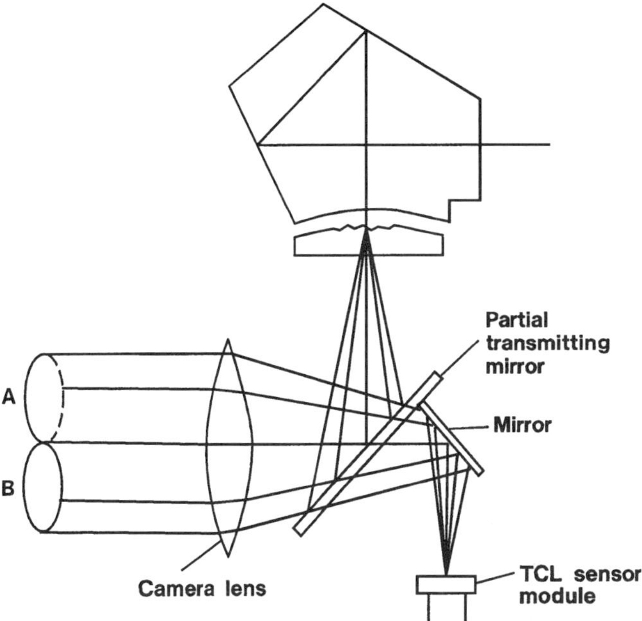

Figure 6.17 Schematic of Honeywell autofocus concept. Part of the light from the objective is directed to the sensor element located at the same distance as the film plane.

an equal amount of light reaches each of the two detectors, as depicted in the middle ray trace. Here, for the sake of simplicity, we are looking at the light from only one point in the object field. When the focus position precedes the correct focus, one gets the situation depicted in the upper ray trace. For this situation, light from the upper half of objective reaches the A string, whereas light from the lower half tends toward the B string. If the focus is beyond the correct focus, the opposite behavior is observed as shown in the lower ray trace. A schematic of what the output from the respective detector would see is shown in the inset. The lack of overlap of the outputs would indicate that the object is not in focus. The nice thing about this method is that depending on the phase of the output curves, that is, the A string leading or lagging, indicates which way the objective lens should move to attain focus.

The initial lens array used for this device was a molded plastic array manufactured by USPL [14]. It was subsequently replaced with SMILE lens array for a number of reasons [15]. The cross talk was less because of the opaque surround to the lenses, the uniformity and positioning of the lenses were better, and the glass lenses were more securely attachable to the detector die. The array in this

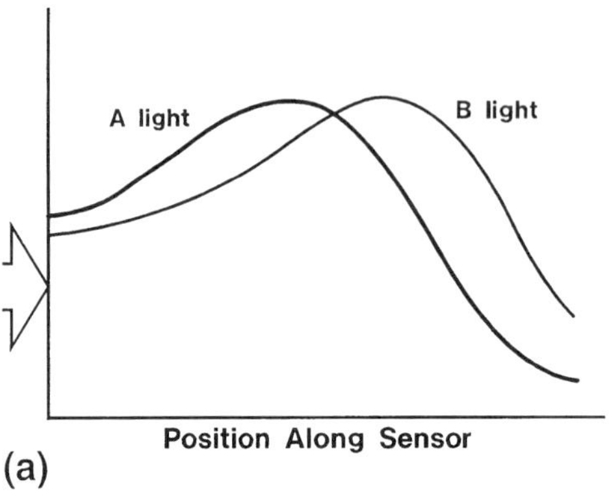

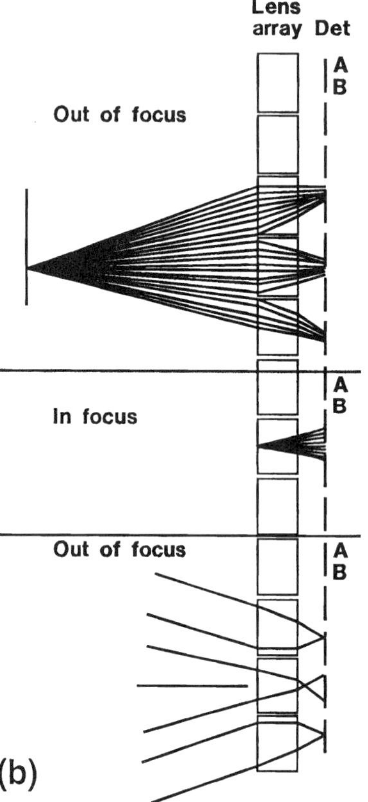

Figure 6.18   (a) Basis for autofocus module. If an object is in focus, the light is focused at the lens array plane and the light is split equally by each lenslet onto the split detectors behind the array. (b) If an object is out of focus, either too close, or too far, one set of detectors sees more light than the other. The relative output of the two sets of detectors is indicated by the diagram. When they coincide, the object is in focus.

case was made up of a row of 26 planoconvex microlenses, 160-μm-diameter lenses on 195-mm centers with a focal length of 400 μm.

6.3 TWO-DIMENSIONAL ARRAYS

There are a number of newer applications that require 2D arrays of lenses. The technology of producing 2D arrays does not differ in any substantial way from that to make linear arrays except that maintaining the precision of location is compounded by the added dimension. In other words, 2D microlens arrays can be made by molding GRIN, diffractive, etc.; however, when such issues as fill factor and uniformity are important, then differences in the method of fabrication can make a difference.

6.3.1 Shack-Hartman Wavefront Analyzer

This is an astronomical telescope application where an incoming wavefront is sampled by means of a $N \times N$ array of detectors, such that it can provide a basis for correcting disturbances that appear over time [16]. The prime example is that where atmospheric disturbance is to be removed from a ground-based astronomical system [17]. There are certain design conditions for optimum utilization of the device which bear on the optical properties of the microlens array. In particular they involve the interplay between the wavelength of the light and the focal length of the microlens, to the optimum subimage to grid and the spot size. Artzner [17] has shown that an optimum ratio of subimage size to the grid spacing should be much larger than $\lambda/4\delta$, where δ is the sag of the microlens. The net

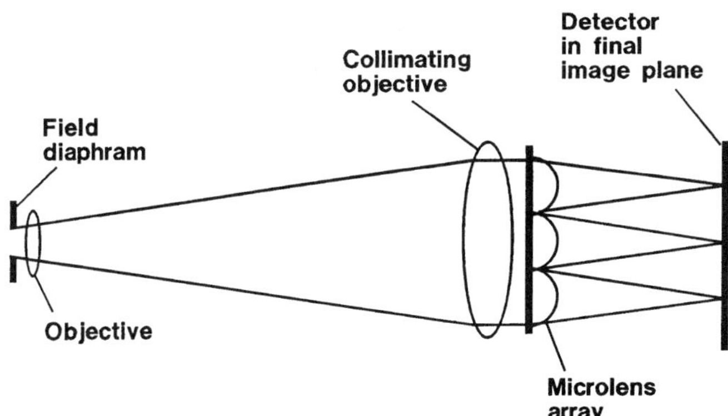

Figure 6.19 Optical diagram of a Shack-Hartman wavefront analyzer. The lens array images to an array of photodetectors.

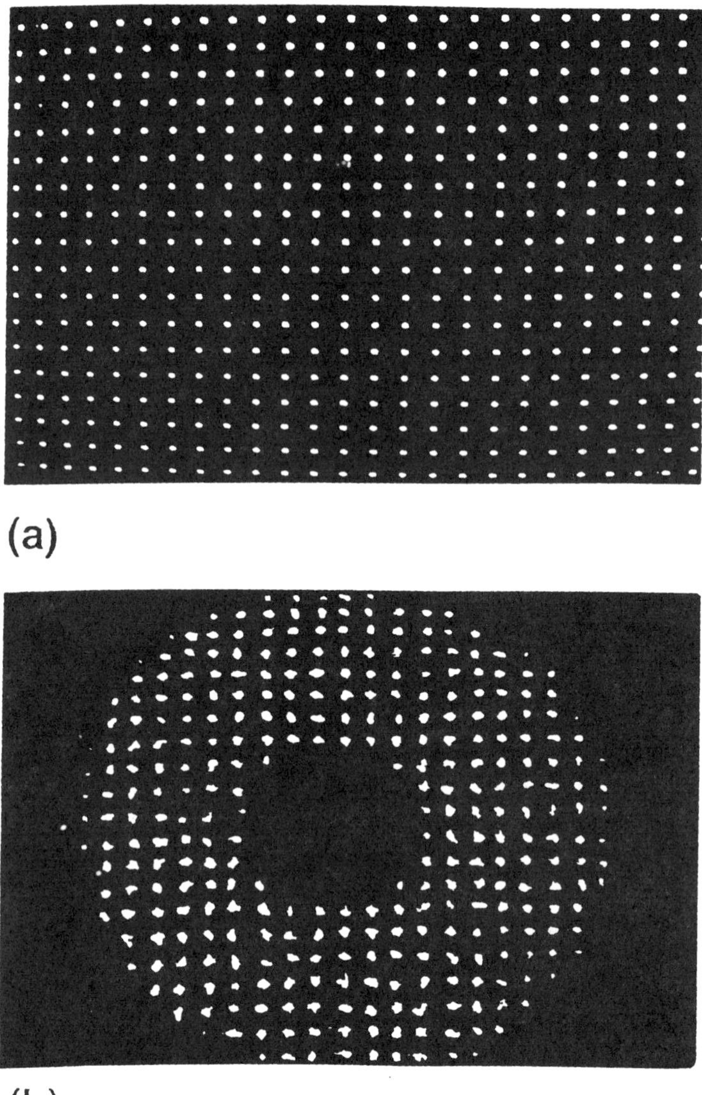

(a)

(b)

Figure 6.20 Typical pattern from a 22×22 Shack-Hartman array with a pitch of 0.185 mm and a focal length of 7 mm. The upper photo is the reference grid for an artificial point source, and the lower is the actual sensing image from a 4.2-m telescope. (From Ref. 16.)

result is that the sag of the microlenses will be small irrespective of the period and focal length of the microlenses.

A sample pattern taken from Ref. 16 is shown in Figure 6.20 for a 22 × 22 array with a pitch of 0.185 mm and a lens focal length of 7 mm. Figure 6.20a is that of the reference grid for an artificial point source (Figure 6.19). Figure 6.20b is that of an actual sensing image from a 4.2-m telescope. Different sub-images have different shapes corresponding to different instantaneous atmospheric wavefront distortions.

This lens array has a sag of about 1.5 μm. The smallness of the sag represents one of the main challenges to the lens preparation for this application. One way to offset this is to reduce the power of the lens by near index matching the lens with a suitable liquid.

6.3.2 Parallel Processing (VCSELS)

With the advancement of GaAs technology there has emerged a new area which is based on a vertical fabrication architecture, that is, structures that are fabricated such that their output is normal to the face of the growth. An important example of this is vertical-channel surface-emitting laser, or VCSEL, for short [18–21]. A schematic representation of such a device is shown in Figure 6.21, which is taken from Ref. [18]. The ability to deposit precise thin layers of varying composition allows one to create not only the active lasing junction, but also the ''mirrors'' which in the case shown are distributed feedback gratings. This type of vertical architecture distributed over two dimensions is the ideal way to imagine a truly parallel system where each laser can be individually addressed. The vertical architecture is not limited to lasers. One can also construct amplifiers, detectors, and switches. Indeed, if optical computers ever become a reality it may very well be based on this type of technology.

The ability to couple light in and out of these devices provides a real challenge because the lateral dimension of these devices is typically in the range of 5–50 μm. Clearly, the smaller the better in order to maximize the areal density of the device, but the lenses must also be smaller. In the lower portion of Figure 6.21, we see a schematic representation of how the microlens array would be fitted to the VCSEL to couple light out. Many of the methods of making lens arrays will not work for lens diameters below 50 μm. What does work at the smaller diameters are the melted photoresist method and the diffractive element. The performance of diffractive lenses for this application have been reported down to a diameter of 43 μm.

6.3.3 Laser Eye Protection

One of the more interesting applications proposed for microlens arrays is where they would be utilized as an element of a broad band laser protection goggle.

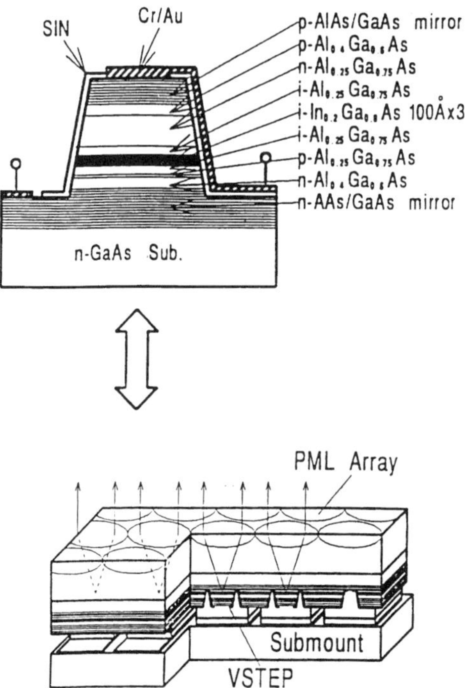

Figure 6.21 Representation of a VCSEL (vertical-cavity surface-emitting laser) and how a lens array would be positioned to couple out the light. (From Ref. 18.)

With lasers present everywhere on the battlefield, the need for eye protection became a real need. Moreover, the protection had to be more than for a fixed wavelength since many lasers are purposely randomly tuned to avoid jamming.

The optical construction is represented in Figure 6.22. It is essentially a 1 × telescope [22]. The parallel light from the object is brought to a focus at the midplane of the lens and then emerges as a parallel beam. Since the intention is to use it as a goggle for human use, the images has to be erect. This is accomplished by using a second identical lens array. The active element is placed at the first focus and has the property that it blocks the light when the intensity reaches a desired threshold. Suffice it to say here that the nature of this limiter material is crucial in determining the usefulness of the device. The example we will use later on is a liquid crystal that changes from a transmitting to a scattering state when a certain intensity is incident. The purpose of the lens arrangement in Figure 6.22 is to provide this high intensity at the focal spot. The impracticality of this concept in using conventional lenses is the total thickness of the device.

a)

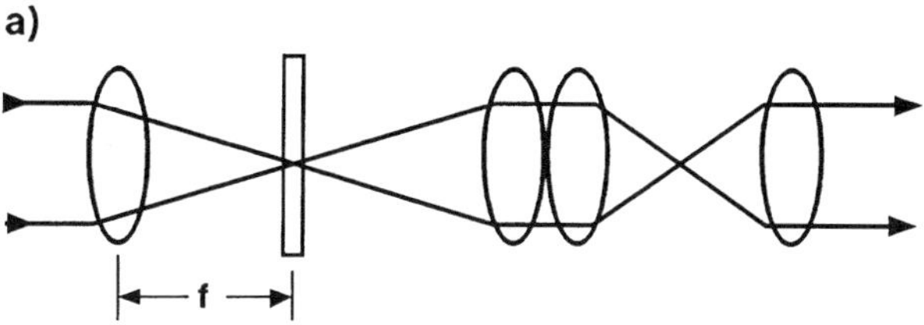

b)

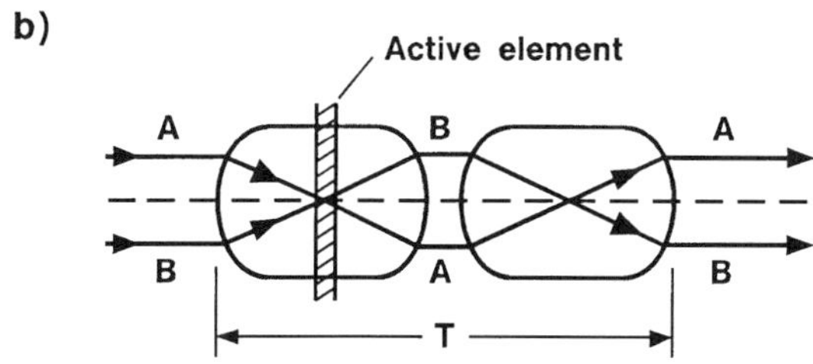

Figure 6.22 Concept of laser eye protection. Light is focused at a midplane containing the active switch material that would shut down at high intensity. Second array is to provide an erect image.

If f is the focal length of the lens corresponding to the required NA to produce the necessary spot size at the active plane to effect the switching, then the total thickness of the goggle could be no less than $4f$. For a conventional lens to cover the eye without restriction, the lens diameter should be at least 3 cm. Assuming that an NA of at least 0.25 is required, this would yield a total thickness of 24 cm.

The microlens approach is to replace the single element with an $N \times N$ array of microlenses operating in the same $1 \times$ mode, with a comparable NA. Since the lens diameter is smaller, the focal length will be correspondingly smaller, and consequently the overall thickness will be smaller. The expression for this condition is given by Eq. (5.6) with t being infinite, which yields the focal length as the lens thickness divided by twice the refractive index.

As an example, a lens array made up of 400-μm-diameter lenses on 480-μm centers was fabricated with an NA of 0.24 and an effective focal length of 0.83 mm. From the expression for the total thickness, this would yield something less than 5 mm. A ray trace of the operation of an array of such elements is shown in Figure 6.23.

There are some trade-offs with the microlens approach. For example, the active material can be switched by either peak fluence or intensity. In either case the energy delivered to the active plane is per unit area. The individual microlens

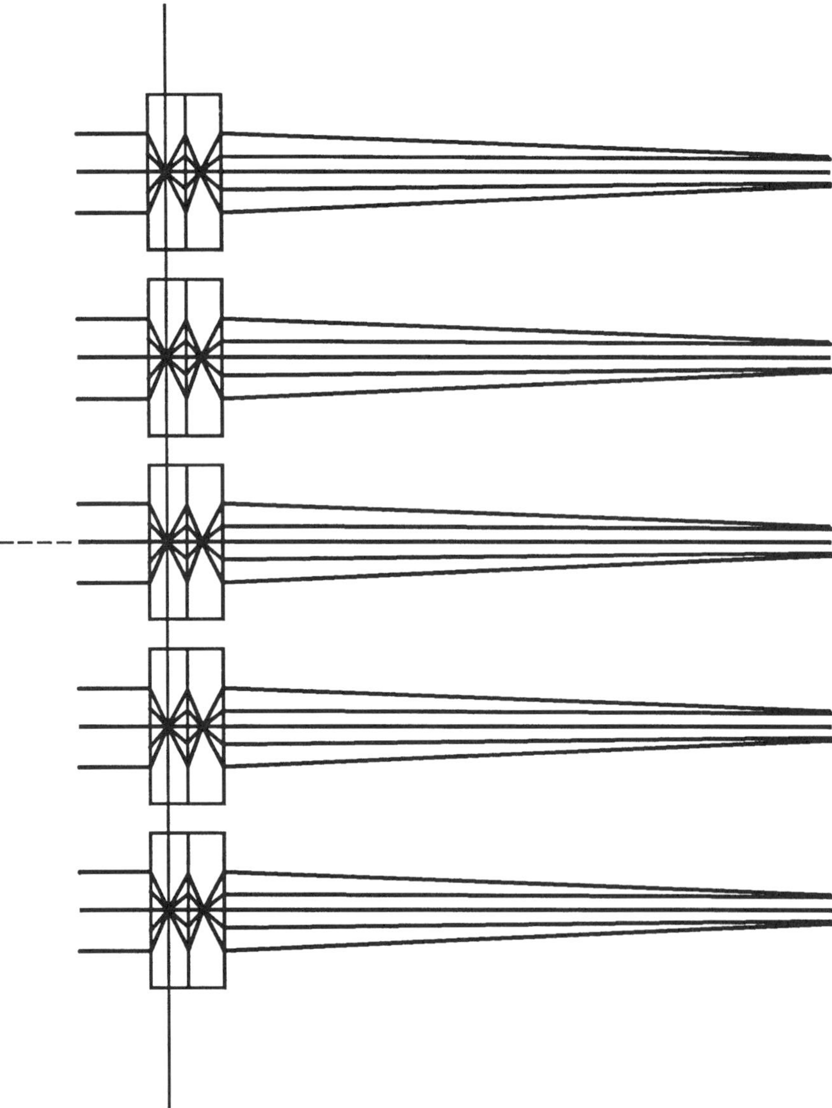

Figure 6.23 Computed ray trace of lens array acting as a one-to-one telelescope.

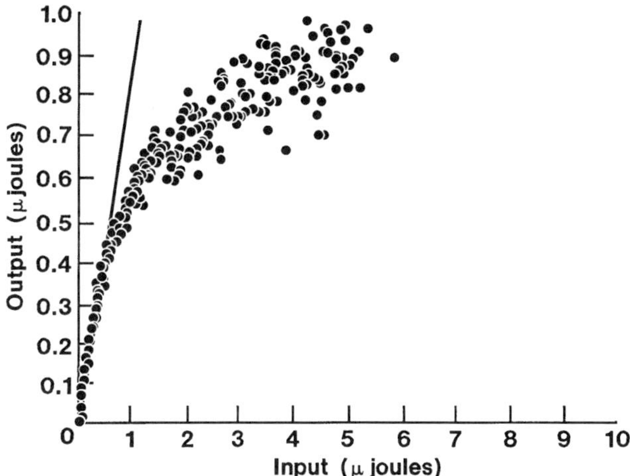

Figure 6.24 Actual optical clamping data obtained by using the optical arrangement shown in Figures 6.25 and 6.26 using a two-dimensional array of 400-μm-diameter lenses imaging onto a nonlinear liquid crystal material.

intercepts less light, yet the intensity must be sufficient to switch the active material. The device has been experimentally demonstrated using the above-mentioned smetic liquid crystal as the active medium REF. The output energy is plotted against the input energy (which would correspond to the ambient laser), as shown in Figure 6.24. The lens used was that described above. The threshold is defined as the point at which the slope deviates from unity, which in this case occurs at about 1 μJ. The maximum level, or so-called clamped value of the output, is the order of 1 μJ and occurs when the ambient energy is greater than 5 μJ. In general, eye safe levels are considered to be below 0.5 μJ.

A serious drawback to this device is the field of view. Unless the lenses could be fabricated with a curve, the vision of the wearer would be too restricted for practical eye use.

6.3.4 Projection LCD-TV

Projection LCD-TV is one approach to large-screen television. The active LCD matrix is contained in a 54 × 75-mm area. The light is projected through the panel onto the wall. One of the major problems with this projection approach is the optical efficiency. A number of attributes contribute to this problem, not the least of which is that it is a display based on polarized light, so one starts with a 50% efficiency. Color filters are another source of light loss. The feature that loses about 60% of the available light is the geometric obstruction on the panel itself produced by the conductors and the transistor gate. One can get an idea of this problem by looking at the structure of a typical panel, as shown in Figure 6.25. The clear area, which is referred to as the aperture ratio, is only about 40%. A microoptic solution to this problem is to use a microlens array to focus the incident light through the open area, as shown in Figure 6.26a,b. In (a), one can see the light lost by the nontransmitting area of the electronics. In (b) one sees how the microlens directs the light through the open ''window'' regions.

For the microlens approach to be viable for this application, it is necessary to produce an array with as close as a 100% fill factor as possible [23–25]. For certain of the fabrication techniques this poses a problem because of the effect that the coming together of the boundaries has on the lens. Related to this is the problem associated with having to make lenses with noncircular boundaries. We shall see this for the GRIN and the SMILE lenses.

Referring to Figure 6.26, the focal length of the lens should be close to the thickness of the cover glass. The standard cover glass thickness is 1.1 mm, but thickness as thin as 0.4 mm has been reported. Typical lens diameters are of the

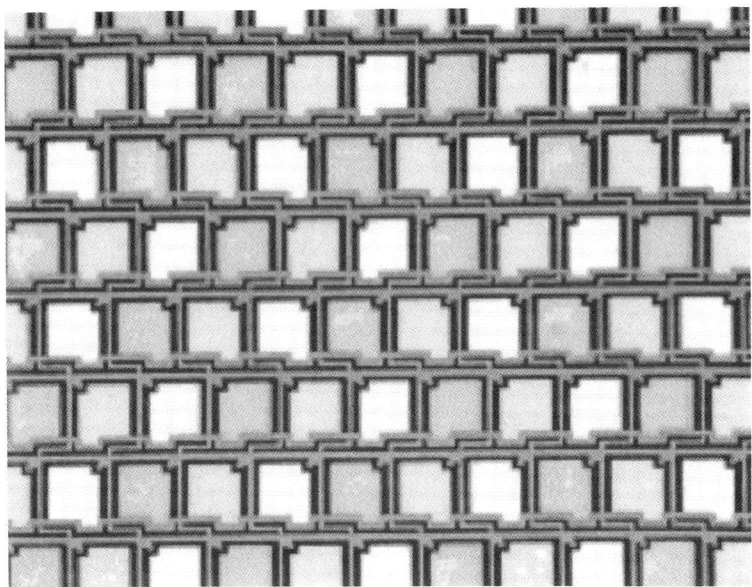

Figure 6.25 Photograph of a typical projection LCD TV panel. The pitch is approximately 100 μm.

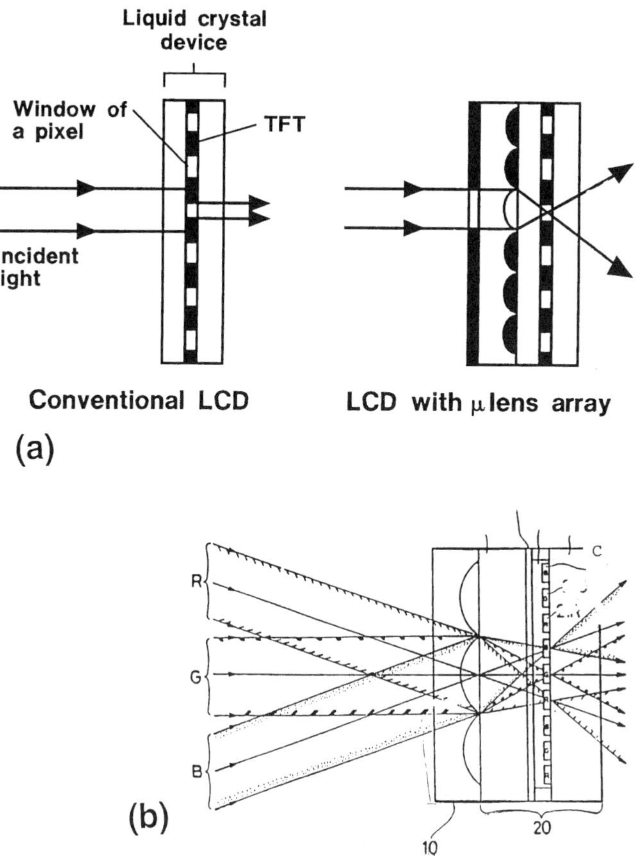

Figure 6.26 Two concepts to utilize lens arrays to improve brightness. The upper diagram (a) shows how the lens array would focus the light through the open areas of the LC panel shown in Figure 6.25. The lower diagram (b) shows how the lens array can direct each color light to a separate pixel, obviating the need for a color dot array.

100-μm order, so that the numerical aperture of the lenses is quite low. The problem this will produce is that this will limit the ultimate spot size. For the present pixel size shown in Figure 6.25, the rough diameter of the clear opening is of the order of 60μm. For an NA of 0.05, the diffraction limited spot size for the red beam would be 15 μm. The problem is that the projection lamp that is used is effectively an extended source, so that the light is not parallel and diffraction-limited performance cannot be achieved. The situation is indicated in Figure

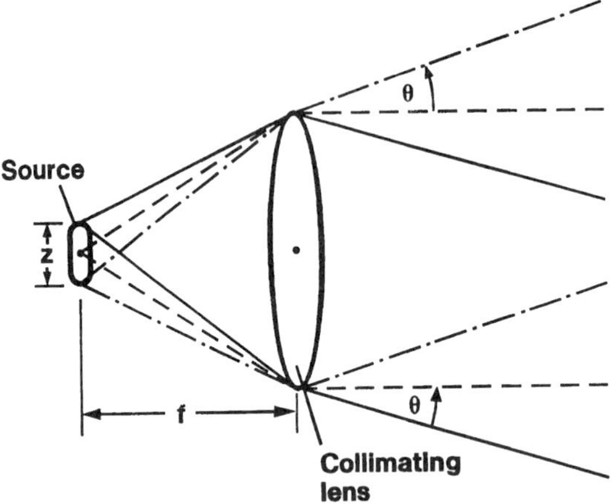

Figure 6.27 Consequence of extended source on the beam spread, and ultimately on the size of the spot that can be imaged.

6.27. The beam spread comes from the off-axis points of the extended source. One can show that the spread angle θ is related to the object height through the simple relation

$$\tan \theta = \frac{z}{2f} \tag{6.3}$$

Here f is the focal length of the lens. The spot size that will form will simply be the product of this angle with the focal length of the lens. For a 5° beam spread the minimum spot size would be 45 μm.

The spot-size issue is even further complicated by the physical layout of the pixels. The pixel separation in the x direction is not the same as in the y direction. In some designs the rows are staggered, presenting an elongated hexagonal layout. What this means is that in order to fill space, one has to resort to lens shapes that are not symmetric. As an example of this we will use the SMILE lens array [23]. The pixel pattern is shown in Figure 6.28a. The x direction pitch is 160 μm and the y direction pitch is 190 μm. If one uses the pattern of 150-μm-diameter circular lenses, Figure 6.28b, only 58% of the area would be covered, but the focus spot is well defined, as shown in Figure 6.28c. If one resorts to oval-shaped lenses to effect better coverage, one suffers the consequence of having an anamorphic behavior, as shown in Figure 6.29a. What this means is that the lens has a different focal length in the long and short directions of the lens. We show this

(a)

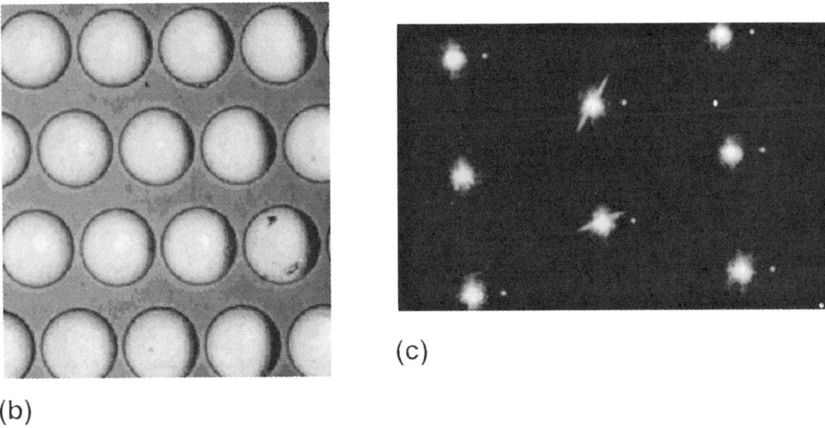

(b)

(c)

Figure 6.28 Array of circularly shaped lenses and the images formed to go through the LC pattern shown.

in the images below that pattern. The best coverage design is that of lenses with a hexagonal shape, which is shown in Figure 6.29b. There is still an anamorphic effect present which has an adverse effect on the spot size in that the image is elongated.

The planar GRIN approach [24,25] suffers from a similar problem. One would start with a patterned set of openings to allow the ion exchange to be accomplished. The pattern would be set by the x and y pitch. The length of time of the

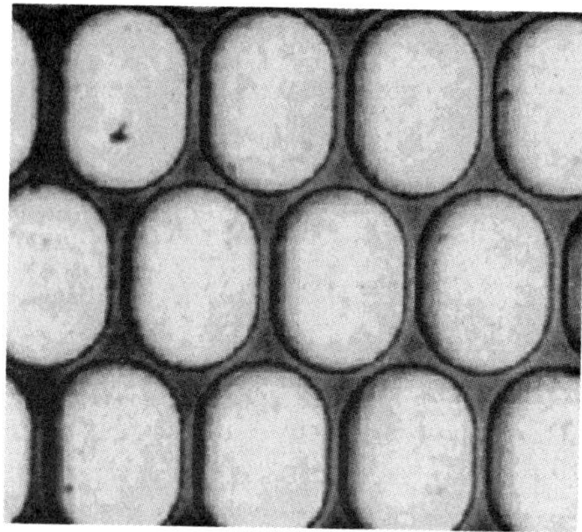

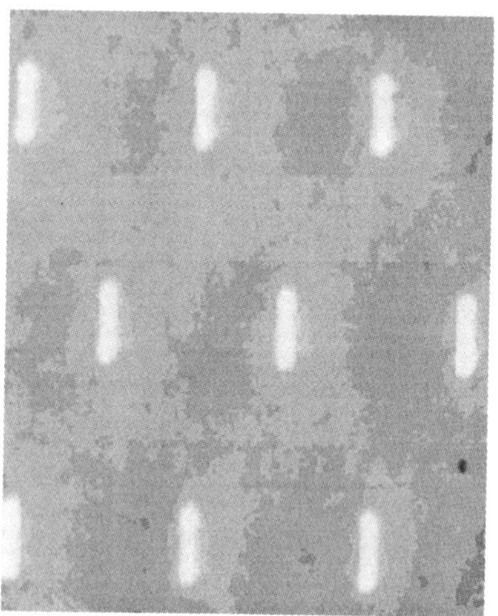 

(a)

Figure 6.29 Array of oval-shaped lenses and hexagonal-shaped lenses and the images formed. Note the lenses are anamorphic. For the oval-shaped lenses the image in each image plane is shown.

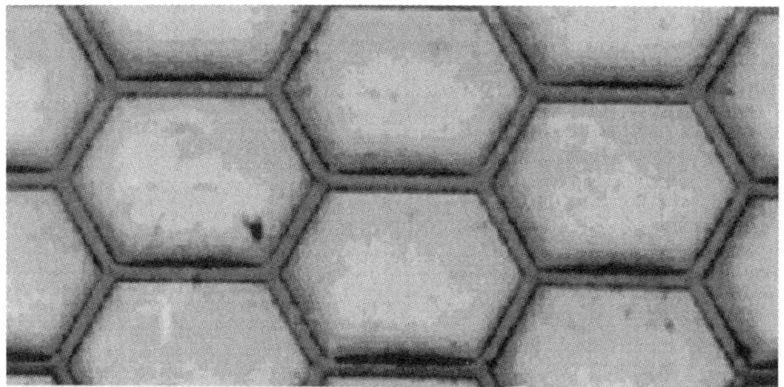

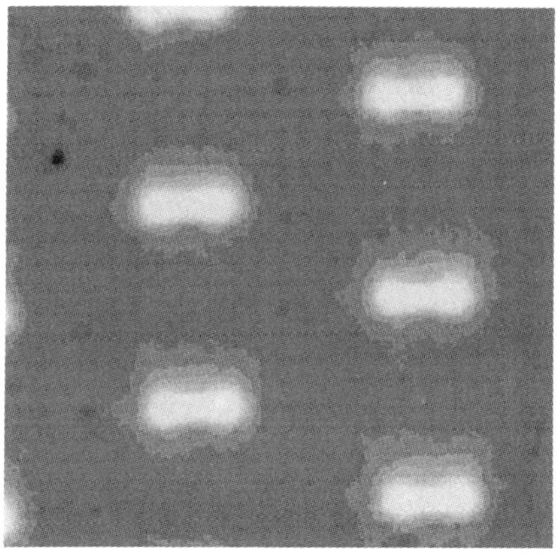

(b)

exchange would dictate the lens shape. This is schematically shown in Figure 6.30. The planar GRIN lens suffers similar asymmetry problems, as demonstrated by the images formed (Figure 6.31).

The ideal method of fabricating a lens array for this application is the diffractive approach (see the discussion in Chapter 4) because it decouples the shape of the lens from the performance. For example, one can almost seamlessly arrange the lenses in any configuration and still maintain the spherical-like lens imaging performance. Moreover, it is a low NA application so the resolution of the required fabrication is not an issue. A full 100% fill-factor layout is shown in Figure 6.32a. The fact that the diffraction efficiency varies with wavelength is easily remedied by adjusting the focal length of each lens in the triad for the RGB wavelengths. The actual function of a diffractive array to increase the efficiency of a LCD projection panel is shown in Figure 6.31b.

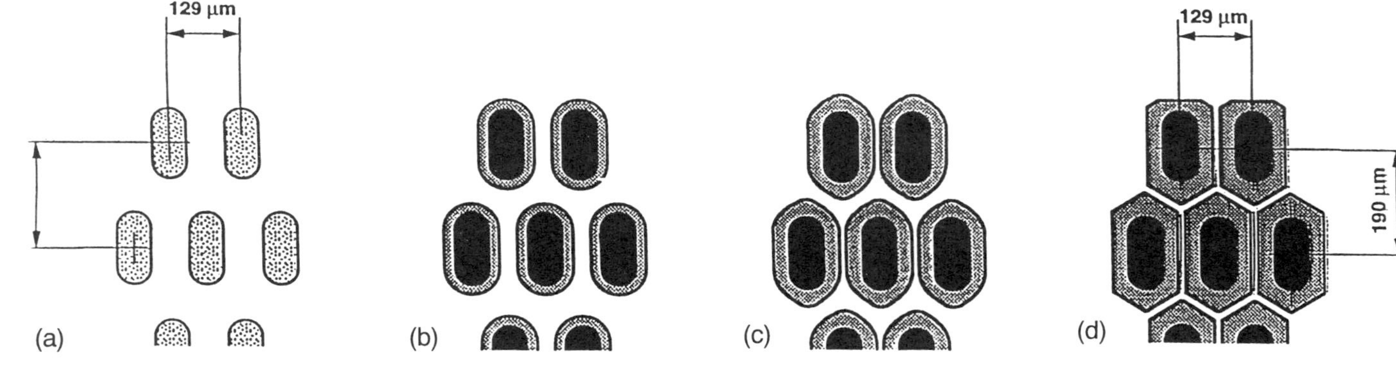

Figure 6.30 Formation of planar GRIN array for the same application. (a) Patterned substrate; (b) separated; (c) partially fused; and (d) closed packing structure (honeycomb shaped). The index region grows as the ion-exchange time through the aperture increases. Finally, when fronts meet, a hexagonal pattern results.

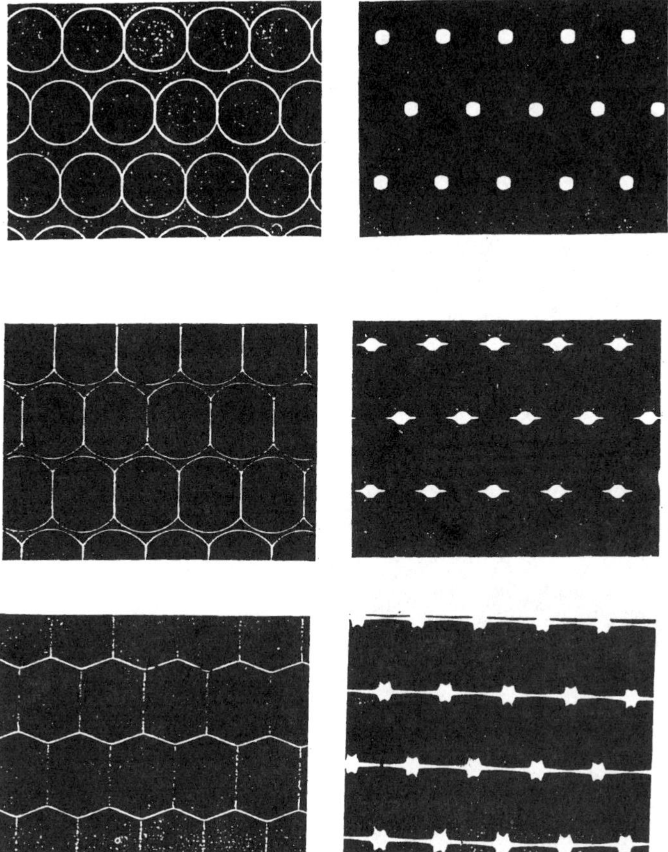

Figure 6.31 Lens shapes of GRIN lens arrays and corresponding images. (From Ref. 25.)

There have been reported efficiencies for the three lens fabrication methods, calculated as the percentage increase in the light through the array with the lens array in place, relative to none. They range from 50 to 90%.

6.3.5 3-D Images

People have always been fascinated by the ability to produce a three-dimensional image. Holography is by far the best-known example of how this can be accomplished. However, there is an older concept which does not rely on coherent optical properties at all. The method is called integral photography.

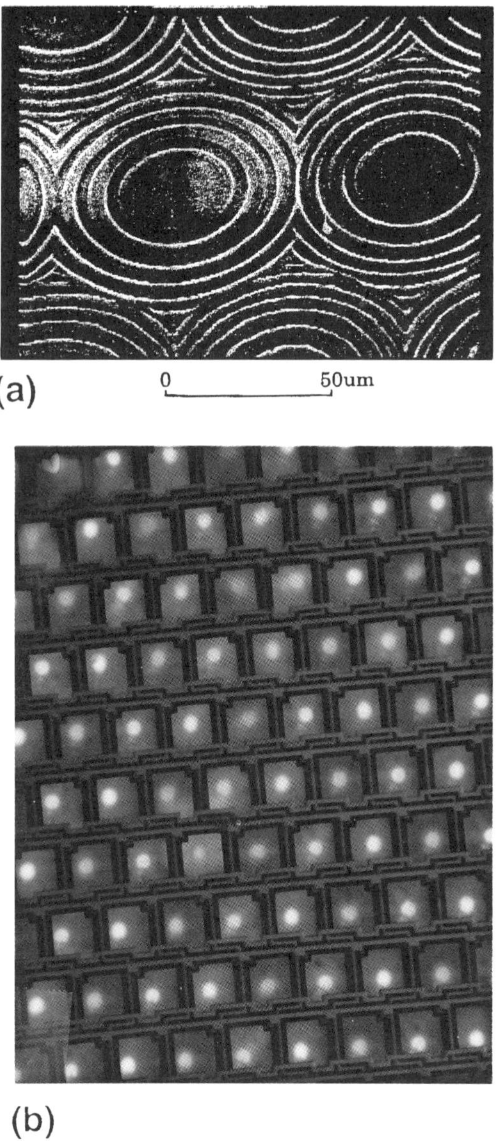

(a)

(b)

Figure 6.32 Diffractive element lens array for the same application, along with the images.

Integral photography consists of using a fly's eye array of lenses to record multiple images on a conventional photographic plate, as shown in Figure 6.33a [26–30]. Consider the recording from a point P on the object. Each lens in the array produces an image of point P at approximately the focal length of the microlenses. Putting it slightly differently, each lens forms an image with a different perspective. The photographic plate is developed and a positive is formed. The developed positive is registered behind the lens array and illuminated as shown in Figure 6.33b. Each point on the image emits a spherical wave that is collimated by the fly's eye lens. All the beams reconverge to the original position

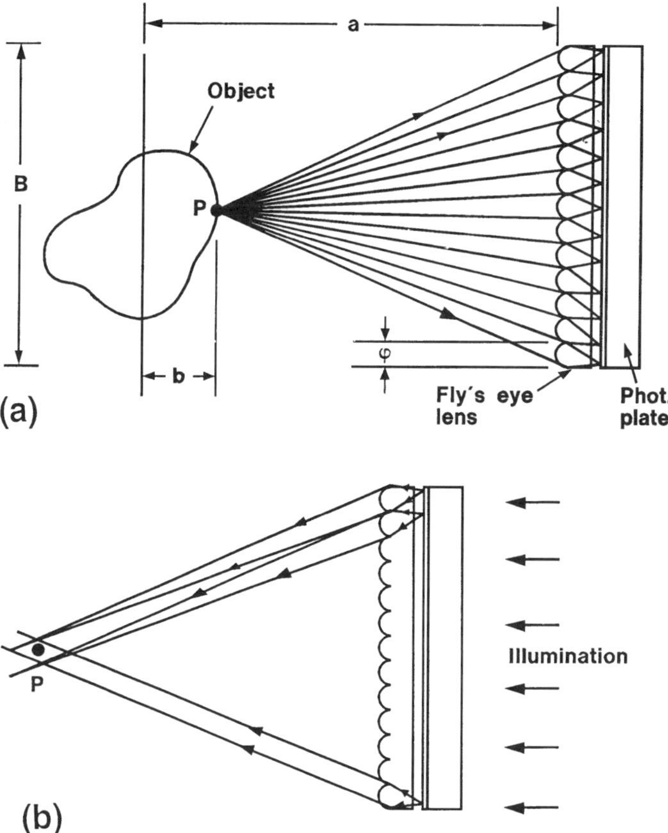

Figure 6.33 Concept of integral photography. Multiple images are formed on emulsion behind the lens array. Negative is developed and light is sent through the negative to the same lens array form a real 3D image.

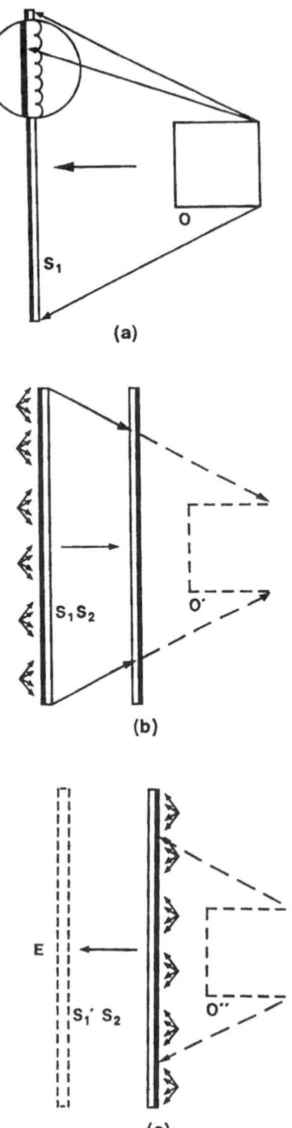

Figure 6.34 Scheme to produce desirable virtual orthoscopic image rather than real pseudoscopic image. The method is to make a second recording from the first as shown in (b). Illumination of this image from the right forms a virtual orthoscopic image when viewed from the left.

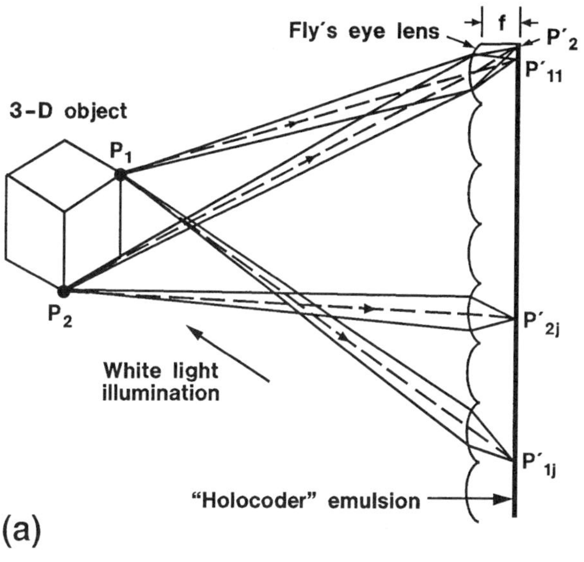

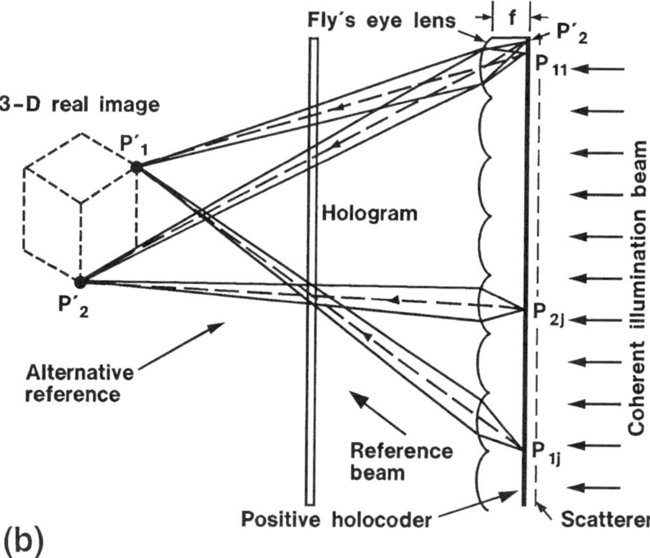

Figure 6.35 Another scheme to produce an orthoscopic image. In this case a hologram is formed in the imaging beam as shown. The first image is formed in the normal way. Viewing the developed hologram will show a virtual orthoscopic image.

of the object point. The image so formed is real and psuedoscopic, the same as would occur for a holographic image. To obtain a more practical concept where the image would be virtual and orthoscopic, a second recording has to be made. This can be done in one of two ways. The first is to make another integral photograph in the converging reconstructed beam as shown in Figure 6.34a [31]. Then one makes a second exposure through the original as shown in Figure 6.34b. Finally, the desired orthoscopic virtual image is obtained by illuminating the second exposure through the lens array as indicated in Figure 6.34c. The disadvantage of this is in the reduction in the quality of the image.

A second method is from a hologram of the imaging forming wavefront, as shown in Figure 6.35 [28]. The problem here is that one has to introduce a different holographic process into mix.

Burckhardt [26] has analyzed the resolution of this method along with an estimation of the ideal microlens parameters to obtain this condition. His analysis shows that the optimum microlens diameter ϕ is given by the expression $1.24a \sqrt{\lambda/b}$ where, referring to Figure 6.32a, a is the object distance, b is half of the object depth, and λ is the wavelength of light. For typical values of the parameters, $a = 50$ cm and $b = 5$ cm, and the estimate of the optimum lens diameter was 2 mm. This could easily be less than 1 mm for other choices of the parameters, or with somewhat less resolution at the values used. The expression for the maximum field angle was also estimated as ϕ/c, where c is distance between the lens array and the film plane, which is essentially the focal length of the microlens. For the above parameters, it is 10 mm.

The lens array requirements for this application are not particularly stringent, and would seem ideally suited for a molded plastic, or even glass, fabrication process. One would want a 100% fill factor which should be readily doable for a spherical lens figure on a square format.

REFERENCES

1. K. Iga, Y. Kokubun, and M. Oikawa, *Fundamentals of Optics*, Academic Press, New York, 1984.
2. K. Nishizawa and M. Oikawa, Microoptic research in Japan, *SPIE*, vol. 1751, *Miniature and Microoptics*, p. 54 (1992).
3. M. Oikawa, H. Imanishi, and T. Kishimoto, High NA planar microlens for LD array, SPIE vol. 1751, *Miniature and Microoptics*, p. 246 (1992), and references contained within.
3a. A. Yariv, *Optical Electronics*, Holt-Sanders, 1985.
4. N. F. Borrelli, private data.
5. T. G. McDonald, R. M. Boysel, and J. R. Sampsell, Deformable mirror based 4 × 4 fiber optic cross-bar switch, *OFC paper* WM–22, July 22–26, San Francisco, 1990.

6. L. J. Hornbeck, Deformable mirror spatial light modulator, *Proc. SPIE*, vol. 1150, paper #6 (1989).

7. R. M. Boysel, T. G. McDonald and J. B. Sampsell, A linear torsion hinged deformable mirror device for optical switching, *Tech. Digest, Topical Mtg. Photonic Switch*, p. 183, Optical Society of America, Washington, DC, 1989.

8. J. M. Florence, Joint-transform correlator systems using deformable mirror, *Opt. Lett. 14*, 341 (1989).

9. D. R. Collins, J. B. Sampsell, L. J. Hornbeck, P. A. Penz, and M. T. Gately, Deformable mirror device spatial modulator, *Appl. Opt. 28*, 4900 (1989).

10. N. F. Borrelli and D. L. Morse, Microlens array produced by a photolytic technique, *Appl. Opt. 27*, 476 (1988).

11. N. F. Borrelli, private data.

12. N. F. Borrelli, private data.

13. N. Stauffer and D. Wilwerding, *Scientific Honeyweller 3*, 1–10 (1982).

14. *Handbook of Plastic Optics*, US Precision Lens Inc., Cincinnati, OH, 1983.

15. N. F. Borrelli, Optoelectronic interconnects using microlens arrays formed in glass, Proc. 43rd Electronics & Components Conference, Orlando, FL (1993).

16. P. C. Montgomery and J. P. Fillard, *Proc. SPIE*, vol. 175, *Miniature and Microoptics*, p. 76 (1992).

17. G. Artzner, Microlens arrays for Schack-Hartmann wavefront sensor, *Opt. Eng. 31*(6), 1311, (1992).

18. S. Kawai, S. Araki, K. Kasahara, and K. Kubota, Optical interconnections using microlens arrays for parallel processing, *Proc. SPIE*, vol. 1751, *Miniature and Microoptics*, p. 255 (1992).

19. M. Oikawa, E. Okuda, K. Hamanaka, and H. Nemoto, Integrated planar microlens and its applications, *Proc. SPIE*, vol. 1898, p. 3 (1988).

20. K. Kasahara, Y. Tashiro, H. Hamao, M. Sugimoto, and T. Yanase, Double heterostructure optoelectronic switch. *Appl. Phys. Lett. 52*, 679–681 (1988).

21. T. Numai, M. Sugimoto, I. Ogura, H. Kosaka, and K. Kasahara, Surface emitting laser operation in VSTEPs, *Appl. Phys. Lett. 31*, 1250–1252 (1991).

22. Optical Shields, Inc./Corning, Inc. Final report to DARPA (1990).

23. N. F. Borrelli, Efficiency of microlens arrays for projection LCD, *Proc. 44th Electronic & Components Conf.*, Washington, DC, May 1–4 (1994).

24. M. Hijikigawa, Society for Information and Display, *SID Digest*, p. 265, (1992).

25. H. Hamada et al., *SID Digest*, paper 239, p. 269 (1992).

26. C. B. Burckhardt, Optimum parameters and resolution of integral photography, *J. Opt. Soc. Am. 58*(1), 71 (1968).

27. G. Lippmann, *J. Phys. Theorique et Appliquee 7*, 821 (1908).

28. R. V. Pole, 3-D imagery and holograms of objects illuminated in white light, *Appl. Phys. Lett. 10*(1), 20 (1967).

29. R. L deMontebello, Wide angle integral photography, *Proc. SPIE*, vol. 120, p. 73 (1977).

30. L. P. Dudley, A new development in autostereoscopic photography, *J. SMPTE 79*, 687 (1970).

31. C. B. Burckhardt, Information and inversion of pseudoscopic images, *Appl. Opt. 7*(3), 628 (1968).

Part III
OTHER OPTICAL ELEMENTS

7

Gratings

7.1 INTRODUCTION

In Chapter 1 we mentioned that the area of *microoptics* constitutes a much larger topic that includes devices and applications beyond the intended scope of this book. Our main objective was to concentrate on fabrication and application of the important subset of microoptics; namely, microlenses. A nonexhaustive listing of microoptical elements of importance is given in Table 7.1 along with an important application area. However, many of the methods of lens fabrication described in the previous chapters are equally applicable to making some of these microoptical elements. Two important structures for which this is true are gratings and planar waveguides. The implementation of the former leads to devices that can be used to redirect light, select out specific wavelengths, or both. The fabrication of the latter in various configurations leads to optical devices such as demultiplexers, directional couplers, and splitters [1,2].

The fabrication of planar optical waveguides is a topic that is amply and well covered in the literature and other texts and monographs [1,2]. The common elements of the fabrication methods covered in the previous chapters are the photolithographic patterning followed by etching, and the use of ion exchange. Although one of the main driving forces for microoptic elements and devices is integrated optics, it is not possible to do justice to this topic here. Rather, we will deal with two of the other important microoptic elements: gratings and the elements that make up optical isolators. These two topics are not often discussed in detail in the microoptic context. Gratings fit naturally into the context of this

219

Table 7.1 Listing of Microoptical Elements

Element	Example	Application
Polarizers	Dielectric/metal stack (Lam-ipol[a])	Optical isolator[b]
Polarization separator[c]	Wedge of birefringent crystal	Polarization-independent isolator
Waveplates	Slab of birefringent material	Isolators
Magnetooptic elements[d]	Slab or thin film of Fe–garnet crystal	Isolators
Filters	Dielectric stack	Demultiplexers
Gratings	Fiber Bragg	Filters/Demux
Waveguides	Etched planar in SiO_2	Planar integrated optics

[a] Tradename of Sumitomo Osaka Cement Co. It is a way to make a wire-grid type polarizer for the NIR by alternating very thin layers of metal and dielectric plates, and then slicing in cross-section.

[b] Optical isolators are devices that allow light to pass in the forward direction, but not in the reverse direction. They are used in such equipment as solid-state lasers and optical amplifiers. They are made up of a series of components.

[c] Element that separates the two polarizations into two paths to provide polarization-independent isolation.

[d] Faraday rotators: Theses are the essential element in the optical isolation because they provide the nonreciprocity property.

book because many of the methods used to make gratings are similar to those described previously in the fabrication of microlenses. The elements that go into optical isolators do not share the commonality of fabrication, but they do share the same overall area of application and are used in conjunction with other microoptical elements, such as microlenses. Because there have been few, if any, reviews of the elements that go into optical isolators, it was deemed worthwhile to include some discussion in this book.

The descriptions given in the following Chapters 8 and 9 will be somewhat brief, compared with those given in the previous chapters, because much of the material properties and processes have already been discussed in the previous chapters. The layout of the chapters will be the following: First we will give a very brief review of the physics of the optical element, to give some context to the subsequent descriptions (diffraction grating and optical isolators). Then we will list the ways these elements can be fabricated within the microoptics format, and finally, we will discuss some of the most important devices in which these elements are used. We will deal with gratings in this chapter and optical isolators in Chapter 8.

7.2 TYPES OF DIFFRACTION GRATINGS

In Section 4.1.3 we discussed the physical basis of diffraction gratings from the viewpoint of a surface relief pattern providing the periodic phase difference. This is certainly one primary way that diffraction gratings can be made. However, there is a much more general description of diffraction gratings of which the surface relief type is a special case. Actually, the underlying parameter of significance in a grating is the phase difference. This pattern can be produced in several ways, which we express as follows.

$$\Delta\phi(x) = \sum_m \Delta\phi_{0m} \cos\left(\frac{2\pi mx}{\Lambda}\right) \tag{7.1}$$

Here, Λ represents the grating period, the phase ϕ is expressed in waves, $\phi = nZ/\lambda$ where n is the refractive index and Z is the thickness. One can see the phase difference can be created either by letting Z vary with x in a periodic way, which would be the surface relief type, or by letting the refractive index vary in a similar way. The way we have written Eq. (7.1) amounts to a Fourier series; thus we are allowing a variety of functional forms of the way the phase can vary through a given period.

Another useful way to classify gratings is by the characterization: *thick* and *thin*. The methods of making gratings (discussed later on) will clearly depend on what type of grating one wants to make; hence, this classification applies equally to the fabrication [3,4]. The distinction originates because the phase change of Eq. (7.1) can be accomplished with a small refractive index over a very long path. From a light-propagation standpoint, this represents a much different case and, as a matter of fact, is closer to the way one would deal with a waveguiding structure (see Ref. 2 in which this is referred to as the *coupled mode treatment* of gratings). On the other hand, if the grating were very thin, then small changes in the path length have little effect on the diffraction efficiency. Such a change might occur by a differing angle of incidence. This is in contrast to the *thick* example, in which very small changes in the path length have a very large effect on the diffraction efficiency. So much so, that in the thick limit the efficient diffraction occurs at only one angle of incidence. This angle is called the *Bragg angle*. A number of gratings are schematically shown in Figure 7.1.

Magnusson and Gaylord [3] have developed a mathematical description of this thick and thin classification that is of considerable help in determining the ultimate behavior of the grating. The parameter of consequence is called the Q *parameter* and is defined as

$$Q = \frac{2\pi Z\lambda}{n_0 \Lambda^2 \cos\Theta} \tag{7.2}$$

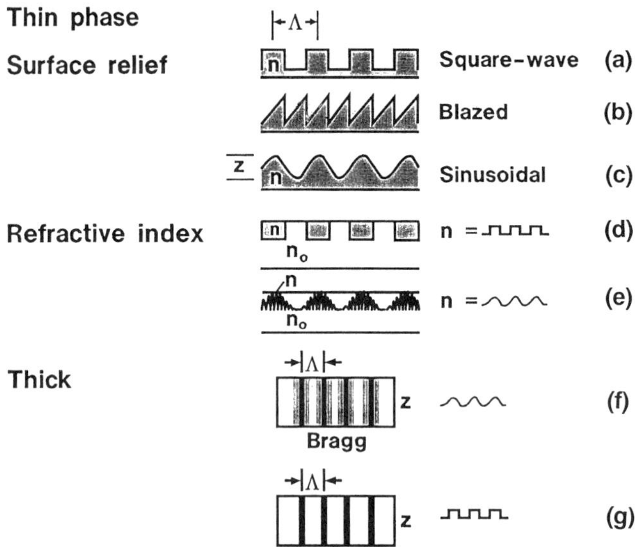

Figure 7.1 Grating types: Examples of thin gratings: (a–c) surface relief gratings, with the geometric shape as shown; (d, e) thin phase gratings where the variation in phase derives from the index change rather than surface relief; (f, g) thick gratings stemming from the periodic refractive index throughout the depth.

Where λ is the wavelength of light, Z is the thickness of the grating, Λ is the grating period, and Θ is the angle that the light beam makes with the grating normal. The reader is referred to Figure 7.2 for the definition of the terms.

7.2.1 Thin Gratings (Q, Small)

From the Magnusson and Gaylord [3,5] analyses, in the limit of small Q, that is less than 1, as defined in Eq. (7.2), one can express the behavior of various thin phase gratings in terms of their diffraction efficiency. The expression for the three most common thin gratings are listed in Table 7.2. Generally, these gratings produce multiordered diffraction patterns. For example, for the sinusoidal grating, such as would be produced by the interference of two monochromatic beams (holographic-grating pattern), the expression for the diffraction efficiency is,

$$\eta(i\text{th order}) = J_i^2 \left(\frac{2\pi n_1 Z}{\lambda \cos \Theta} \right) \tag{7.3}$$

where J_i is the ith order Bessel function, and the other terms are as defined in

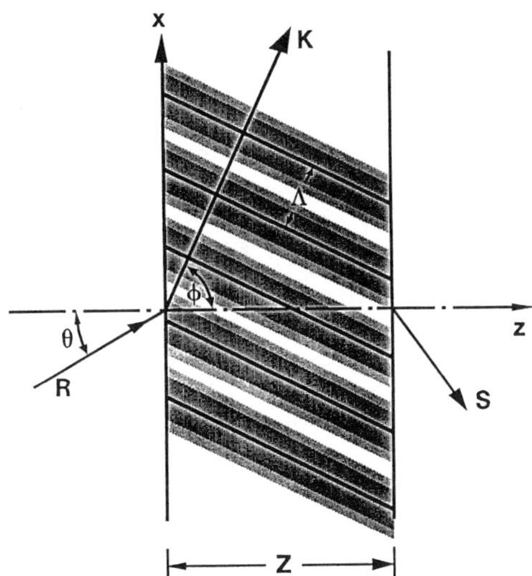

Figure 7.2 Model of thick hologram with slanted fringes: The spatial modulation of the refractive index in indicated by the shaded regions. The grating has period Λ, and grating vector **K**, as shown. The incidence angles θ and the grating angle ϕ are defined.

the foregoing. An exception to the multiordered behavior can be achieved with the use of a blazed grating, which is a special case of the sawtooth index pattern shown in Table 7.1. In this example, as shown in Eq. (4.16) in different notation, the diffraction efficiency is given by the expression,

$$\eta(\iota\text{th order}) = \frac{\sin^2[\pi(\gamma - 1)]}{[\pi(\gamma - 1)]^2} \tag{7.4}$$

Table 7.2 Diffraction Efficiency of Thin Phase Gratings

Type	Efficiency	Comment
Sinusoidal	$J_j^2(2\gamma)$	j = order
Square	$\cos^2(\pi\gamma/2)$	0th order
	0	Even order
	$(2/j\pi)^2 \sin^2(\pi\gamma/2)$	Odd order
Sawtooth	$\sin^2(\pi\gamma)/[\pi(\gamma + j)]^2$	j = order

$\gamma = \pi n_1 d/\lambda$

where, $\gamma = \pi n_1 Z/\lambda$ The difference in the notation between Eqs. (4.16) and (7.4) is because for a surface relief grating (see Fig. 7.1a,b,c), the phase difference is fixed by the depth of the groove. In other words, the phase shift is expressed by the term $2\pi(n_0 - 1)Z/\lambda$. Whereas, for the thin phase grating (see Fig. 7.1d,e) the phase shift is determined by the refractive index difference, $2\pi n_1 Z/\lambda$. For the surface relief grating one ensures the optimum efficiency by making the depth correct for the argument of Eq. (4.16) to be zero.

An important point in the fabrication of these thin gratings is that quite different methods are required. For the surface relief type, methods such as or etching or molding of the surface of the substrate, are appropriate. On the other hand, for the thin phase gratings, the photosensitive process whereby the refractive index is locally ordered is the method of choice. We will go into much more detail about the methods when we describe the fabrication methods.

7.2.2 Thick Gratings

For the parameter Q, as defined in Eq. (7.2), when its value is large, $Q > 10$, the diffraction behavior changes from a multiordered phenomenon to that of a single-order one when the Bragg condition is met; namely,

$$\cos(\phi - \Theta) = \frac{(\lambda/n_0)}{2\Lambda} \tag{7.5}$$

where Θ is the angle of incidence, ϕ is the angle that the grating vector makes with the surface normal, and n_0 is the refractive index of the grating medium (see Fig. 7.2).

For a holographic grating (sinusoidal index profile), the diffraction efficiency is given by the expression,

$$\eta = \sin^2\left(\frac{\pi n_1 Z}{\lambda \cos \Theta_0}\right) \tag{7.6}$$

Clearly, from the equation, this type of grating can reach 100% efficiency when the argument of the sine reaches 90°. Kogelnik [4] has shown by coupled mode analysis of the behavior of thick gratings that the Bragg condition corresponds to a momentum-conserving relation between the wave vectors of the input beam, the diffracted beam, and the grating vector. Thick gratings have various properties that make them particularly interesting from an optical applications standpoint. In particular, the sensitivity of the thick grating to small changes in wavelength and angle provides unique applications that will be discussed later. Their behavior

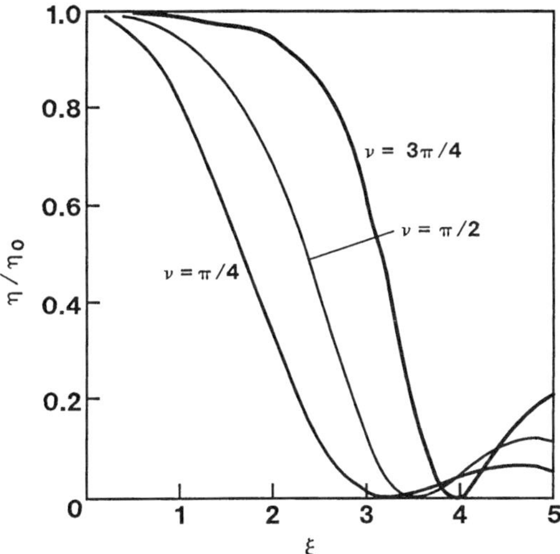

Figure 7.3 Representative behavior of the falloff in diffraction efficiency with deviation from the Bragg condition. The *x*-axis can be thought of as the deviation in the angle or the wavelength from the optimum. The curves are given for gratings of three strengths, as indicated by the phase angle. (From Ref. 4.)

is shown in Figure 7.3. Although the ν and ξ parameters are somewhat obscure, one can think of ν as the measure of the index modulation n_1, and ξ as a measure of the deviation from the Bragg condition of either the incident angle or the wavelength. Here, the diffraction efficiency is normalized to the value at $\xi = 0$, which corresponds to $\sin^2 \nu$.

These types of gratings are invariably formed in solids by a photosensitive technique, usually interfering coherent beams provide the required exposure. This will be covered in some detail when we discuss the actual fabrication process. There is sufficient literature on all aspects of holography [6,7] that it is not considered necessary to deal with it here other than how it relates the actual fabrication. Our main concern is its usefulness as a grating in the field of microoptics and with the various materials and methods by which they can be made. In a subsequent section, we will list and discuss the types of devices that use microgratings.

7.3 FABRICATION OF MICROGRATINGS

In the following sections we will describe the various methods employed for making gratings on the microoptic scale. The dimension here is roughly 1 mm, or less, in spatial extent. From the foregoing discussion, the methods will be broken down into those applicable for the fabrication of thin surface relief gratings and those appropriate for thick holographic-type gratings.

7.3.1 Surface Relief Gratings

The methods to provide finely spaced patterns in surface relief of a material are similar to, if not substantially identical, with those for making microlenses, described in the previous chapters. Nonetheless, we will review the methods with specific attention to the aspects that affect the grating performance.

Photolithography

Photolithography is the most straightforward method used to make surface relief gratings, although it is not necessarily the best one. This will depend on such factors as determined by its ultimate use. The photolithographic method is used when high-resolution gratings are required. We have discussed the basic method in Section 4.2 for the fabrication of diffractive lenses, and for convenience it is outlined again in Figure 7.4. The photoresist is spun on a suitable substrate and exposed through a mask that contains the appropriate spatial pattern. After development, the grating pattern is in the resist layer. The grating pattern is then produced in the substrate by a suitable etching technique, such as reactive ion etching (RIE).

In general, the gratings require high resolution—spacing of the order of 0.5

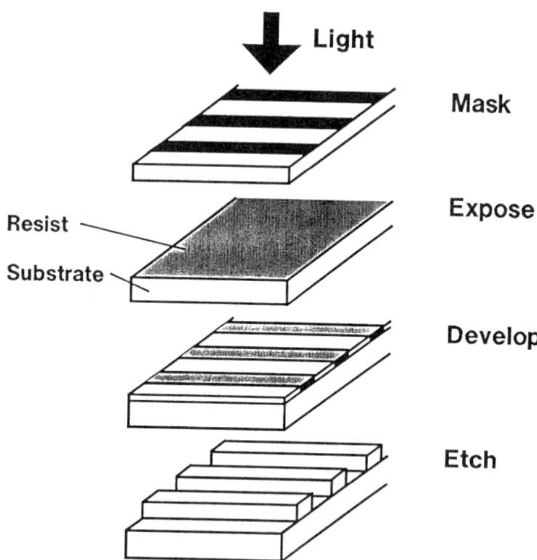

Figure 7.4 Representation of the classic photolithographic production of a surface relief grating.

μm—thus, the mask must be fabricated with the highest-resolution technique that is available. There are two methods that can be used to prepare such masks. The first uses standard e-beam lithography writing onto resist that covers a metal film. After development of the resist, followed by dissolution of the exposed metal, a pattern of metal on glass is achieved. The type of relief pattern that will be produced by the mask can be varied to some extent by the e-beam—writing methods. For example, as mentioned in Section 4.2.1, one can vary the transmittance through the pitch of the pattern, by half-tone techniques that approximate transmittance functions that are sawtooth, or sinusoidal. The former pattern would allow a blazed grating to be made, and the latter a pattern approximating a holographic grating.

The second method is the holographic method for which the interference of two mutually coherent beams is used to expose the resist (Fig. 7.5). This method produces a sinusoidal pattern that may be desirable. Another advantage of the holographic exposure method is that it is capable of a much larger area of exposure. The e-beam system operates in a step and repeat mode; hence, stitching errors are possible.

Recently, a third type of mask called a *phase mask* has found its niche in the area of exposing fiber Bragg gratings. We will discuss this important application later in this chapter. A phase mask is itself a diffraction grating that, in turn, is used as a mask to produce gratings in photosensitive materials. The grating is made in a way, usually using the holographic exposure method [8], such that it maximizes the $+1$ and -1 orders and suppresses the 0th order. It is then used as shown in Figure 7.6. The advantage of this type of mask is that it is essentially

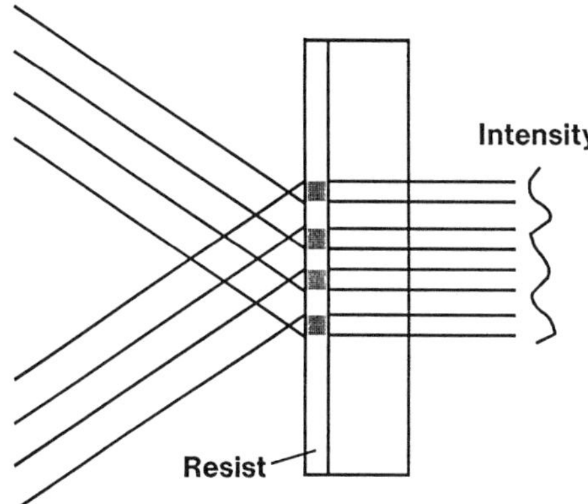

Figure 7.5   Representation of the interference method of exposure to form a sinusoidal grating in a photosensitive film.

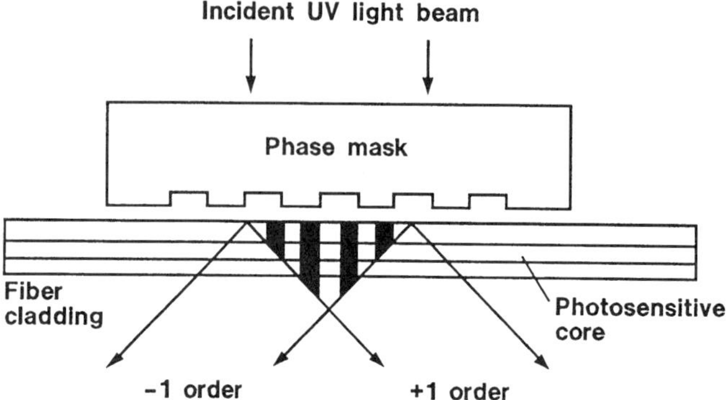

Figure 7.6 Diagram showing the use of a phase-mask to expose a sinusoidal grating.

equivalent to a holographic-type exposure, but requires very little coherence length of the laser. In the conventional holographic exposure, the coherence length of the laser sets the allowable path length difference. The excimer lasers are not very satisfactory coherent sources, with coherence lengths of less than 100 μm. With the exposure geometry shown in Figure 7.6, this is not a problem because the beam is essentially split at the grating–sample interface.

Molding

The molding of diffraction gratings strongly depends on the nature of the material requirements. For most applications the requirement is for a robust material, such as silica. To mold into glass requires the use of a master. The master would contain the opposite relief pattern, as shown in Figure 7.7. In Chapter 2 we dis-

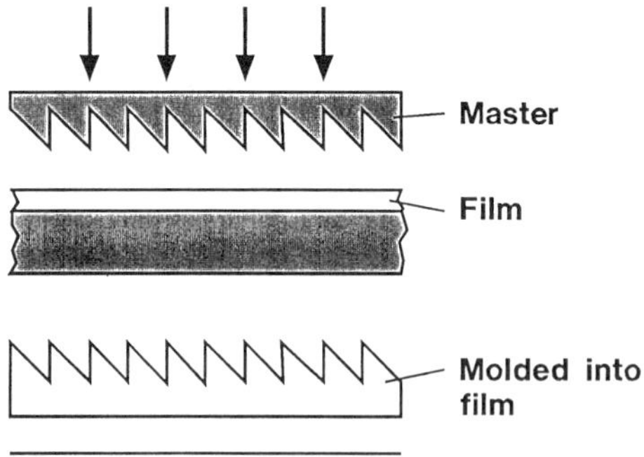

Figure 7.7 Schematic representation of the molding of a surface relief pattern.

cussed the molding of aspheric lenses. This same method can be applied to the molding of diffraction gratings. Here, a special low-temperature glass was developed that was compatible with the molds [9]. For the grating one can actually use a silica master. In Figure 7.8A and B we show an AFM surface profile of the silica master and the resulting pattern in the glass.

Another similar approach is to mold into a soft layer deposited, or otherwise produced, on the surface of a hard material. This layer could be a polymer, or a sol–gel-derived layer. Figure 7.9 shows the pattern produced in a sol–gel layer (mixture of polymer and TEOS-derived glass) produced by the same master as used to make the grating shown in Figure 7.8.

7.3.2 Photosensitive Films

The type of grating we have described is a phase grating, of which the surface relief type is the important, but nonetheless, special example. The more general approach is to apply a thin photosensitive film to a hard substrate. *Photosensitive* means that on exposure to light the film will undergo some change, structural or otherwise, that will result in a change in refractive index. There are examples of this type of material, the most common being photographic emulsion. For example, in Kodak 649-F, the conventional use of a silver-based emulsion is to produce an image through a development process that reduces silver halide to silver. However, by special development conditions, one can bleach the silver and end up with an almost pure refractive index change [10]. This is how holography was performed in the early days. The problem is the durability of the resulting material. A related process makes use of what is called *dichromated gelatin* [11]. This material can be prepared on glass, and when exposed to light, cross-linking of the gelatin molecules occurs. Much like photoresist, this causes changes in solubility or volume, which lead to phase changes in proportion to the exposure. More recently, families of photopolymers have been developed that undergo structural changes, such as polymerization on exposure to light, which lead to refractive index changes.

The thickness of these layers can vary from a few tenths of a micrometer to tens of micrometers. The thickness can bridge the gap between the thick and thin regimen discussed earlier.

7.3.3 Thick Gratings

The fabrication of thick gratings follows the method by which one would make a plane-wave hologram. This was schematically shown in Figure 7.5 and we expand on it in Figure 7.10. The coherent phase difference between the two beams **U** and **R** produces an interference pattern, such that the intensity is given by the following expression.

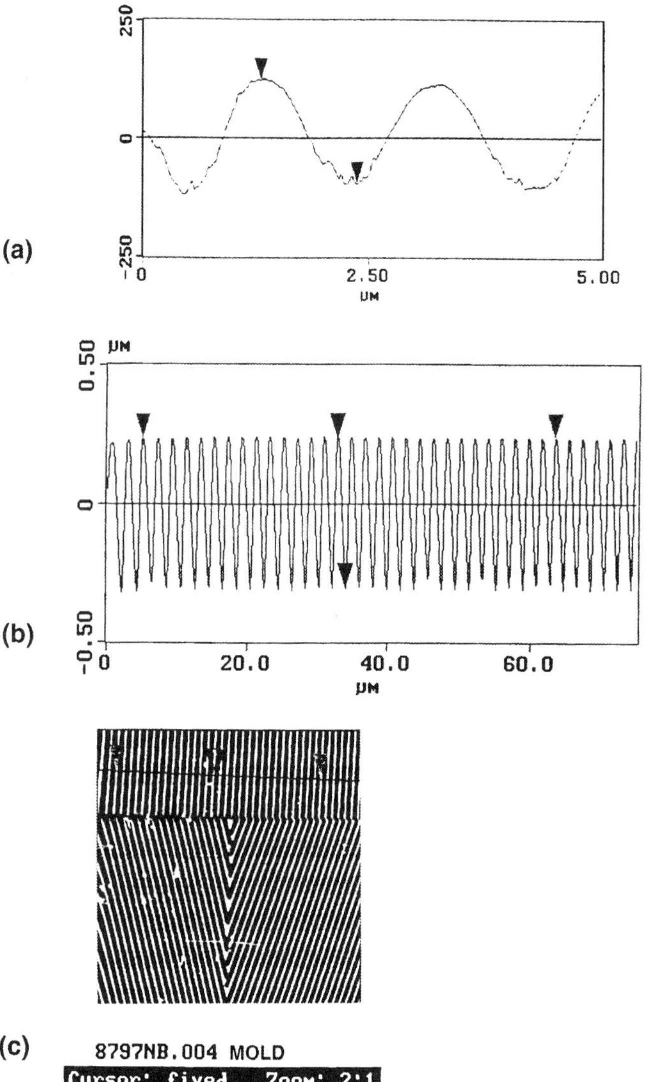

Figure 7.8   (a) AFM pattern of the surface of a molded grating embossed into a special low-temperature glass; (b) the actual mold pattern is shown; The important point to note is the depth of the pattern. For the mold the maximum depth was 556 nm and the molded part yielded 543 nm.

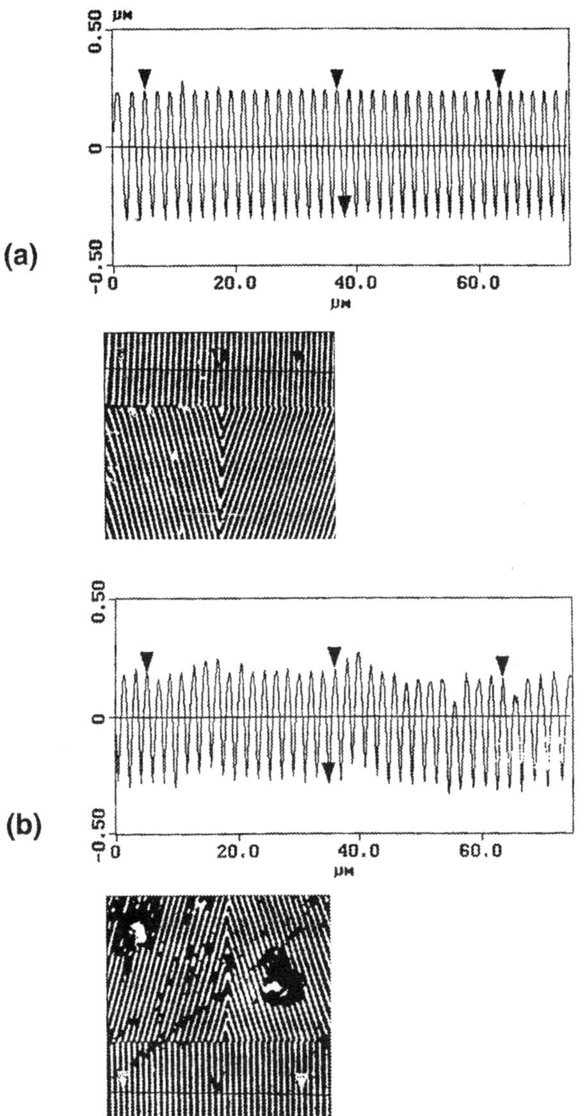

Figure 7.9 AFM pattern of the surface of a molded grating embossed into a polymer/ sol–gel composite. (a) the silica mold and (b) the sample. Note that the depth here is 493 nm, compared with the mold depth of 556 nm.

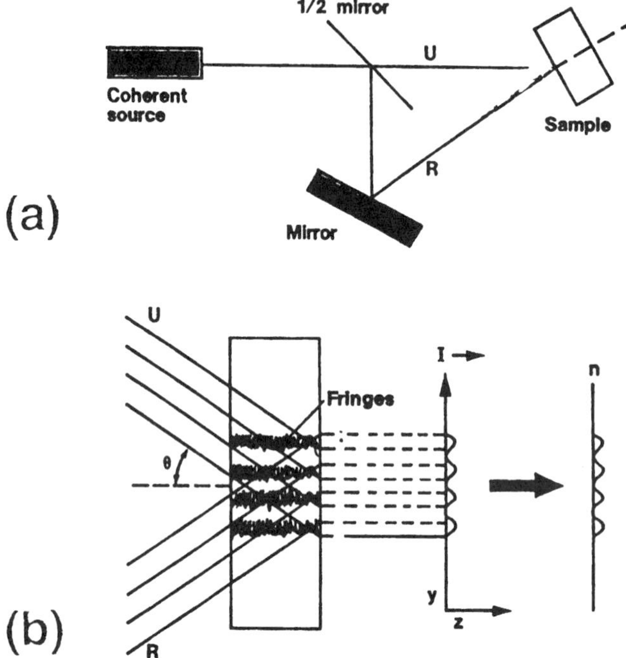

Figure 7.10 Standard plane wave holographic exposure process: (a) The coherent beam is split in two and is made incident onto the photosensitive body at equal angles to produce (b) an unslanted fringe pattern. The interference of the two coherent beams produces a sinusoidal intensity pattern.

$$I = (\mathbf{U} + \mathbf{R})(\mathbf{U} + \mathbf{R})^* \tag{7.7}$$

Where the asterisk stands for the complex conjugate. In the simple plane wave, $\mathbf{U}$ and $\mathbf{R}$ differ in phase only by a value of δ. This means one can write Eq. (7.7) as

$$I = E_0^2\{1 + \exp[i\delta(y)][1 + [-i\delta(y)]]\} = I_0\cos^2\left[\frac{\delta(y)}{2}\right] \tag{7.8}$$

This sinusoidal intensity pattern is incident on and through a photosensitive material. The material responds to the exposure in a way such that a permanent refractive index variation is subsequently established. (This property of permanence is important because the applications in this book are to make robust optical elements. This is not to say that there is a whole other technology that is built on being able to make and erase gratings with light or heat.) This need not be a direct response, but may require a further step, as we shall see, such as heating. It is important to make a distinction between the nature of the initial change that occurs in the body and what ultimately happens. For example, the initial state of the material may absorb the light, promoting a structural change that will eventually be a pure refractive index grating. To the extent that absorption persists, the diffraction efficiency decreases.

As a result of the photoprocess induced by the intensity variation, a phase function is produced.

$$t(y) = \exp[2\pi\iota\phi(y)] \tag{7.9}$$

Where the phase function $\phi(y)$ is given by the product of Eq. 7.8 with the function that represents the manner in which the refractive index develops with intensity. For the simple plane case, this just means that a sinusoidal phase grating has been produced, the depth of which will depend on the specific material response.

Photosensitive Materials

There are several examples of *photorefractive* materials; that is, materials that when exposed to light can undergo some degree of *permanent* change leading to a refractive index change. We have made a fairly representative list of the materials that have been reported to have such an effect in Table 7.3. The mechanism for the production of the refractive index change is quite distinct for the various materials. We shall very briefly review each of the materials in Table 7.3.

Ferroelectric Single Crystals. $LiNbO_3$ and $BaTiO_3$ can be considered the "original" photorefractive materials, and even today to many, the photorefractive effect refers exclusively to the effect observed in these materials. There is much literature on this and related materials [12–14]; thus, we will touch only briefly on the phenomenon. $LiNbO_3$ is a pseudocubic single crystal of the perovskite family. The important properties of this class of materials are that it is ferroelectric, and it is photoconductive. The ferroelectric materials are highly polar, and

Table 7.3 List of Photorefractive Materials

Ferroelectric single crystals
Polymer guest/host
Photosensitive polymers
Photosensitive glasses
Porous glass
UV-photosensitivity (fiber Bragg grating)
As_2S_3

undergo a spontaneous polarization at a temperature termed the *Curie tempera-ture*. Because of this highly polar structure, these materials possess a large elec-trooptic effect, which is expressible as,

$$n(\text{par}) - n(\text{per}) = \Delta n = rE \tag{7.10}$$

Where $n(\text{par})$ is the refractive index parallel to the applied electric field, and $n(\text{per})$ is the perpendicular. The photoconductive property derives from defect centers and impurities from which electrons can be promoted into the conduction by exposure to light with sufficient energy. The level of defects required to pro-duce the photorefractive effect is roughly parts per million (ppm).

The induced photorefractive effect occurs in the following sequence of steps [15,16]. On exposure, the mobile electrons are promoted from the defect into the conduction band, leaving behind a positively charged site. Eventually, the charge drifts into the unexposed region where it becomes permanently trapped. An elec-tric field is developed between the region of negative charge (dark) and the posi-tive charge (bright). This strong internal field, estimated to be 10^4 V/cm, induces a refractive index change through the electrooptic effect [see Eq. (7.10)].

The production of the diffraction grating follows the standard process indi-cated in Figure 7.10. The light source is in a spectral range sufficient to provide the excitation of the electrons out of the defects, and this depends on the specific crystal and the nature of the defects. The band gap of most of these crystals is in the region of 3–4 eV; however, the defects' absorption that lead to the photoelectrons extends through the visible portion of the spectrum. The depen-dence of the index change on the exposure is shown in Figure 7.11 [17]. It is characteristic that this effect requires only small intensities to produce significant index changes. This means that the production of photocarriers is a very efficient process which, when coupled with a large electrooptic effect, produces the large index change.

Because electrons are trapped in the dark regions, it is possible to thermally stimulate them out of the traps. Subsequently, they can recombine with the trapped holes. The energy depth of the trapped electrons determines the thermal stability of the photorefractive effect. In other words, one can thermally erase gratings.

As we will see in the applications section, $LiNbO_3$ is a versatile substrate in the planar-guided wave application. It provides a platform for much functionality; the ability to form gratings is one important example.

Polymer Guest–Host.   In the past few years an all-organic version of the photo-refractive effect has evolved [18,19]. The basic idea is to mix into a polymer host, various organic molecules that will function as the required agents to produce the refractive index change. An example is as follows: One dissolves into the polymer matrix a molecule that is photoreducible; that is it can promote a hole into the

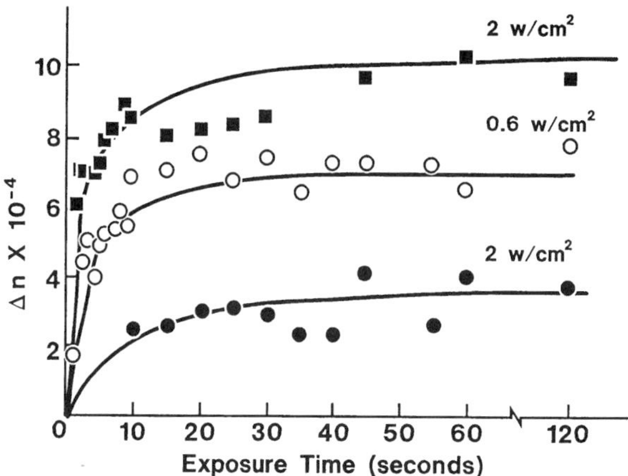

Figure 7.11 Induced refractive index change in LiNbO$_3$ as a function of time for three exposure intensities. (From Ref. 17.)

polymer. Examples of good charge generators are donor–acceptor charge transfer complexes, examples of which are shown in Figure 7.12. The required next element is a molecule to provide charge transport. The charge-transporting function is usually provided by a charge transfer agent that is in sufficient concentration to permit hopping of the hole or electrons. These are chosen from the class of molecules such as carbazoles, hydrozones, and aryl amines. The next molecule

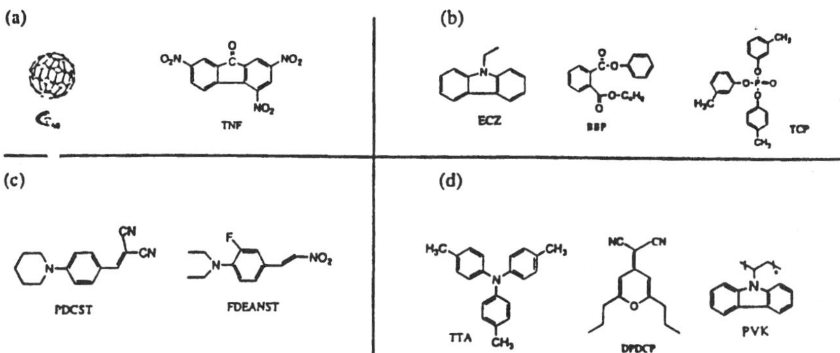

Figure 7.12 Listing of some of the molecules used for the organic photorefractive effect: (a) Photosensitizers to produce the carriers; (b) plasticizers to control the stiffness of the polymer; (c) NLO chromophores to produce the electrooptic effect; and (d) transporting molecules and polymers to produce the conduction. (From ref. 18.)

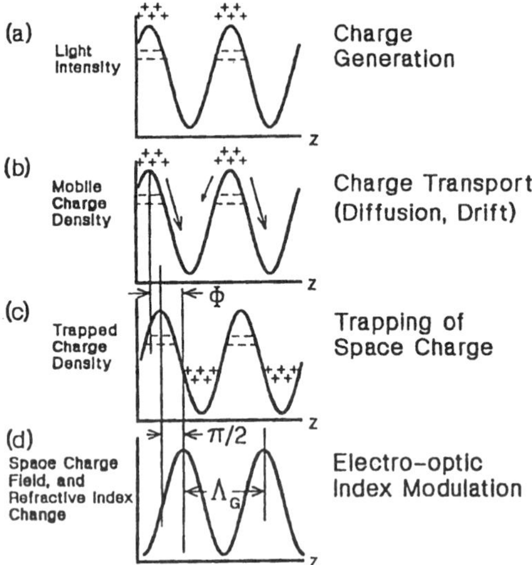

Figure 7.13 Schematic of the photorefractive process: (a) Carriers are formed in the illuminated area; (b) charge is separated either by drift or the bias field; (c) mobile carriers are trapped in the dark regions; and (d) the space charge field provides the electrooptic effect. (From Ref. 18.)

needed is one that provides relatively deep traps for the photogenerated mobile species. These trapping sites are intrinsic to the polymer configuration.

With these three elements, one can envision the evolutionary formation of a space charge region, as shown in Figure 7.13. Unfortunately, the natural fast diffusion or large drift mobility that occurs in the inorganic version is not present in the organic version, so that a bias voltage is maintained both during exposure and after. This bias field is close to 90 V/μm [19].

The final ingredient in the mix is the molecule that will provide the electrooptic effect as a result of the internal space charge field. Here one chooses to include a high concentration of a molecule with a large optical nonlinearity. These nonlinear optical chromophores (NLO) are molecules that have structures that involve a large dipole moment change with applied field. The symmetry-breaking step that is required here is that the sample be *poled*. This means that all of the mix is now complete and the resulting refractive index change is achievable.

The exposure for grating production in these materials is the standard plane-wave holographic exposure (see Fig. 7.10). One significant difference between the organic version and the inorganic one is that a larger Δn is achievable in the organic material [19].

Photosensitive Polymers.   We include here all of the polymeric materials that undergo structural changes when exposed to light, usually followed by some thermal treatment. Conventional photoresist is a good example of this class. The dichromated gelatin mentioned in the foregoing section is another example. The reason they are included here is that these layers can be thick enough that the grating can exhibit thick grating behavior. The gratings are made as shown in Figure 7.10, with the specific exposure conditions determined by the particular photosensitive material.

Photosensitive Glass.   In section 3.4.3, we described the composition of certain special glasses, and how the photosensitive development of a second microphase could be used to alter the refractive index. Reviewing the mechanism that was described in the previous section, the UV light promotes an electron from a donor, such as Ce^{+3}, that ultimately becomes trapped. The thermal treatment that follows the exposure frees the electron from the trap and reacts with the mobile Ag^{+1} ions that are contained in the glass. The reduced silver agglomerates to such a size to nucleate a phase separation of an NaF phase. If one ''overexposes'' the sample, thereby creating many nuclei, then the growth of the individual particles is limited to a size sufficiently small to keep the glass transparent. The refractive index increases where the crystal is produced because the fluoride is removed from the surrounding glass.

The exposure conditions required to produce index changes are shown in Figure 7.14 [20]. One can see that the exposure level at 337 nm is quite high, and that a high-temperature thermal treatment is required.

In this material, and for any photosensitive material for which the photorefrac-

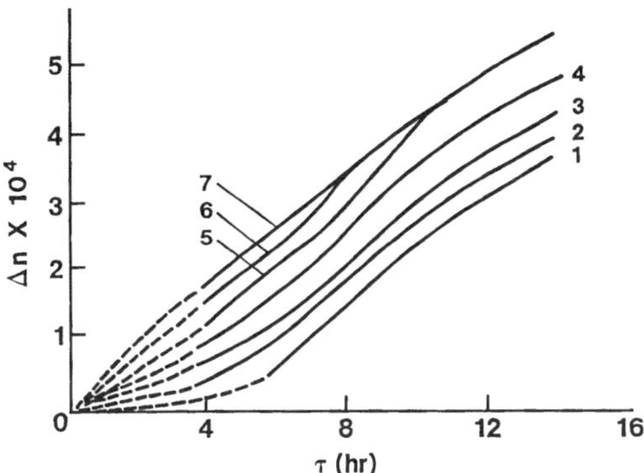

Figure 7.14   Induced index change versus time developed in photosensitive glass: Temperature of thermal development was 500°C. Numbers over graph signify the exposure level (J/cm^2). (From Ref. 20.)

tive process is initiated by an absorption process, the depth of penetration of the excitation and, hence, the value of the induced refractive index change, is limited. Consequently, the thickness of the grating can be adjusted by the choice of the excitation wavelength. For photosensitive glasses, the peak of the excitation is at 307 nm. However, by using an exposure wavelength of 337 nm, which is on the tail of the excitation peak, one can obtain a deeper grating, at the expense of a corresponding longer exposure time.

Doped Porous Glass.   As with the photosensitive glasses (see Section 3.4.3.), permanent refractive index changes can be produced in porous glass that has been impregnated with photosensitive materials, we make a distinction between the two different ways the porous medium can be used to this end.

In the first method, a photosensitive organometallic compound is loaded into the porous glass through vapor loading or through the use of a solvent that is subsequently evaporated off. Here, the action of the light is to alter the physiosorbed molecule in such a way that it bonds more strongly to the surface of the porous glass. Subsequent heating oxidizes the metal, and the result is an oxide phase, contained within the pores, that results in an index of the refractive change [21,22]. An example of holographic formation is shown in Figure 7.15. Multiple gratings were formed in a porous glass sample impregnated with an organometallic titanium compound. The exposure arrangement is also shown. The exposure was accomplished with a 10-mW He–Ne laser with exposure times of minutes. The sample after exposure was heated to 600°C. The diffraction efficiencies of gratings prepared in this way were over 80%. A wide range of organometallic compounds can be used for this purpose [20].

The second method [23,24] involves the infiltration of the porous structure with a photosensitive material such as a photoresist. The photoinduced refractive index change derives from the resist in a manner in keeping with the composite nature of the structure [see Chapter 3, Eq. (3.50)]. There are several important steps that can be used to modulate the index in the resist-impregnated porous glass. These involve the various changes that the resist undergoes. The inherent effect is from the structural changes in the resist, such as the density and volume. However, one can proceed further and use the solubility difference between the unexposed and unexposed resist to enhance the contrast. Moreover, one can use the retained resist as a protection when the glass is reexposed to a leaching bath. After the resist is removed, the index pattern is purely structural in that the exposure pattern is embedded in the pore structure. In Table 7.4, we reproduce the refractive index (RI) changes achieved after the various stages [23].

UV Photosensitivity in SiO_2–GeO_2 Waveguides.   One of the more interesting phenomena that has been uncovered over the last few years is what has come to be called *fiber Bragg gratings*. This name has arisen because of its wide application as a Bragg grating in optical fiber devices as we will see in the later section.

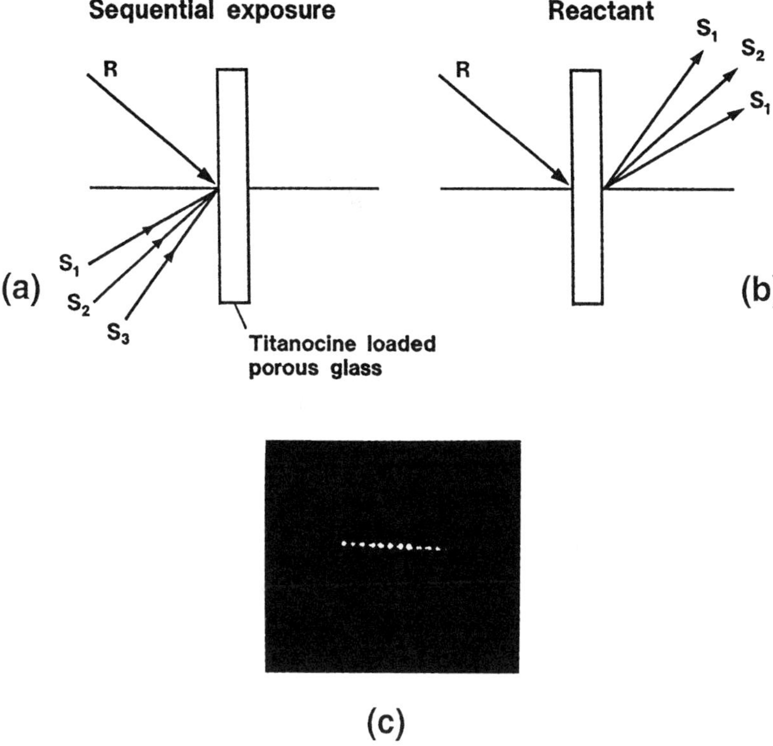

Figure 7.15 Diffracted output from gratings made in titanocene-loaded porous glass: The exposure was $1\,\text{W/cm}^2$ for 5 min. Glass was heated to $600°\text{C}$ for 1 hr after exposure. The exposure arrangement is shown. Eleven gratings were exposed about $2°$ apart. The reconstruction was with the single reference beam, as shown.

Table 7.4 Induced Index Change After Stages of Porous Glass Composite Treatment

Stage	Exposed resist	n Resist removal	Releaching
RI	0.01	0.03	0.15

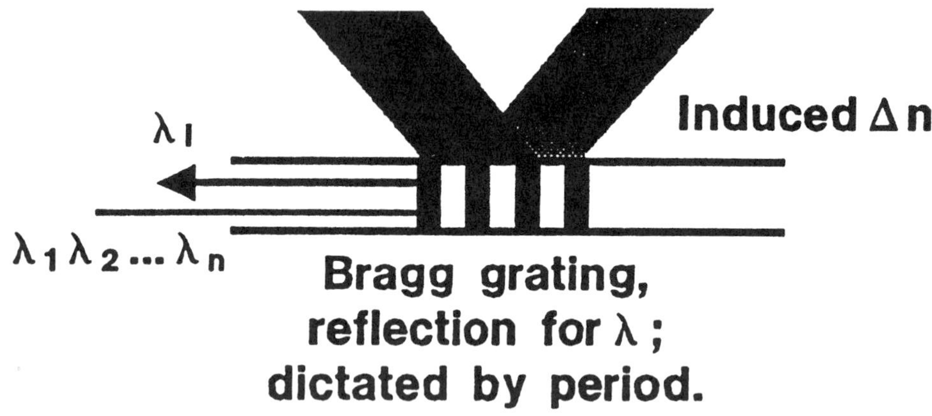

Figure 7.16 Schematic drawing of the exposure arrangement to develop a fiber Bragg grating. Two interfering beams are exposing the core of a single-mode fiber through the cladding from the side. Typical exposure of SiO_2–GeO_2 fiber is 100 mJ/cm^2 for 10–20 min at 10 Hz with 15 nsec pulses from a 248 nm excimer laser.

What the term really means is that in certain binary silica systems a UV photosensitivity exists. In particular, the binary system SiO_2–GeO_2 is of most interest because it has the largest effect and is the basis for most of the optical fiber in use today.

The amount of work on this effect could fill a chapter in itself, perhaps even a book. Consequently, here we can only briefly describe the nature of the effect and its proposed origin. The original report of an induced index change of the order of 10^{-5} in a single-mode GeO_2–SiO_2 optical fiber is attributed to Hill et al. [25]. The initial experiment used green light into the core of a single-mode fiber. After prolonged exposure there appeared a weak reflection indicating that some sort of grating was forming as a consequence of the green exposure. However, the real interest began when it was reported that one could induce the change much more efficiently using deep UV radiation. Moreover, the exposure could be made from the side of the fiber passing through the transparent silica cladding [26]. This is the so-called side-writing technique (Fig. 7.16). With reference to the Bragg condition given by Eq. (7.5), the condition corresponds to $\Theta = \phi = 90$ degrees. Measured reflection efficiencies obtained with the side-written exposure geometry were now consistent with that predicted from a Bragg grating with an induced refractive index change of greater than 10^{-4}.

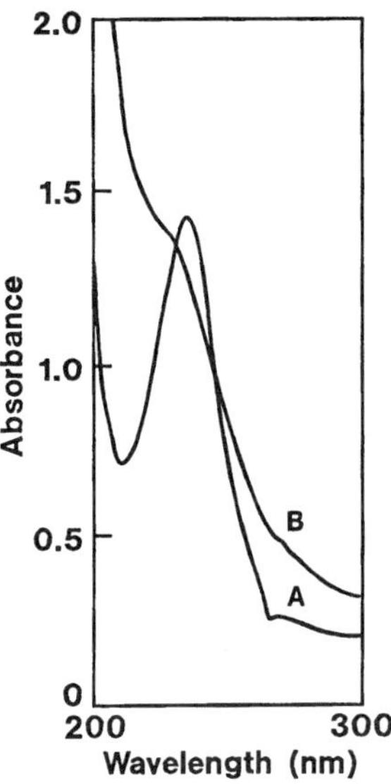

Figure 7.17 An UV absorption spectrum of 0.5-mm–thick section of 93% SiO_2–7% GeO_2 waveguide core-cane blank showing the characteristic defect absorption feature at 240 nm (A) before and (B) after exposure to 248 nm excimer radiation.

Although there is no total agreement on the exact mechanism for the induced index change, there is a consensus on certain aspects. First, there is a Ge defect that has a distinct optical absorption at 240 nm. This absorption band is bleached when exposed to light within its absorption spectrum. Finally, new absorption bands, extending to the higher energies are created as a consequence of the disappearance of the 240-nm absorption. This is shown in Figure 7.17, measured on a 0.5-mm–thick slice from a preform. (The single-mode waveguide is made by redrawing the preform into fiber.) There is a correlation between the existence of this absorption center and the subsequent ability to produce a grating in the fiber. The defect center has been attributed to an oxygen deficiency [27]. In support of this are the data that show the strength of the 240-nm–absorption band as a function of the oxygen partial pressure used in consolidation of the soot preform. This is the so-called outside process (Fig. 7.18). There are other processes used to prepare waveguide blanks for which it is harder to pinpoint the conditions that favor the formation of this oxygen-deficient center.

The mechanism for the photoinduced refractive index change, and hence, the grating, is given by the following explanation. The 240-nm–absorption band is bleached, which corresponds to the destruction of the Ge–oxygen-deficient cen-

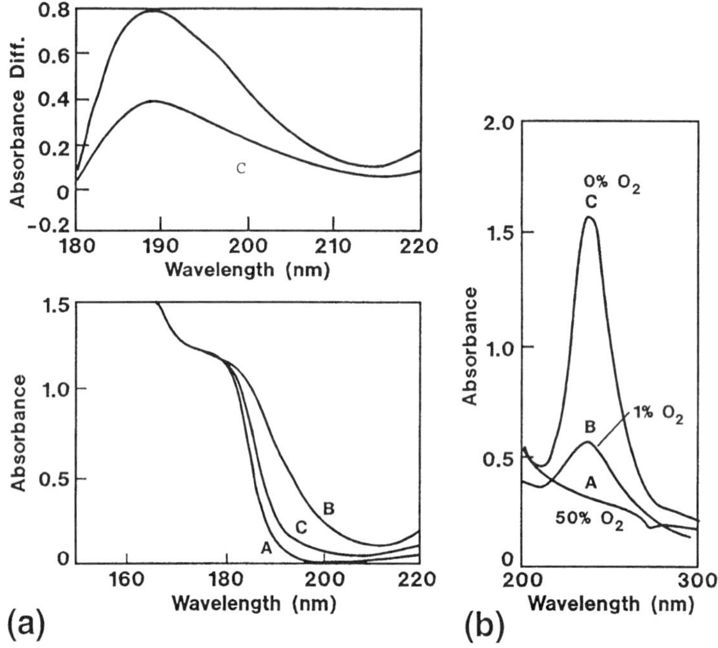

Figure 7.18 Ultraviolet absorption spectra showing the dependence of the 240-nm–absorption band on the oxygen partial pressure during consolidation of the porous preform. Samples are 0.5-mm slices of the consolidated blank: curve A represents sample from consolidation in 50% O_2/50% He, whereas that of curve B is of 1% O_2, and C that of <1 ppm O_2. Both VUV and near UV regions are shown, as well as the difference curves for the VUV.

ter. The electrons and holes produced in the process become trapped and produce new strong absorption features in the vacuum UV portion of the spectrum. An example of this phenomenon is shown in the absorption spectrum, before and after exposure in Figure 7.19. This is the result of the exposure at 248 nm (100 mJ/cm², 20 Hz for 10 min) of a slice of a waveguide blank containing 40% GeO_2. One can clearly see the disappearance of the 240-nm–absorption band as a consequence of the 248-nm exposure, and the increase in the absorption at shorter wavelengths. One can measure a significant number of paramagnetic species after exposure, corresponding to trapped holes and electrons. The most significant is the Ge E′ center, which is a hole trapped on a three-bonded Ge (Fig. 7.20). Also shown is a possible origin of this center; namely, the Ge–Ge bond, which when broken by the action of the light, produces the E′ center. The electron

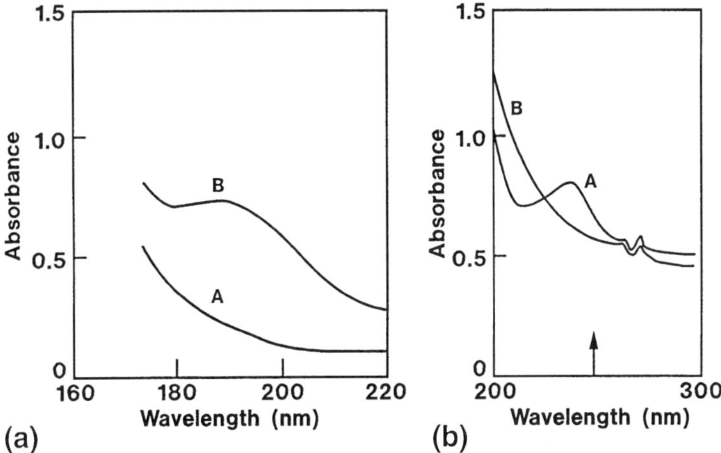

Figure 7.19 Typical absorption changes optically induced in 0.5-mm slice of core-cane 40% GeO_2 sample by 248 nm exposure (100 mJ/cm^2, 10 Hz, 10 min): (a) before exposure; (b) after exposure,

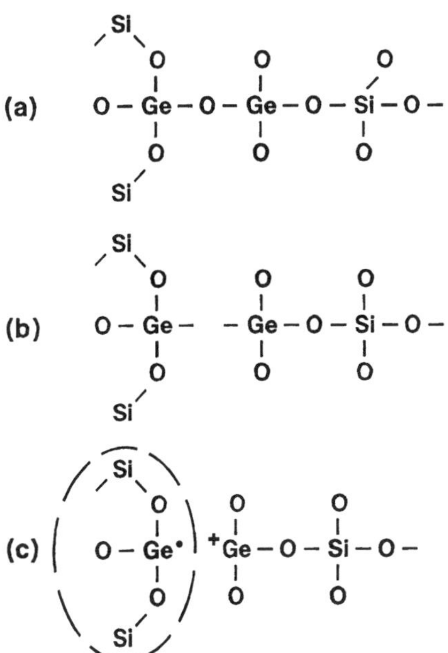

Figure 7.20 Schematic representation of the possible defect structure in SiO_2–GeO_2: (a) A 2-D representation of the Si–Ge network; (b) the oxygen vacancy; and (c) the way the defect would interact with light.

promoted out of this bond is also trapped, and corresponding electron spin resonance (ESR) signals are also observed [28].

Whatever the specific details of the reaction are, the refractive index change at wavelength λ is proposed to come about through the Kramers-Kronig relation.

$$\Delta n(\lambda) = \int \frac{n(\lambda')\alpha(\lambda') \, d\lambda'}{(\lambda - \lambda')} \tag{7.11}$$

This is just the mathematical way of saying that if significant absorption changes are produced in the deep UV portion of the spectrum, then the refractive index will be influenced at all longer wavelengths.

A much larger photoinduced refractive index change could be produced in GeO_2–SiO_2 compositions by impregnating the fiber with molecular hydrogen before the exposure [29,30]. This impregnation is carried at room temperature, or slightly above to ensure that no reaction occurs. The pressure of the hydrogen is high to allow the order of 10^{20} molecules per cubic centimeter to be dissolved. The mechanism is thought to involve the direct attack of the Si–O–Ge bond with molecular hydrogen in the presence of the radiation [28], leading to GeH and SiOH species. Both of these are observed after exposure, as shown in Figure 7.21. Another possibility is the formation of SiOH and a $=Ge^{+2}$ center [27]. The increase in the absorption in the deep UV as a consequence of the exposure is large, as shown in Figure 7.22. Note that the magnitude of the photoinduced absorption in the hydrogen-loaded preform slices is independent of the initial defect absorption at 240 nm. As a matter of fact, in the hydrogen-loaded example, the excitation is equally effective at 193 nm. The explanation for this resides in the suggestion that the process here is dominated by a two–photon-initiated exciton formation that, on being trapped, reacts with H_2. We give a brief summary of the phenomenon in Table 7.5 in its various manifestations.

The exposure method is generally to place a phase mask, corresponding to the period that will yield the desired peak wavelength reflectivity, in contact with the single-mode fiber over the region where the fiber is to be formed. The exposure conditions are somewhat different, but in general, a pulsed UV source is used. This is typically an excimer laser, KrF or ArF. There is some indication that 193-nm exposure is better than 248 nm [31]. The reason is unclear, although the higher energy would favor processes dependent on a two-photon process. The use of the phase mask relaxes the requirement of long coherence length that would be needed for the conventional holographic exposure. Nonetheless, one could also use the more coherent excimer-pumped tunable dye lasers as the source. The exposure fluence is in the vicinity of 100–200 mJ/cm^2, operating at 10 Hz for tens of minutes. It is also possible to use a doubled cw Ar ion laser. Here, although the peak power is low, the average power is comparable with the

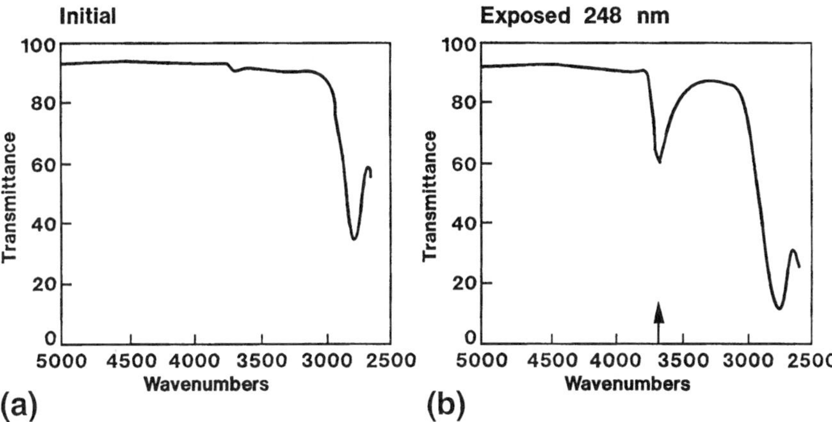

Figure 7.21 Near infrared transmission spectra of hydrogen-loaded core-cane blanks, before and after 248-nm exposure. Note the increase in the absorption at 3670 cm^{-1}, which is attributed to the OH stretching mode.

pulsed sources for which duty cycles are only 10^{-9} [32]. There are also reports of single-shot production of gratings aimed at producing gratings ''off the draw'' [33]. Here the peak fluence is much higher, and it is likely than the gratings are forming from a contribution other than the one just described. The oxygen defect that produces the optical absorption at 240 nm is necessary to couple the light. The major effect may be thermal. Successive pulses produce large swings in the measured efficiency, indicating a different mechanism than pure photochemistry [33]. Nonetheless, the grating is stable, once made.

One last point of practical significance is the thermal stability of these gratings. The thermal stability is characteristic of a material with a distribution of trap depths. Erdogan et al. [34] have analyzed the dependence on temperature. To obtain long-term stability, the gratings are heated to somewhere near 300°C to remove the unstable portion, leaving the long, slowly decaying part of the distribution.

As$_2$S$_3$. Refractive index changes can be readily induced in As$_2$S$_3$ glasses and films [35,36] by exposure to light near the bandgap. More appropriately, one might include the family of glasses As-Ge-S-Se [36]. The origin of the optically induced changes is not fully understood, although it has been the subject of much study [34]. Two proposed explanations are based on light-induced structural changes, sketched in Figure 7.23. The one mechanism is a double-well configuration in which the sulfur ion can be optically excited to the other site, or in the

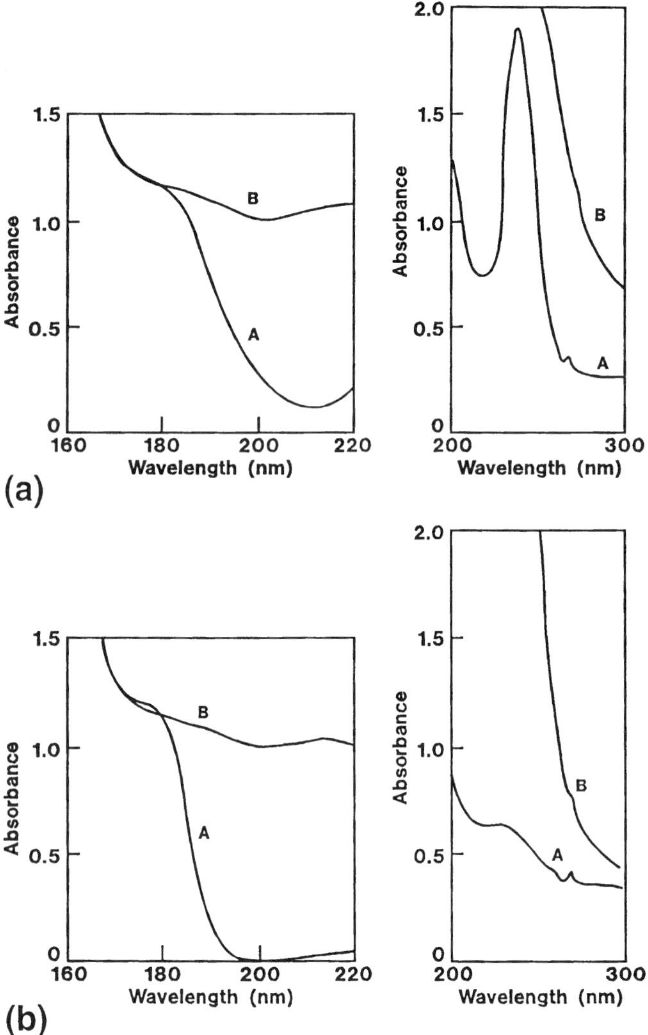

Figure 7.22 UV and VUV absorption spectra of hydrogen-loaded 7% GeO_2–93% SiO_2 core-cane samples, curve A before and curve B after 248 nm exposure. (a) Curves for a sample with a high concentration of oxygen vacancy defects, as indicated by the strength of the 240-nm–absorption band; (b) spectra for a sample with little, or no oxygen defects. In both cases the amount of induced absorption is equivalent, irrespective of the initial difference in the oxygen vacancy concentration.

Table 7.5 Relative Comparison of Mechanisms

Initial state	Defect	Reaction	Products	$\Delta n \times 10^{-4}$
As made[a]	$=$Ge:, $=$Ge–Ge$=$	Bleaching of Ge–Ge	Ge$'$, Ge–1,2	5
H$_2$ low temp[b]	$=$Ge–O–Si	Bond scission	GeH, SiOH	40
	$=$Ge–O–Si	Bond scission	$=$Ge^{+2}	

[a] Number of defects and mix varies considerably with method of preparation and composition.
[b] Samples are usually impregnated at $<150°C$ at pressures >100 atm.

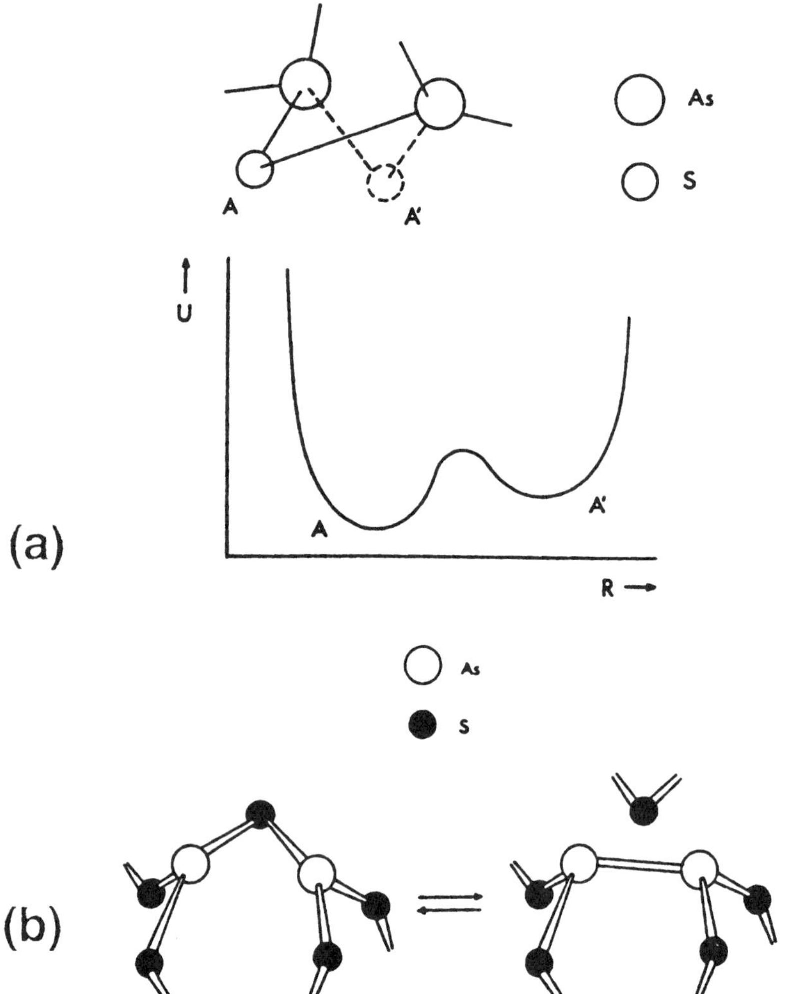

Figure 7.23 Schematic representation of two possible mechanisms leading to light-induced refractive index changes in As_2S_3. (a) A double-well mechanism is proposed; (b) a bond breakage and re-formation is considered. Each of these would lead to a structural change that would cause the index to be altered. (From Ref. 36.)

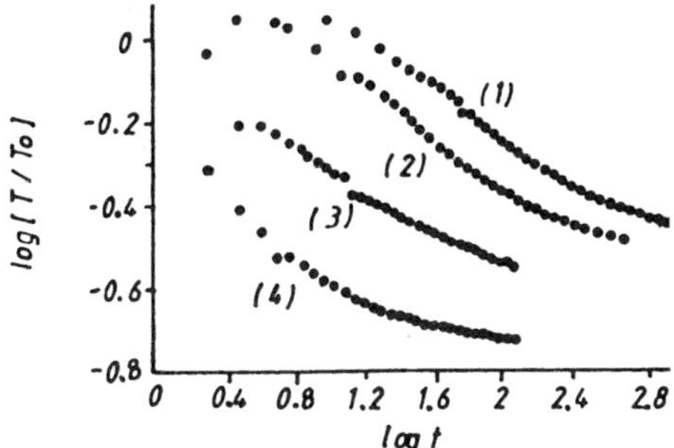

Figure 7.24 Log of measured transmission change in As_2S_3 film at 514 nm as a function of time for the powers of (1), 0.4; (2), 0.9; (3), 3.5; (4), 20 W/cm^2. (From Ref. 37.)

mechanism the proposed optically induced change is where a As–As bond is formed. The exposure also produces optical absorption on the edge, referred to as *photodarkening* (Fig. 7.24). Here the absorption on the absorption edge is monitored in time as a function of exposure [37].

In addition to uniform refractive index changes, there are also reports of photoinduced optical anisotropy [36]. Here, linearly polarized light is used to produce some kind of defect alignment rendering the sample birefringent. An example of this effect is shown in Figure 7.25 where a linearly polarized He–Ne laser was used to produce the birefringent pattern viewed under crossed polarizers.

Holograms have been written in As_2S_3 films. First-order diffraction efficiencies of over 80% have been reported in a 10-μm–thick film using a He–Ne writing beam [38]. The exposure was 15-m W He–Ne laser onto a 1-mm^2 spot for 10 sec.

7.4 APPLICATIONS

The applications of diffraction gratings of small dimension can be broadly classified into two areas: optical interconnects and optical waveguide components and devices. The grating utilization for the two areas is quite distinct. In the optical interconnect application the grating is primarily used to redirect the light. In the optical waveguide application, the grating is primarily used to separate different wavelengths and to modify propagation characteristics. We will cover a few of the representative applications in the following sections.

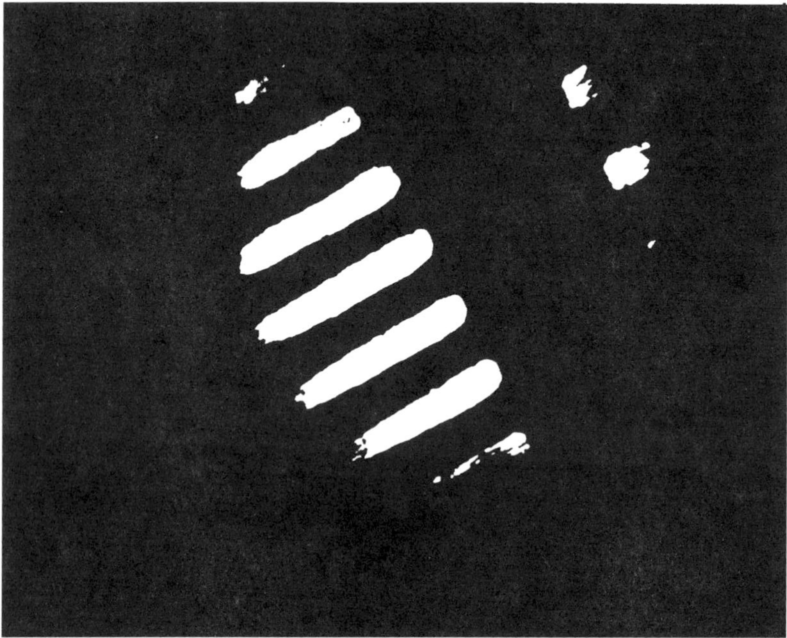

Figure 7.25 Photomicrograph of induced birefringence in As_2S_3 glass: The exposure was polarized with an He–Ne laser at 633 nm, 15 mW, 1.5 mm beam, for 15 min. Picture was taken under crossed polarizers with the exposure polarization direction oriented 45° relative to the polarizer direction.

7.4.1 Optical Interconnects

The optical interconnect application has to do with the transfer of optical signals in the three-dimensional (3-D) spatial domain [39–41]. For example, the connection between chips or boards could be accomplished faster optically than electronically. The electrical signal is converted to light by a LED, and transmitted as light to a detector at some other spatial location. The advantage of this is based on the assumption that the electronic method is limited by the RC time, which becomes more an issue as dimensions of the circuit becomes smaller.

A couple of simple functions are shown in Figure 7.26. Figure 7.26A shows the way light from one place on the chip might be communicated to another position on the same chip. The diffraction grating would be a thick or phase holographic type for high efficiency. Figure 7.26B shows the communication

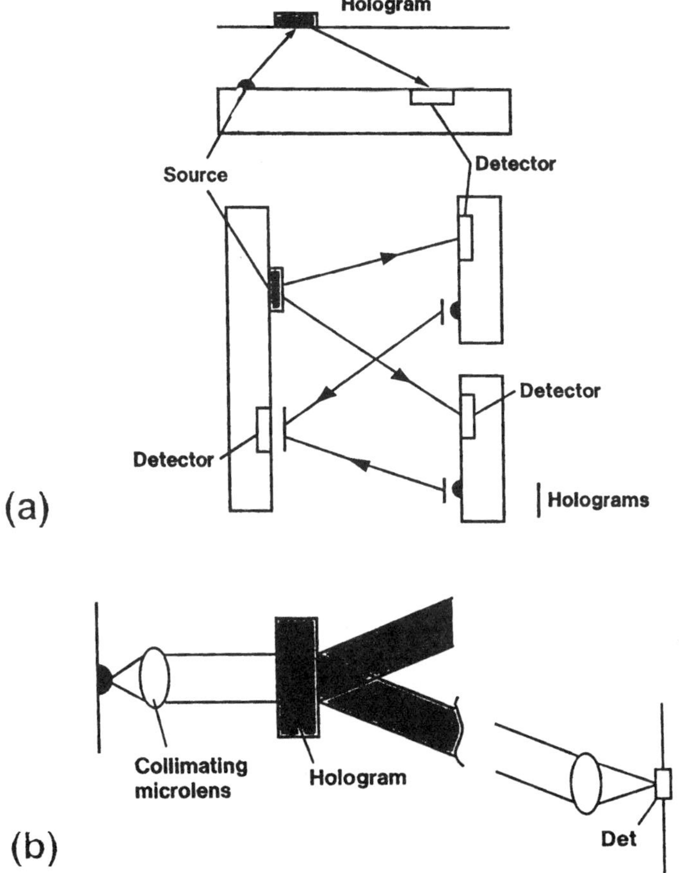

Figure 7.26 Schematic drawings of the way gratings can be used to optically connect places on (a) the same chip, and (b) on separate chips or boards.

between boards. In this case there are both fan-out type gratings, one beam to N, and fan-in gratings, n beams to one.

In this function, the grating element is used in combination with other microoptic elements. For example, the output from the source is collimated for the diffraction grating to be effective, similarly a lens would be required to focus the light onto the detector.

A wonderful example of the power of diffractive optics is in the ability to

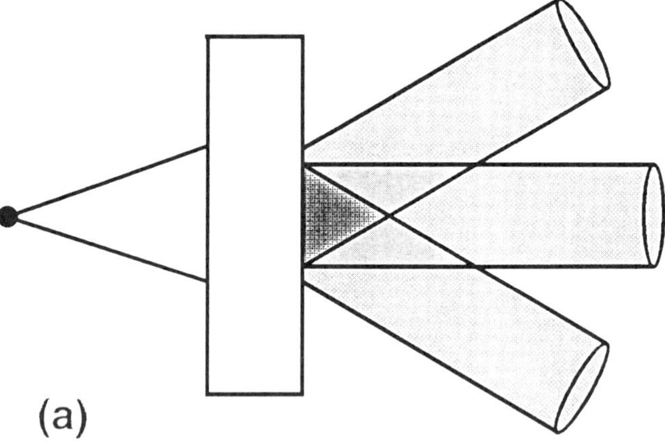

(a)

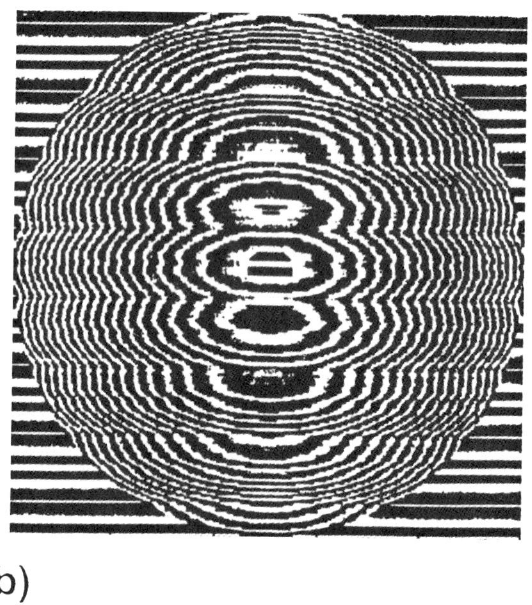

(b)

Figure 7.27   Schematic of a diffractive element that combines (a) collimating lens and fan-out function; (b) a photograph of the actual diffractive element. (From Ref. 39.)

combine optical functionality. For example in the foregoing application, there are two elements required, one to collimate the beam from the source and the other to diffract it with the grating to form a fan-out pattern. Kostuk et al. [39] have designed a diffractive element that combines both functions. We schematically show the combined function in Figure 7.27A and a picture of the eight-level diffractive element in Figure 7.27B.

A somewhat different, but related, grating function has to do with the beam tracking of a compact disk (CD) player [42]. The optical device is shown in Figure 7.28. There are actually two gratings that are used. The first is a simple thin phase diffraction grating placed right after the laser diode as shown. Its function is to split the beam into three, a strong zeroth order, and two weaker first-order beams. The three beams then pass through a segmented holographic-type grating that focuses them onto the disk. After reflection, the beams are directed differentially by the different segments onto the five-element photodetector array. Comparison of the signal from the detector array leads to a discernment of the tracking error.

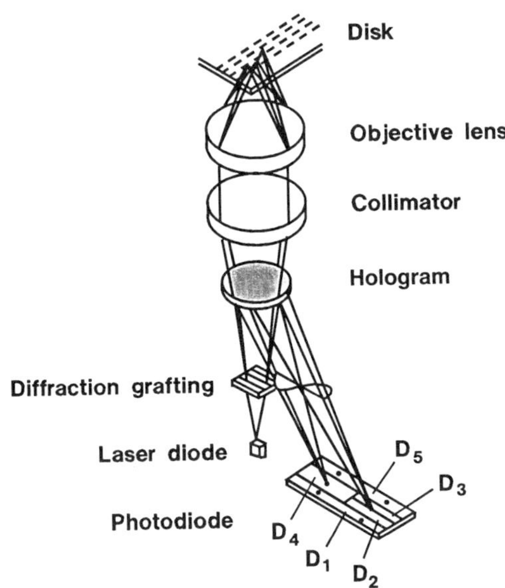

Figure 7.28 Schematic of the read element of a CD player showing the position of the two gratings. The function of the holographic grating is to diffract the return beam onto a detector array to provide tracking information. (From Ref. 40.)

7.4.2 Optical Waveguide Components

The fiber Bragg grating is by far the most important example of the use of a grating structure in the optical fiber area, in particular in wavelength division multiplexing (WDM). Its function is to select a wavelength from a number being carried in the fiber and to reflect it backward where it can be coupled out. Before we address the specific applications it will be useful to present the phenomenon in terms of mode coupling. With this background one will be able to see how gratings can be used for a wide range of applications.

The general formulation is written in terms of the perturbing influence of a periodic refractive index pattern (grating) on the mode propagation in a given waveguiding structure. It can be shown (1) that the coupling between two modes, represented by form of Eq. (7.12) will occur if the condition expressed by Eq. (7.13) exists between their propagation constants.

$$E_k = A_{0k} \exp[2\pi j\, \beta_k(z)] \tag{7.12}$$

$$\beta_k - \beta_m = p \left(\frac{2\pi}{\Lambda} \right) \tag{7.13}$$

Here, p is an integer and Λ is the grating period. One represents the grating as variation in the dielectric constant of light guiding medium,

$$\epsilon = \epsilon_0 + \Delta\epsilon \tag{7.14}$$

For the cases with which we will be dealing, the variation in ϵ will be in the propagation direction; that is, $\epsilon = \epsilon_0 + \Delta\epsilon(z)$. Starting with Maxwell's equations and representations in the form of Eq. (7.12), together with making all the standard assumptions [1], one obtains the standard coupled mode equations.

$$\frac{dA_1}{dz} = \sum_j \kappa_{1m} A_m \exp(-j\Delta\beta_{1m}z)$$

$$\frac{dA_2}{dz} = \sum_j \kappa_{2m} A_m \exp(-j\Delta\beta_{2m}z) \tag{7.15}$$

$$\frac{dA_l}{dz} = \sum \kappa_{lm} A_m \exp(-j\Delta\beta_{lm}z)$$

where the κ_{lm} are the coupling constants that are defined by the overlap of the lth and mth mode, in the following way:

$$\kappa_{lm} = \iint E_l{}^*(x,\, y)\, \Delta\epsilon(x,\, y)\, E_m(x,\, y)\, dxdy \tag{7.16}$$

and $\Delta\beta$ is defined by

$$\Delta\beta_{lm} = \beta_l - \beta_m + p\left(\frac{2\pi}{\Lambda}\right) \tag{7.17}$$

There are two applications that can now be discussed in terms of this development.

The first is the use of the fiber Bragg grating as a highly selective filter in WDM systems [26]. In this example, the fiber is carrying many wavelengths and the grating is fabricated at the appropriate spacing to reflect one. This is schematically shown in Figure 7.29. The performance of an actual fiber grating made in the manner described in the foregoing is shown in Figure 7.30, both in reflection and transmission. This represents a rather simple example in that one has to concern oneself with only a single forward and backward mode, which reduces Eq. (7.15) to just two terms $A_1(\beta)$ and $A_2(-\beta)$. One can make $\Delta\beta = 0$ by choosing $\beta_1 = \pi/\Lambda = -\beta_m$, according to Eq. (7.17). The solution for the reflectivity—that is, $A_2(z)/A_1(0)$—is simply given by,

$$R = \tan h^2(\kappa z) \tag{7.18}$$

If one assumes a sinusoidal form for the axial refractive index, that is, Eq. (7.14) is written as

$$n(z) - n_0 = \Delta n \cos (2\pi z/\Lambda) \tag{7.19}$$

then κ in Eq. (7.18) is simply given by the quantity $(\beta/2)\,\Delta n$. For this geometry the Bragg condition, [see Eq. (7.5)] gives $\Lambda = (\lambda_0/2n_0)$, thus one can see that

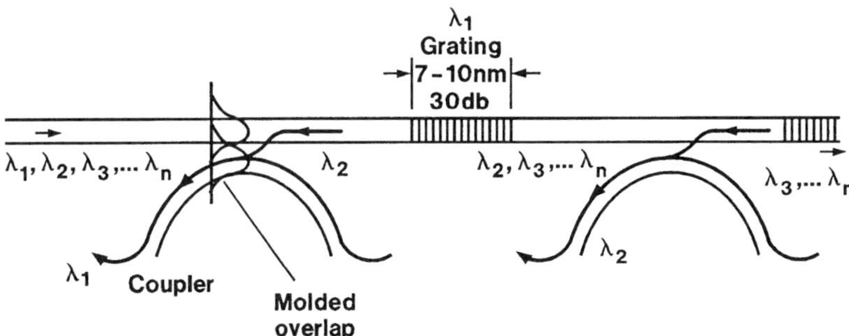

Figure 7.29 Schematic of the way fiber Bragg gratings would be used to remove selected wavelengths traveling in a WDM system. The trunk carries all of the wavelengths. After encountering a grating made to highly reflect one of the wavelengths, it is picked up by the directional coupler and carried out.

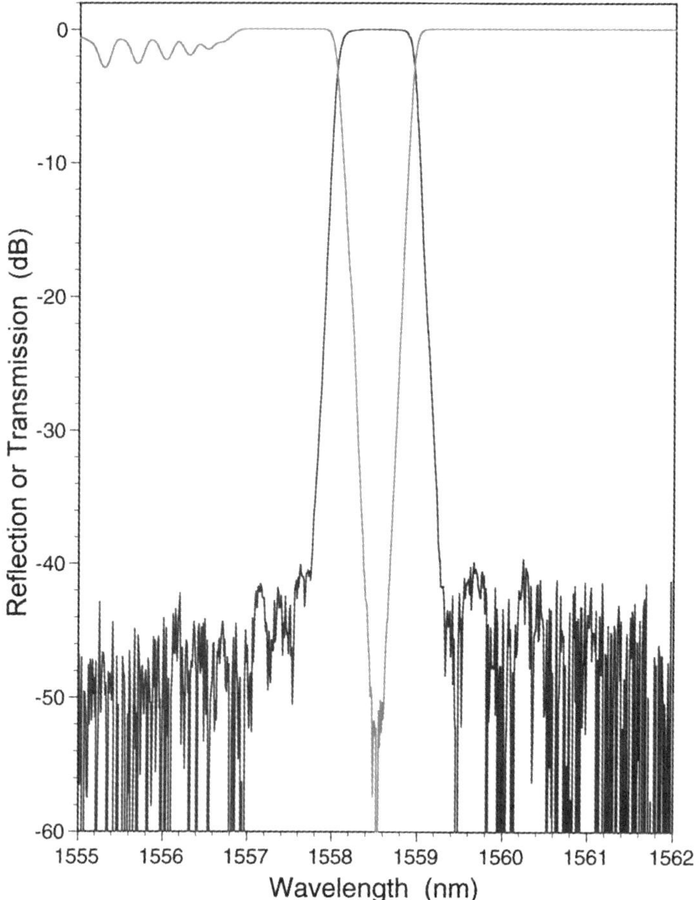

Figure 7.30 Actual transmission (red) and reflection (blue) curves versus wavelength of a fiber Bragg grating. In this example the GeO_2 content was approximately 10 mol%, and the fiber was hydrogen loaded before exposure to 240 nm radiation.

$\kappa = \pi \Delta n / \lambda_0$. Theoretically, a peak-to-valley index change of 2×10^{-4} (corresponds to a $\Delta n = 1 \times 10^{-4}$) would produce a 96% reflection (14 dB) for a 1-cm–long grating. The actual specifications for this application are quite stringent not only in terms of the magnitude of the reflectivity, but also the flatness of the reflectivity versus wavelength. To achieve this flatness, very deep gratings are needed (large Δn). A problem that is encountered in the grating formation stems from the fact that one is using gaussian-shaped beams, so that Eq. (7.19) has a gaussian envelope. To compensate for this, there are several tricks that could be termed *apodization*. One way is to pull the sample away from the mask. In this way the diverging beams provide a uniform exposure on the ends to raise the average refractive index. There are also such things as apodized masks [8].

There is another way one can obtain an estimate of the induced refractive index change. This comes from the measured shift in the peak reflection wavelength as the grating develops. From the Bragg condition one sees that the peak wavelength is just the product of the grating spacing and the refractive index of the medium in which the grating resides. Here, in which the refractive index change is induced by the laser exposure, the average index is also changing. One can see from Eq. (7.19) that the average index change is Δn, so that the peak reflection wavelength of the grating would shift to longer wavelengths by the amount $\lambda \Delta n$. For the foregoing example, one should see a shift of 0.15 nm for a grating made for 1500 nm. In general, the measured wavelength shift indicates a refractive index change in excess of what is determined from the reflectivity. This occurs because the actual contrast in the material produced by the interfering beams is coupled with the explicit dependence of the refractive index with the exposure intensity. It should be clear that, to the extent that the intensity pattern produced by the interfering beams does not produce high contrast, for whatever reason, then the average index will increase and the Δn will decrease. This will produce a larger wavelength shift, larger n_{avg}, yet a lower reflectivity, smaller Δn. Generally speaking, the hydrogen-impregnated samples yield a better agreement between the two index change numbers.

The next example for the use of the fiber Bragg gratings is for selective wavelength filtering [43,44]. For example, in a Er-doped fiber amplifier, the gain may not be flat over some wavelength interval. Thus, one designs a grating such that it couples light from the core mode to lossy cladding modes over the appropriate wavelength interval.

The loss should exhibit a wavelength dependence over the interval to compensate for the variation in gain. To accomplish this, one generally uses forward coupling; that is, one couples the core mode to lossy cladding modes traveling in the same direction. This means that the βs are all positive. From Eq. (7.17) one can see that the grating spacing will now be large. Because of this, these gratings are often called long-period gratings. Typically the grating period is 0.4 mm to effectively couple the core mode to lossy cladding modes. The modes one is trying to couple are those that have some overlap with the core mode, consistent with having a nonzero integral of Eq. (7.17) and, at the same time, extend out into the cladding sufficiently to access the surface so that they can escape. A schematic of this application is shown in Figure 7.31. The coupling constants are smaller here because the modal overlap is not large. This re-

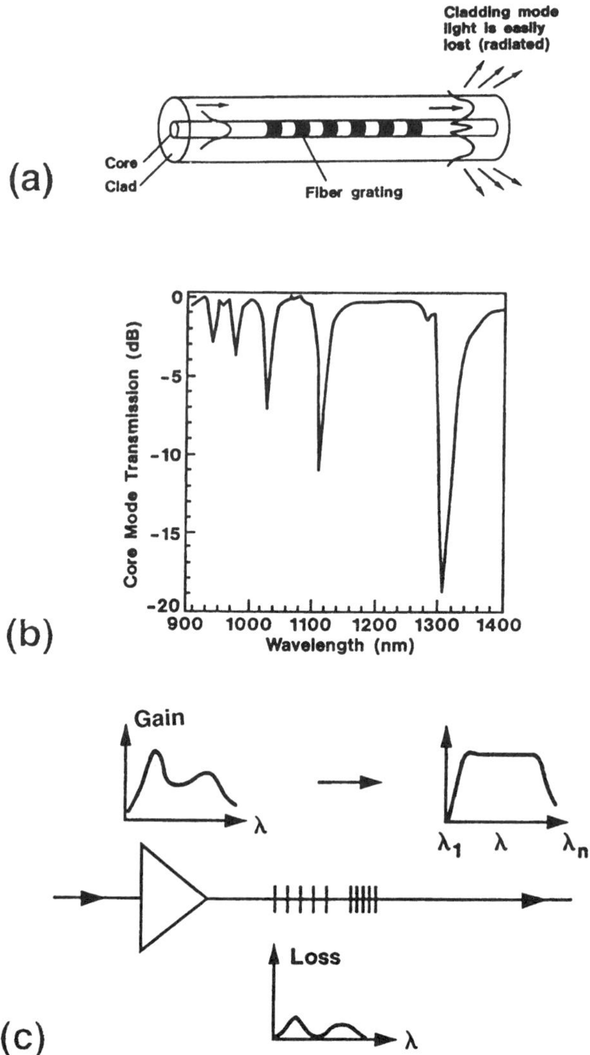

Figure 7.31 Representation of a long-period grating used for passive gain equalization: (a) The modal profile of the core mode and a typical lossy cladding mode with sufficient overlap; (b) the transmittance versus wavelength, indicating the wavelength of the various modes that are being coupled out; (c) an idealized representation of the gain-flattening function. (From Ref. 44.)

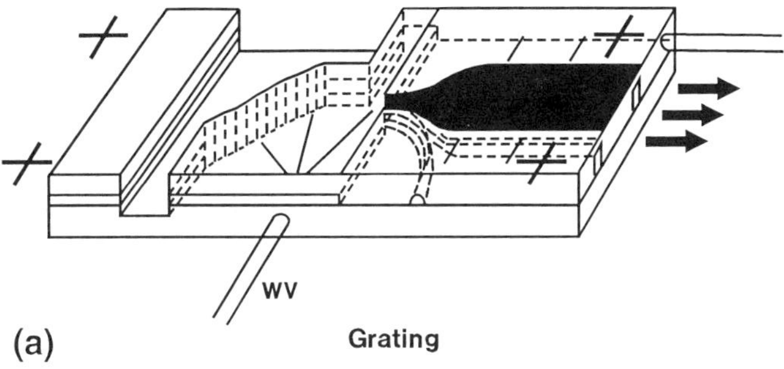

Figure 7.32 (a) Schematic of planar grating-based wavelength demultiplexer made by photolithography and reactive ion etching; (b) SEM photomicrograph of the grating pattern in the SiO_2.

quires a larger value of Δn to make κ a reasonable number, as indicated by Eq. (7.17).

Another possible application is the use of fiber gratings for dispersion compensation [45]. The width of a pulse propagating down a fiber broadens in time as a consequence of wavelength dispersion. This just means that the band of wavelengths that are contained within the time pulse do not travel at the same speed and, as a result, spread apart. To compensate for this, one could design a series of gratings that couple wavelengths into the cladding for some distance where their speed is different to compensate for the speed difference in the core mode. This is most efficiently done with a long-period *chirped grating*. A chirped grating is one for which the pitch is a function of the distance. With appropriate design of the chirp, each wavelength can be coupled to a cladding mode accordingly to provide compensation.

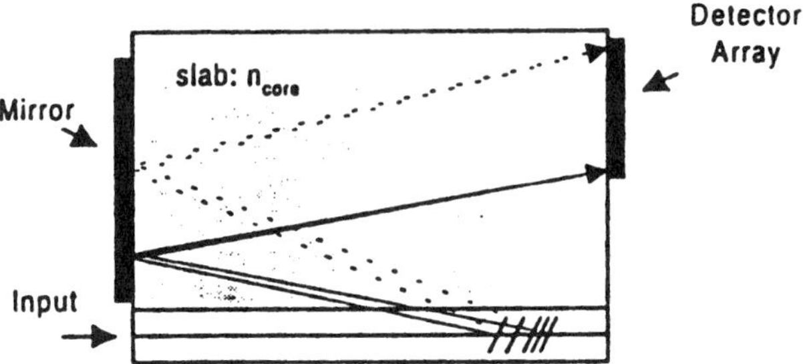

Figure 7.33 Schematic drawing of the demultiplexer based on a photorefractive Bragg grating made in a planar waveguide format. Multiple wavelengths enter the planar guide, then encounter a chirped Bragg grating. The diffraction angle is a function of the wavelength, as shown. (From Ref. 46.)

Finally, a more conventional grating, although hardly in a conventional way, is one that uses demultiplex light in the manner shown in Figure 7.32A. This is a planar device by which the output of a waveguide-carrying multiple wavelengths is directed at a grating. The function of the grating is to separate the wavelengths into separate paths, ultimately coupled back out into separate waveguides. The grating is etched into a planar material from the top using photolithography and RIE. A scanning electron micrograph (SEM) of a portion of the etched grating made in a CVD-prepared silica film in shown in Figure 7.32B [46]. The reported performance here was for a individual channel bandwidth of 0.45 nm, the crosstalk (isolation between channels) was 15 dB. Issues concerning wall verticality and edge feature sharpness contribute to the overall loss of the device, which is polarization-sensitive and more than 20 dB.

One can also use the same idea using planar photorefractive Bragg gratings, as shown in Figure 7.33. Here the entire structure is in the planar waveguide framework. The multiwavelengths guided in encounter the slanted chirped Bragg grating, sending each wavelength out at a different angle. The reported performance of this grating is for a 50 GHz spacing, the isolation is 9 dB [46].

REFERENCES

1. A. Yariv and P. Yeh, *Optical Waves in Crystals*, Wiley-Interscience, New York, 1984.
2. L. D. Hutcheson, ed., *Integrated Optical Circuits and Components*, Marcel Dekker, New York, 1987.

3. R. Magnusson and T. K. Gaylord, Diffraction regimes of transmission gratings, *J. Opt. Soc. Am., 68*(6), 809–814 (1978).

4. H. Kogelnik, Coupled wave theory for thick hologram gratings, *Bell Tech. J., 48*(9), 02909–02947 (1969).

5. R. Magnusson and T. K. Gaylord, Diffraction efficiencies of thin phase gratings with arbitrary grating shape, *J. Opt. Soc. Am., 68*(6), 806–809 (1978).

6. H. M. Smith, *Principles of Holography*, John Wiley & Sons, New York, 1975.

7. R. J. Collier, C. B. Burckhardt, and L. H. Lin, *Optical Holography*, Academic Press, New York, 1971.

8. Lasiris Inc. Quebec Canada, H4R 2K3, product brochure.

9. B. Dahmani, Fountainebleau Research Center, Corning Inc, private data.

10. H. M. Smith, ed., *Topics in Applied Physics, Holographic Recording Materials*, Springer-Verlag, Berlin, 1977.

11. R. G. Brandes, E. E. Francois, and T. A. Shankoff, Preparartion of dichromated gelatin films for holography, *Appl. Opt. 8*(11), 2346–2348 (1969).

12. A. Ashkin, G. D. Boyd, J. M. Dziedzic, R. G. Smith, A. A, Ballman, J. J. Levinstein, and K. Nassau, Optically-induced refractive index inhomogenieties in $LiNbO_3$ and $LiNTaO_3$, *Appl. Phys. Lett., 9*(1), 72–74 (1966).

13. R. V. Schmidt and I. P. Kaminow, *Appl. Phys. Lett., 25*, 458 (1974).

14. S. K. Korotky and R. C. Alferness, $Ti:LiNbO_3$ integrated optic technology, *Integrated Optical Circuits and Comp.* (L. D. Hutcheson, ed.). Marcel Dekker, New York 1987, Chap. 6, and extensive references contained.

15. D. M. Pepper, J. Feinberg, and N. Kukhtarev, The photorefractive effect, *Sci. Am.*, Oct, 62–74 (1990).

16. P. Gunter and J.-P. Huignard, *Photorefractive Materials and Their Applications*, Vols. 1 and 2, Springer, Berlin, 1988, 1989.

17. F. S. Chen, Optically induced changes in refractive index in $LiNbO_3$, *J. Appl. Phys., 40*, 3389–3396 (1969).

18. E. Moerner, A. Grunnet-Jepsen, and C. L. Thompson, Photorefractive polymers, *Annu. Rev. Mater. Sci., 27*, 585–623 (1997).

19. K. Meerholz, B. L. Volodin, Sadalphon, B. Kippelen, and N. Peyghambarian, A photorefractive polymer with high gain and diffraction efficiency, *Nature, 371*, 497–500 (1994).

20. V. V. Astakhova, N. V. Nikonorov, E. L. Panysheva, V. V. Savvin, I. V. Tunimanova, and V. A. Tsekhomskii, A study on the kinetics of photoinduced crystallization in polychromatic glass, *Sov. J. Glass Phys. Chem., 18*, 152–156 (1992).

21. N. F. Borrelli and D. L. Morse, Photosensitive impregnated porous glass, *Appl. Phys. Lett., 43*, 992–993, (1983).

22. N. F. Borrelli, M. D. Cotter, and J. C. Luong, Photochemical method to produce waveguiding in glass, *IEEE J. Quantum Electronics, QE-26*, 896 (1986).

23. S. A. Kuchinskii, V. I. Sukhanov, and M. V. Khazova, The principles of hologram formation in capillary composites, *Laser Phys. 3*, 1114–1123 (1993).

24. V. L. Sukhanov, Porous glass as a storage medium, *Opt. Appl. 24*(1,2), 13–26 (1994).

25. K. O. Hill, Y. Fuji, D. C. Johnson, and B. S. Kawasaki, Photosensitivity in optical fiber waveguides, *Appl. Phys. Lett. 32*, 647–649 (1978).

26. G. Meltz, W. W. Morey, and W. H. Glenn, Formation of Bragg gratings in optical fibers by a transverse holographic method, *Opt. Lett., 14*, 823–825 (1989).

27. V. B. Neustruev, Colour centers in germanosilicate glass and glass fibers, *J. Condens, Matter, 6*, 6901–6936 (1994).

28. T. E. Tsai, C. G. Askins, and E. J. Friebele, Pulse energy dependence of defect generation in Bragg grating optical fiber materials, *Mat. Res. Soc. Symp., 224*, 47–52 (1992).

29. P. J. Lemaire, R. M. Atkins, V. Mizrahi, and W. A. Reed, High pressure H_2 loading as a technique for achieving ultrahigh UV photosensitivity and thermal sensitivity in GeO_2 doped optical fibers, *Electr. Lett. 29*, 1191–1193 (1993).

30. R. M. Atkins, P. J. Lemaire, T. Erdogan, and V. Mizrahi, Mechanisms of enhanced UV photosensitivity via hydrogen loading in germanosilicate glasses, *Electr. Lett., 29*, 1234–1235 (1993).

31. J. Albert, B. Maio, F. Bilodeau, D. C. Johnson, K. O. Hill, Y. Hibino, and M. Kawachi, Photosensitivity in Ge-doped silica optical waveguides and fibers with 193 nm light from an ArF laser, *Opt. Lett., 19*, 387–389 (1994).

32. D. P. Hand and P. St. J. Russell, Photoinduced refractive-index changes in germanosilicate fibers, *Opt. Lett., 15*, 102–104 (1990).

33. L. Dong, J. L. Archambault, L. Reekie, P. Russell, and D. N. Payne, Single pulse Bragg gratings written during fiber drawing, *Electr. Lett. 29*, 1577–1578 (1993).

34. T. Erdogan, V. Mizrahi, P. J. Memaire, and D. Monroe, Decay of UV-induced fiber Bragg gratings, *J. Appl. Phys., 76*, 73–80 (1994).

35. J. P. De Neufville, S. C. Moss, and S. R. Ovshinsky, Photostructural transformations in amorphous As_2S_3 and As_2Se_3, *J. Non Cryst. Sol., 13*, 191–223 (1973).

36. A. E. Owen, A. P. Firth, and P. J. S. Ewen, Photo-induced structural and physico-chemical changes in amorphous chalcogenide semiconductors, *Phil. Mag. B 52*, 347–362 (1985).

37. M. Bertolotti, F. Michelotti, V. Chumash, P. Cherbari, M. Popescu, and S. Zamfira, The kinetics of the laser induced structural changes in As_2S_3 amorphous films, *J. Non-Cryst. Sol., 192, 193*, 657–660 (1995).

38. S. A. Keneman, Holographic storage in arsenic trisulfide thin films, *Appl. Phys. Lett., 19*, 205–207 (1971).

39. R. K. Kostuk, D. L. Ramsey, and T.-J. Kim, Connection cube modules for optical backplanes and fiber networks, *Appl. Opt., 36*, 4722–4728 (1997).

40. S. Sinzinger and J. Jahns, Integrated micro-optical imaging system with a high interconnection capacity fabricated in planar optics, *Appl. Opt. 36*, 4729–4735 (1997).

41. S. H. Song, J- S Joung, and E.-H. Lee, Beam array combination with planar integrated optics for three dimensional multistage interconnection networks, *Appl. Opt., 30*, 5728–5731 (1997).

42. Y. Kurata and T. Ishikawa, CD pickup using a holographic optical element, *Int J. Jpn. Soc. Prec. Eng., 25*(2), 89–92 (1991).

43. A. M. Vengsarkar, J. R. Pedrazzani, J. B. Judkins, P. J. Lemaire, N. S. Bergano, and C. R. Davidson, Long-period fiber-grating-based gain equalizers, *Opt. Lett., 21*, 336–338 (1996).

44. P. F. Wysocki, J. B. Judkins, R. P. Espindola, M. Andrejco, and A. M. Vengsarkar,

Broad-band erbium-doped fiber amplifier flattened beyond 40 nm using long-period grating filter, *IEEE Phot. Tech. Lett., 9,* 1343–1345 (1997).

45. R. I. Laming, M. Ibsen, M. J. Cole, M. N. Zervas, K. E. Ennser, and V. Gusmeroli, Dispersion compensation gratings, 1197 OSA Technical Digest Series 17, Bragg Gratings, Photosensitivity, and Poling in Glass Fibers and Waveguides: Applications and Fundamentals, 1997, pp. 271–273.

46. K. Liu, F. Tong, and S. W. Boyd, Planar grating wavelength demultiplexer. *SPIE*, Vol. 2024, *Multigigabit Fiber Communication Systems*, 1993, pp. 278–283.

47. C. K. Madsen, J. Wageener, T. A. Strasser, and M. A, Milbrodt, Planar waveguide grating spectrum analyzer, Integrated Photonics Research Conf., Victoria, Canada, March 29, 1998. *Opt. Soc. of Am. Tech. Digest, 4,* 99–101, (1998).

8

Optical Isolators

8.1 BACKGROUND

Small-sized optical isolators are a relatively recent invention brought about by the need for protection of optical devices from stray reflected signals [1,2]. The two most common examples are laser diodes and optical amplifiers. The reflected light destabilizes the output of laser diodes, whereas with optical amplifiers, the reflected light could prematurely ''dump'' the inverted population. The optical isolator package must always be small to fit into optoelectronic device structures; therefore, they require a rather ingenious arrangement of microoptical components.

The elements that make up the optical isolator depend on the specific kind of isolation required. This distinction is usually made by requirements on the state of polarization of the input beam and the source of the reflections. There are essentially three types of isolator constructions that we show in Figure 8.1.

The simplest is shown in Figure 8.1A; it consists of a polarizer aligned with the polarization direction of the optical source, followed by a quarter-wave plate with its optic axis aligned at an angle 45 degrees to the polarizer direction. For this arrangement to work, the backward-directed light would have to come from a direct reflection. The initial light is made circularly polarized by the action of the $\lambda/4$ (90-degree–phase shift). After the reflection, the phase is advanced (or retarded) by 180 degrees, and then it passes through the waveplate again, and a -90-degree shift is produced. The net effect then is a 180-degree–phase shift that corresponds to a 90-degree rotation. This is crossed to the polarizer thereby

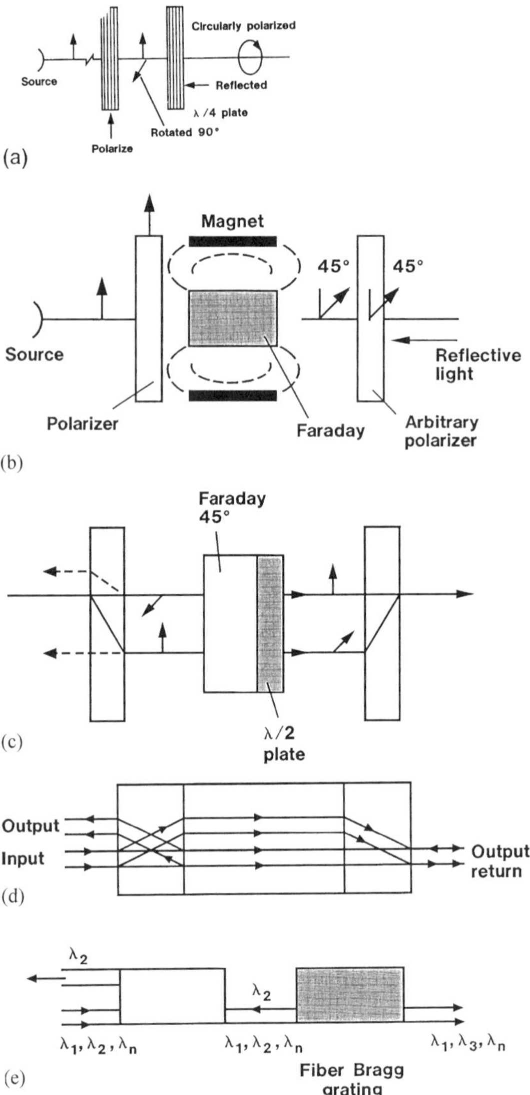

Figure 8.1 Schematic representations of three types of optical isolators. (a) Simple circular polarizer type; (b) more general Faraday effect type where all reflections, irrespective of the source are rejected; (c) polarization independent type, meaning that it rejects backward traveling light irrespective of input polarization (polarization direction through the device are indicated by the arrows); (d) circular device showing how the return beam is captured at the upper port; (e) application of the circulator where it is used to capture the reflected beam, λ_2 in the drawing, from a Bragg grating.

blocking the reflected light. The limitations of this type of isolator are obvious: First, the source must have a fixed polarization direction; second, and much more restrictive, is that the only blocked light is from direct reflections that do not suffer any additional phase shift.

The structure shown in Figure 8.1B is a much more general isolator in that all light returning, regardless of its origin, will be blocked. This is because it uses the Faraday effect that acts on the phase of the light in a nonreciprocal fashion. The *Faraday effect* is the rotation of the axis of polarization of light with an applied longitudinal magnetic field.

$$\Theta = V H L \tag{8.1}$$

where, Θ is the rotation, H is the applied magnetic field in the direction of the light, L is the sample path length in that direction, and V is the constant of proportionality, called the "Verdet" constant. The nonreciprocity means that the rotation will continue in the same sense independent of the direction of the light. This is in contrast to say a half-waveplate that will also rotate the polarization, but in the opposite sense, depending on whether the direction of travel is forward or backward. The origin of this nonreciprocity stems from the fact that the magnetic field direction does not change, and the sense of rotation derives from its direction, not the light direction. In referring to Figure 8.1B, one, sees that the light from the polarized source is rotated 45 degrees by the Faraday element. It then passes another polarizer that is aligned in this direction. Any light returning will be polarized at 45 degrees by this polarizer that rotates the polarization by *an additional 45 degrees* in the same direction. This makes the polarization 90 degrees relative to the initial polarizer; hence, it is blocked. The one major limitation of this type of isolator is that it requires a linear-polarized source. Its major use is for laser diodes in which this criterion is met.

The third type of isolator is the so-called polarization-independent type. This structure removes the limitation of the previous two isolators in that it blocks return light irrespective of the initial state of polarization of the source. This is an important condition because in single-mode optical fibers, the state of polarization is not constant owing to mode coupling between the TM and TE modes. Not only is the state of polarization unknown, but it is changing with time. The structure is shown in Figure 8.1C. The polarization independence derives from the splitting of input beam into the two orthogonal polarizations. This is accomplished by using the different ray paths of the ordinary and extraordinary rays of a birefringent material. The Faraday element is used to provide the nonreciprocity. We have shown in Fig. 8.1c a representation of the polarization condition after each element. This has been done for both the forward and backward direction. It should be clear that the rejection arises by having the polarization of the backward-traveling beams arrive at the input birefringent prism, reversed by 90 degrees from what they were at the beginning.

In the Appendix we will give the Jones matrix form [3] for each of the elements. What this will allow one to do is to quickly determine the state of polarization as a consequence of passing through any sequence of these elements in any order.

A related device is called a *circulator* [4]. It has the same exact structure as the polarization-insensitive isolator; however, its function is to separate and capture the backward beam as shown in Figure 8.1D. A common application is to place it before a reflecting fiber Bragg grating, thus routing the reflected beam out. This is shown in Figure 8.1E.

In the next subsections we will discuss the elements that go into the isolator. For example, we will cover polarizers, waveplates, polarization separators, and Faraday elements. Following that we will discuss actual optical isolator structures.

8.2 POLARIZERS

The familiar polarizing material called Polaroid is not adequate for highly reliable microoptic devices because it is made in a plastic base. Moreover, that limits its use for the telecommunication wavelengths window. There are various inorganic materials that can be used as polarizers in this region. It would be worthwhile to make the following distinctions for the type of polarizers. There are essentially three types of polarizers, (1) *dichroic*, (2) *wire-grid*, and (3) *beam separators*. We schematically represent the three effects in Figure 8.2.

In the dichroic type, one is dealing with a material that absorbs the light strongly in one polarization orientation, and not in the other. The Polaroid is such a material in which the anisotropic absorber is an aligned molecule within the plastic strip. The stretching of the plastic leads to the alignment. We shall see that there is a glass analogue to this approach called Polarcor.™

The wire-grid type relies on reflection to differentiate the respective polarizations. A good example of this is the polarizer used to polarize light in the near infrared (IR) spectrum. It is based on the property that light polarized parallel to the wires is strongly reflected, whereas that perpendicular is not.

A common example of the beam separator type is the birefringent crystal polarizers in which the linear polarized components refract differently as a consequence of the anisotropic refractive index. As a result of this, for light entering at an angle to the optic axis, the two polarizations will see different refractive indices and travel different paths. This is discussed in a later section.

In the following sections we will go through examples of each of these types, together with some discussion of how they are made and their level of performance.

8.2.1 Dichroic Polarizers

The dominant IR polarizer is the ''stretched'' metal–halide glass type. They are made from special alkali aluminoborosilicate glasses that contain Ag and or Cu,

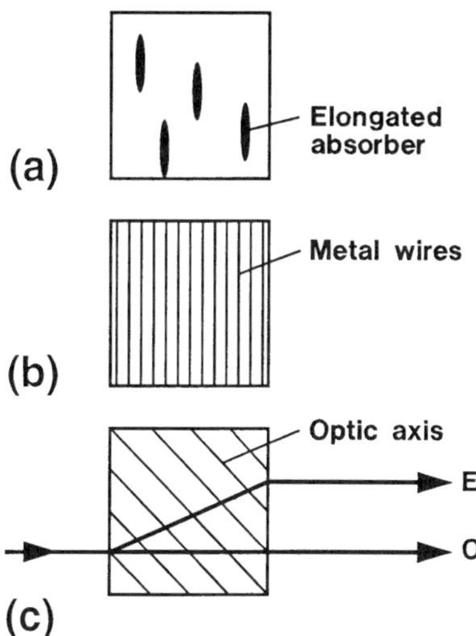

Figure 8.2 Three types of polarizers: dichroic, which works on the principle of an anisotropic absorption; wire-grid that is based on the behavior of the reflection of light from a pattern whose spacing is comparable to the wavelength of light; and birefringent crystals where the polarization components are separated by double refraction.

together with halides, such as Cl and Br [5]. The glasses are subsequently heat-treated to develop a phase-separated droplet metal–halide phase within the glass. The glass is then heated to near its softening temperature and then stretched, which provides elongation of the metal–halide droplets in the process. The schematic of the process is shown in Figure 8-3a, together with an electron photomicrograph of the elongated metal–halide phase (Fig. 8.3b). At this stage the glass is birefringent, but not dichroic. The glass is now in the form of a strip 2–4 cm wide and 3 mm thick. To produce the dichroic property, the stretched glass is treated in an atmosphere of pure hydrogen at 400°C for sufficient time to chemically reduce the metal–halide phase to the metal. The optical dichroism now comes from the anisotropic interaction of the metal needle with the incident radiation. The absorption in the direction parallel and perpendicular to the long direction of the particle can be expressed as Eq. (8.2) [6]:

$$\gamma = \frac{NVk\varepsilon_2}{[(L_l(\varepsilon_1 - 1) + 1]^2 + [L_l\,\varepsilon_2]^2} \tag{8.2}$$

where N is the number of particles per unit volume, V is the particle volume, k is the wavenumber ($2\pi/\lambda$), and ε_1, ε_2, are the real and imaginary parts of the dielectric constant of the metal normalized to the optical dielectric constant of the glass. The L_l refers to the depolarization factors for the respective symmetry

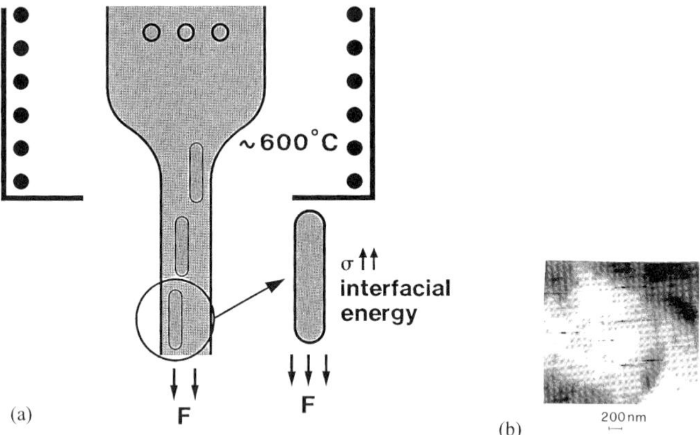

Figure 8.3 Schematic drawing of the method used to make polarizing glass. (a) Glass containing a droplet phase is stretched above the glass softening temperature that produces a force that elongates the droplet. The forces involved are the shear force on the particle as a consequence of the applied stress, and the restoring force related to the interfacial energy. (b) An electron micrograph of the resulting stretched glass.

directions derived from potential theory. The depolarization terms define the local field in the principal directions of the particle.

$$E_{loc} = E_{ext} + 4\pi LP \tag{8.3a}$$

where E_{ext} is the external field. The anisotropy develops as a consequence of the way the external field redistributes (E_{loc}) around the particle. One notes that if the first term in the denominator goes to zero, then a resonant-like behavior is observed. This behavior is assured if the following is true.

$$\varepsilon_1 = \frac{L - 1}{L} \tag{8.3b}$$

When for a given aspect ratio—measure of the major to minor axes—of the particle, the Ls are determined, the term will become zero if at some wavelength ε_1 achieves this value. For free–electron-like metals, such as Ag, Au, and Cu, the real part of the dielectric constants are negative, thus the condition in Eq. (8.3b) is always met for some wavelength. For relatively simple shapes, such as prolate spheriods, the depolarization terms are easily obtained. The transmittances in the stretched direction and perpendicular to that direction for a Polarcor material are shown in Figure 8.4. Contrasts of more than 50 dB are common. Another desirable feature is the breadth of the high-contrast region. This is likely due to a distribution of particle aspect ratios.

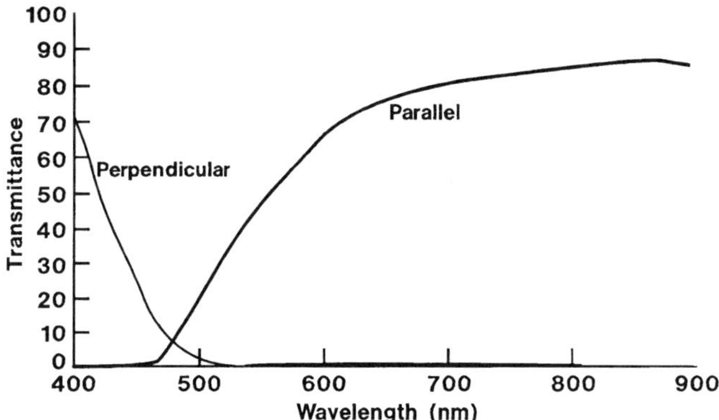

Figure 8.4 Typical transmittance spectra of the polarizing glass in the direction parallel to the stretched direction, and perpendicular to that direction.

8.2.2 Wire-Grid Polarizers

The wire-grid polarizer is based on the fact that, if the spacing of a parallel array of long conducting wires is small enough, then it will reflect radiation for which polarization is in the direction parallel to the wires, and transmit radiation with the polarization perpendicular to the wire array direction. The transmittance and the contrast as a function of the ratio of the wavelength to the wire spacing, where it is assumed that wire width is equal to half the wire spacing, is shown in Figure 8.5 for three different substrate indices, n_0 [7]. The expressions for this case in the two-dimensional limit are the following.

$$T_j = \frac{4n_0 A_j^2}{1 + (1 + n_0)^2 A_j^2} \quad j = 1 \text{ is perpendicular} \tag{8.4}$$

$$j = 2 \text{ is parallel}$$

where,

$$A_1 = \frac{1}{4B}, \; B = (d/\lambda) \left[\frac{0.3466 + 0.25Q}{(1 + 0.25Q) + 0.0039(d/\lambda)^2} \right] \tag{8.5}$$

$$Q = \frac{1}{[1 - (d/\lambda)^2]^{1/2} - 1}$$

One can see from Figure 8.5 that to obtain usable contrast one needs to have the ratio of λ/d be greater than 2 and preferably greater than 8, to obtain a contrast of 20 dB. For the near-infrared wavelengths this means a separation of 200 nm,

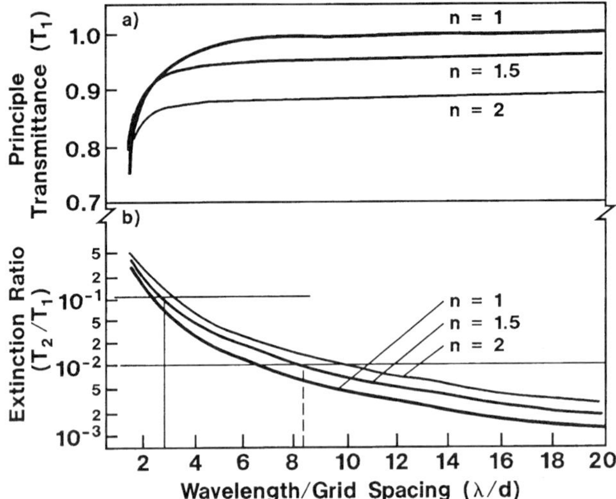

Figure 8.5 Graphs of the computed transmittance and the contrast of a wire-grid polarizer as a function of the grid pitch normalized the wavelength. (From Ref. 7.)

which is about at the present photolithographic limit. The contrast is relatively flat with wavelength because the contrast curve slopes are not steep.

One method of making such devices is the straightforward use of photolithography with which one would pattern the metal lines at the desired separation. This is pushing the limit of resolution of the optical method for the shorter wavelengths. One additional possibility is to pattern a larger period and reduce the separation by a high-temperature redraw process, similar to that described earlier for the Polarcor material.

Another technique is to prepare the structure by depositing alternate layers of a metal and dielectric film of the thickness corresponding to the separation for the desired wavelength. The number of layers are built up to a thickness of the about 100 μm. The wire-grid structure is formed from a cross-sectional slice, as shown in Figure 8.6. This is a process used to produce polarizers under the trade name Lamipol™. The reported extinction ratio is higher than 45 dB [8].

8.2.3 Gratings

Gratings also have a polarization dependence, that is, the diffraction efficiency can be different for light polarized along the grating relative to across the grating [7]. However, most of the large effects are associated with reflection gratings and are angle- and wavelength-sensitive. For the microptic application, a normal incidence transmission grating is preferred, if not required.

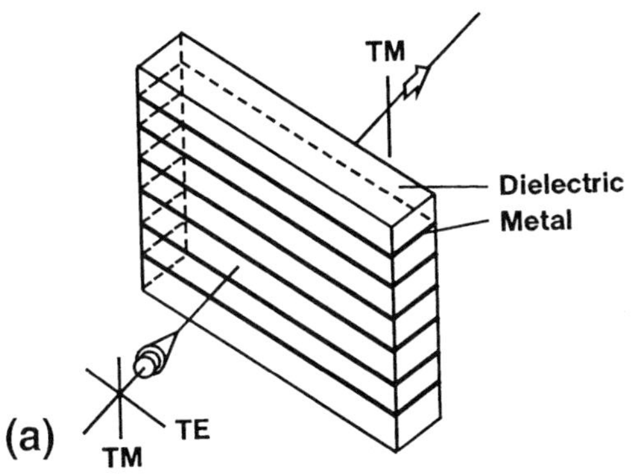

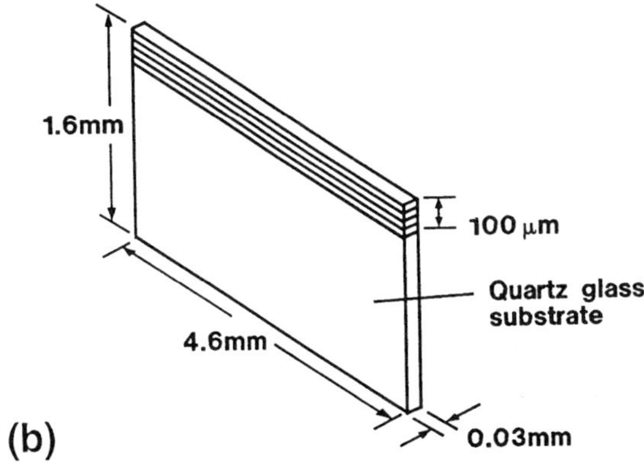

Figure 8.6 Representation of Lamipol™ wire-grid type polarizer. (a) The schematic of the alternate metal dielectric structure and (b) the actual dimensions of the finished element.

An example of a normal incidence transmission grating that produces a high polarizing effect is one that is based on single-crystal $LiNbO_3$. It is based on the interesting property of the induced index change that occurs on ion-exchange of H^+ into $LiNbO_3$. The H^+ ion-exchange is used to increase the refractive index and is a standard way in which waveguiding structures are made in $LiNbO_3$ [9]. The incorporation of the Ti^{+3} into the lattice produces an anisotropic refractive index change, changing more in the crystallographic c-direction than in the basal plane. If a pattern of equally spaced lines were produced by the ion-exchange by the standard photolithographic method, then as a consequence, the grating will have a different efficiency for light polarized along the grating than for the perpendicular. This follows because the phase of the grating, $\phi = 2\pi\Delta nL/\lambda$ will be different for the two directions; $\Delta n(\text{par}) = n_c - n_0$, and $\Delta n(\text{per}) = n_a - n_0$. It actually turns out that $\Delta n(\text{per})$ is negative. The polarizer action occurs by applying a patterned film of a given index of sufficient thickness to exactly balance the $\Delta\phi(\text{per})$, then there will be no phase shift for light polarized perpendicular to the grating direction, and have $\Delta\phi(\text{par}) + \Delta\phi(\text{per})$ for the direction parallel to the grating (Fig. 8.7). The efficiency can be quite high. The limitation is in the spacing of the ion-exchange lines produced that will determine the diffracting angle.

8.2.4 Reflection from Thin Metal Layers

It is clear from the solution of Maxwell's equation at the boundary between a metal and a dielectric that the reflectivity will be different for light for which polarization is perpendicular to the plane of incidence, s-polarized, from that for

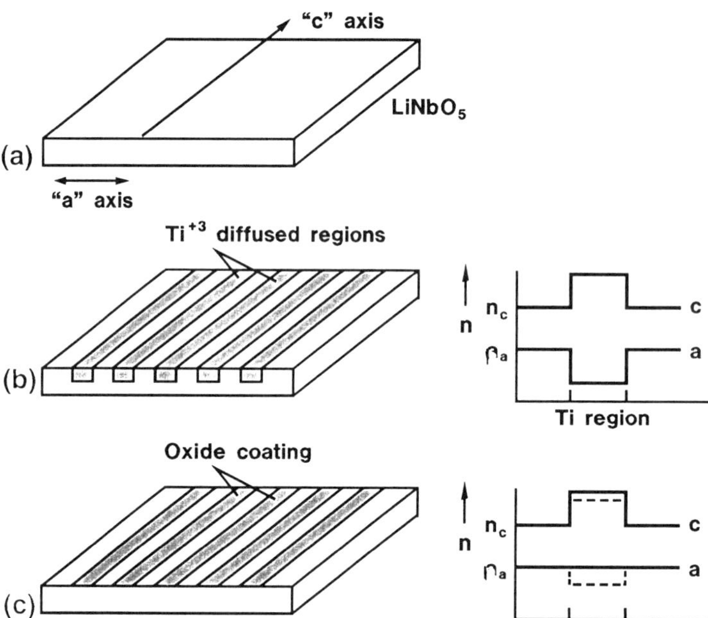

Figure 8.7 Schematic representation of the LiNbO₃ grating polarizer. Initially H⁺ is diffused in the alternating regions. This produces the index profiles shown at the right for the a and c directions. The last step is to give the diffused regions an oxide coating to eliminate the index step in the a direction as shown on the right.

which the polarization lies within the plane, p-polarized [10]. Consequently, it is possible to produce a polarization separation as a result of reflection from a metal–dielectric interface. One of the more dramatic examples of this effect is the coupling to the surface plasmon of the metal [11]. Here, the reflectivity for the s- and p-polarized component can be quite large for a specific wavelength, as shown in Figure 8.8A. The arrangement is that of light traveling in a medium

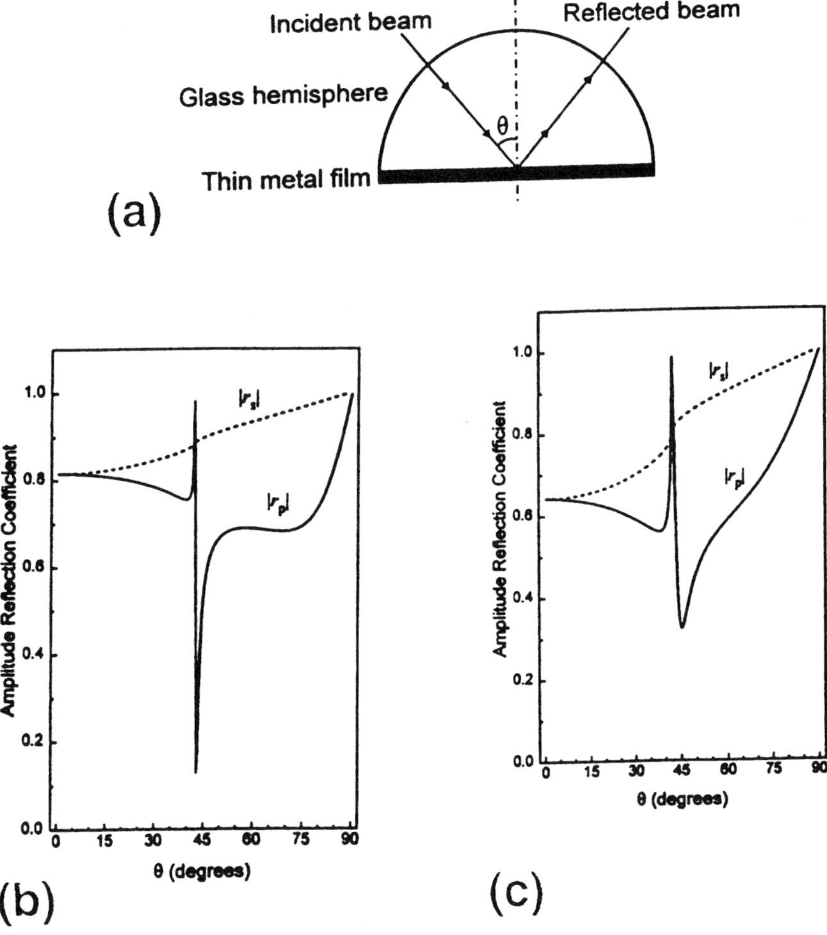

Figure 8.8 Diagrams defining the behavior of the reflection from a thin metal film. In the two graphs the amplitude reflection is shown as a function of incident angle for the *s* and *p* states of the polarizations; this is shown for two different metal layer thicknesses. (From Ref. 11.)

of refractive index *n* making contact with an A1 film an angle θ. The calculated reflectivity as a function of the incident angle for 633 nm is shown in Figure 8.8B,C for two different metal thicknesses. For this arrangement the metal must be very thin to allow the electromagnetic wave to set up the surface plasmon mode. The word *plasmon* is used to represent a quantized inhomogeneous electron density wave.

8.2.5 Optical Waveguide Polarizers

From the solution of Maxwell's equations for single-mode guided waves, one naturally obtains two solutions, usually referred to as the TE and the TM modes. Considering $E = (E_x, E_y, E_z)$ and $H = (H_x, H_y, H_z)$, then z-propagation the TE mode is defined by the nonzero components (H_x, E_y, H_z), whereas the TM mode is defined by (E_x, H_y, E_z). It is clear from this that the polarizations of these two modes are orthogonal. The propagation constant is different for these two modes; however, the difference is usually not large. One can take advantage of this difference to create a polarizer. A straightforward way to accomplish this is to use a Mach-Zehnder configuration. This is shown in Figure 8.9. Directional couplers are guides brought sufficiently close to together to overlap their modal fields for a length long enough to allow that coupling to proceed to whatever extent is desired. There are a few ways this can be accomplished, one of which is to draw down a rod that has two cores to waveguiding dimensions. It is conceptionally easier to imagine this is a planar fabrication, although the operation is the same.

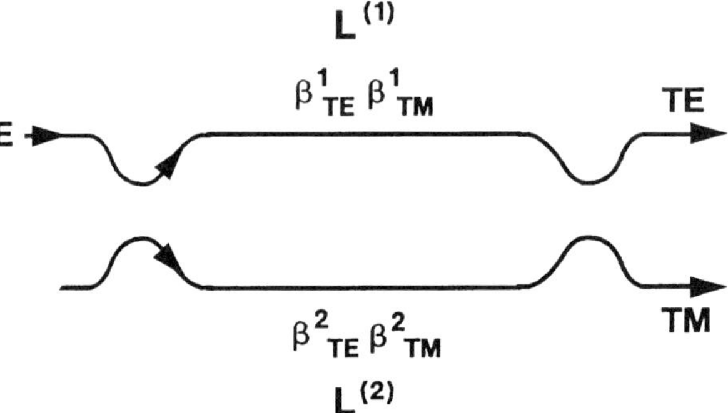

Figure 8.9   Representation of a single mode fiber Mach-Zehnder structure that would separate the input polarization into TE and TM at the output. β^i represents the propagation constant in each arm for the respective polarization. L represents the length of the respective arm. Choosing the correct value for β and L make the phase shift for the TE mode in the upper arm an even multiple of π, and in the lower an odd multiple. The opposite should be true for the TM mode.

One attaches unequal arms to each leg as shown. By unequal is meant that the propagation constants, or lengths, or both are not the same. One then completes the structure with another directional coupler. The idea is the following: For an arbitrary polarization input into one arm, the respective βLs are chosen such that the TM is made to come out one arm and TE out the other. If there is no phase shift between the arms, light entering, say the upper arm, will always exit the lower arm. If one can introduce a π-phase shift, then light in the upper arm will exit the upper arm. If in the upper arm one wants the phase shift for the TE to be an even multiple of π, it will exit the lower arm. Likewise one designs the βL_{TE}, so that this will produce a phase shift of an odd multiple of π; consequently, the TM mode will exit the upper arm. One designs the lower arm to do just the opposite. It should be clear that the guides in the two arms should be reasonably anisotropic to allow for the different phase shifts for the TE and TM components.

8.2.6 Birefringent Crystals

The method of using the difference in the propagation of orthogonally polarized components of a ray in birefringent, single crystals to produce a polarizer is well known. One can understand this behavior referring to the *indicatrix*, which is a way to represent the refractive index for crystals of arbitrary crystallographic symmetry [12]. One writes it in the form,

$$\left(\frac{a}{n_1}\right)^2 + \left(\frac{b}{n_2}\right)^2 + \left(\frac{c}{n_3}\right)^2 = 1 \tag{8.6}$$

where a, b, and c are the principal axes of the crystal, and n_j is the values of the refractive index along this direction. This is the general case of a biaxial crystal. With no loss of generality one can discuss a uniaxial crystal for which a n is equivalent to b, and $n_1 = n_2 = n_0$. Now we have a two-dimensional (2-D) representation that we can show graphically in Figure 8.10A. A ray traveling in any direction other than along the two axes will have a different refraction, depending on the polarization of the ray. Consider the ray shown in Figure 8.6B. One polarization component will see the refractive index n_1, no matter what the ray direction, called the ordinary ray, whereas the other component will see an index somewhere between n_0 and n_3, depending on the exact angle. It is this situation that is represented by the ellipse in Figure 8.10A and represents the condition for the extraordinary ray. The direction this ray will take is determined by the wave normal, as shown in Figure 8.10B. The wave normal is perpendicular to the tangent to the curve at the point P. One sees that the wave normal makes an angle with the ray direction that represents the difference in refraction relative to the ordinary ray, the wave normal of which would always be in the ray direction.

There are several variations of the method, but they are all based on separation of the two rays and then, ultimately, rejecting the extraordinary ray. A simple

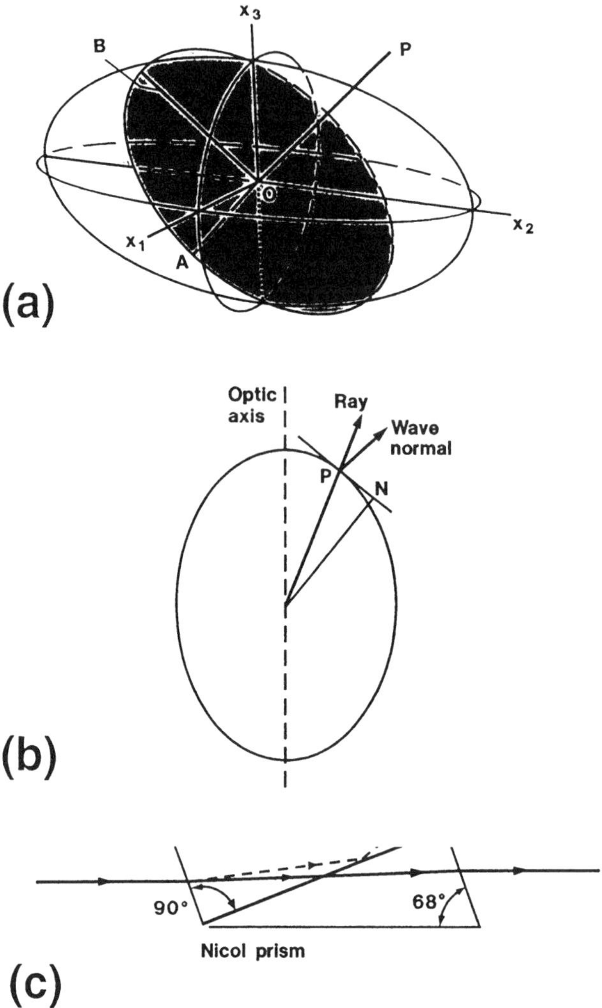

Figure 8.10   The upper figure is a representation of the index ellipsoid with principal axes x_i where the dark region represents the plane normal to the direction of incidence as indicated by the line OP. The middle figure shows that the wave normal does not follow the same direction as the ray normal for the extraordinary which accounts for the double refraction phenomenon. The lower figure shows how one can use this effect to create a polarizer called a Nicol prism. The crystal is cut to make the extraordinary ray totally internally reflected when the ordinary ray is transmitted.

example of this is the so-called Nicol prism. Here the cut of the crystal combined with its orientation relative to the input beam, totally internally reflects the extraordinary ray (see Fig. 8.10C).

These are obviously all derived from single crystals, but there is an artificial way to create birefringence that we discuss later in the beam separator section. The reason for this is that it is used in devices as a beam separator, rather than as a polarizer, although it would be possible to fashion it into a polarizer.

8.3 WAVEPLATES

Waveplates are optical elements that provide specific phase shifts between the orthogonal polarization components of a light beam. In general one can write the expression for a light beam of arbitrary state of polarization relative to an x–y reference frame,

$$\boldsymbol{E} = E_x\,\boldsymbol{x} + E_y e^{-j\delta}\,\boldsymbol{y} \tag{8.7}$$

where $\boldsymbol{x}$ and $\boldsymbol{y}$ are the unit vectors in the given direction, E_x, E_y, the real amplitudes, and δ is a phase shift. If $\delta = 0$ one has linearly polarized light making an angle $\tan^{-1}(E_y/E_x)$ with the x-axis. If $E_x = E_y$ and $\delta = \pi/2$, one would have circularly polarized light. There is a convenient way to present the function of a waveplate using the Jones matrix formalism that we show in the appendix. One can refer to Figure 8.11 to obtain the result of the action of a waveplate of phase shift equal to $\delta = 2\pi\Delta n L/\lambda$, the optic axis of which makes an angle ϕ relative

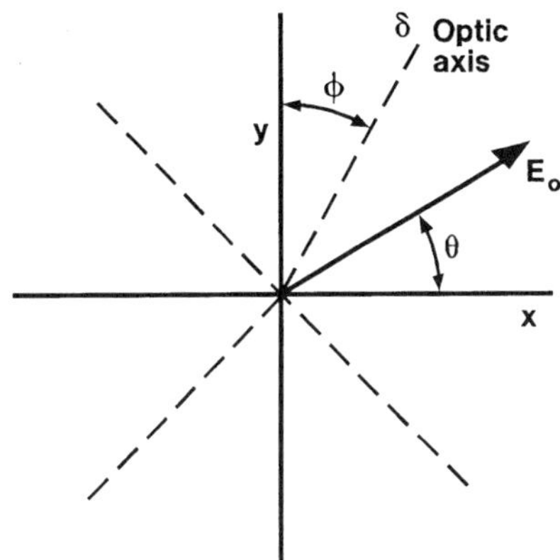

Figure 8.11 Reference system showing a δ phase shift waveplate whose optic axis makes an angle ϕ with the reference axis. The incident beam makes an angle $(90 - \theta)$ with the reference axis.

to the reference axis. An arbitrary polarization can be decomposed into E_x and E_y, which represents the input to the matrix equation.

$$\begin{pmatrix} E_x^{\text{out}} \\ E_y^{\text{out}} \end{pmatrix} = \begin{pmatrix} \cos\phi & -\sin\phi \\ \sin\phi & \cos\phi \end{pmatrix} \begin{pmatrix} e^{j\delta} & 0 \\ 0 & 1 \end{pmatrix} \begin{pmatrix} \cos\phi & \sin\phi \\ -\sin\phi & \cos\phi \end{pmatrix} \begin{pmatrix} E_x^{\text{in}} \\ E_y^{\text{in}} \end{pmatrix} \tag{8.8}$$

One obtains the state of polarization of the beam after having passed through the δ phase-shift waveplate. Some simple examples that can be easily verified from Eq. (8.8) are the following. If $\delta = \pi$ (called a half waveplate, or $\lambda/2$), then for a linearly polarized input, say $E_x = 1$, $E_y = 0$, the result will be linearly polarized beam rotated 2ϕ from the x-direction. Clearly if ϕ were equal to 45 degrees, the effect of the half waveplate would rotate the polarization direction of a linearly polarized beam by 90 degrees. The use of half waveplates to rotate the polarization is the common. Another common use is to create circularly polarized light from a linearly polarized state. This is done by letting $\delta = \pi/4$, and $\phi = 45$.

As we have mentioned in most applications, the role of the waveplate is to produce one state of polarization from another. More commonly, to convert linearly polarized light into another state of polarization, or vice versa. From the discussion in the previous section it is easy to see that birefringent materials provide the simplest way to produce the desired phase shifts. Referring to Figure 8.10A, it is seen that by putting OP in either the x_1 *or* x_2 direction will give rise to the condition that the vibrating component in the x_3 direction will travel at a different uniform speed relative to that in either the x_1 or x_2 directions. In this orientation there is no double refraction.

8.3.1 Single Crystals

The common waveplate materials are made from slices of uniaxial single crystals cut in such a way that they contain the c axis. Crystals such as quartz, calcite, rutile, and mica are often used because of their relatively large birefringence. The phase shift is calculated from the expression,

$$\delta = \frac{2\pi(n_e - n_0)\,L}{\lambda} \tag{8.9}$$

where $(n_e - n_0)$ is the difference in refractive index along the respective optical axes, and L is the thickness of the plate. The refractive indices for the aforementioned crystals are listed in Table 8.1.

Clearly these waveplates are fabricated to a give thickness to produce the desired phase shift at a given wavelength. Because the birefringence is quite large (0.01–0.1) the phase shift is multiple order; that is, $\delta + m2\pi$, m being an integer. There are certain advantages to what are called zero-order waveplates ($m = 0$)

Table 8.1 Birefringence of Common Crystals at 589 nm

Crystal	n_0	n_e
Calcite	1.658	1.486
Quartz	1.544	1.553
Rutile	2.616	2.903
Mica	1.56	1.59
$CaWO_3$	1.92	1.936

because of their reduced sensitivity to wavelength, thickness, and angle of incidence [13]. To make a zero-order plate of materials with large birefringence would make the thickness inpractically thin. One method to make a zero-order waveplate with a reasonable thickness is to glue two pieces together with the optic axis rotated 90 degrees. The difference in the thickness then constitutes the phase shift, which can be zero-order.

8.3.2 Stretched Glass

The stretched metal halide glasses discussed earlier are transparent before the halide phase is chemically reduced to the metal. Consequently, they exhibit birefringence derived from the anisotropic local fields. Similar to the foregoing expression for the absorption cross section for a collection of aligned particles, one can write the expression for the birefringence [14]

$$\Delta n = \left(\frac{V_f}{4n}\right) (\varepsilon - 1) \{[L_1(\varepsilon - 1) + 1]^{-1} - [L_2(\varepsilon - 1) + 1]^{-1}\} \qquad (8.10)$$

where V_f is the total volume fraction of the particles, ε is the ratio of the optical dielectric constant of the particle relative to the medium $(n_p/n_0)^2$, and L_1 and L_2 are the depolarization factors along the symmetry axes of the particle that are dependent only on the aspect ratio of the particle. The phase shift as a function of thickness for various different wavelengths is shown in Figure 8.12 [14].

8.4 BEAM SEPARATORS

In the foregoing section we discussed the use of double refraction for achieving polarization separation. It also can be used to provide a convenient way to separate the incident beam into separate paths. The advantage of this approach is that one can produce an interferometer-type structure necessary to produce an optical isolator that is independent of the input polarization state (e.g., see Fig. 8.1C). Again, there are essentially two ways to bring about this behavior; birefringent single crystals and artificial structures that utilize form birefringence.

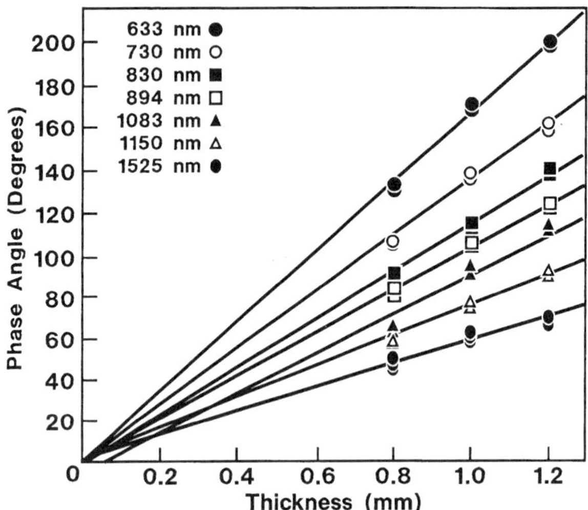

Figure 8.12 Phase shift versus sample thickness for stretched halide glasses. Results are from a number of measurement wavelengths.

8.4.1 Single Crystals

We have already discussed the aspect of double refraction in uniaxial single crystals relative to the polarization directions. The phenomenon of refraction is more complicated to deal with especially when the angle of incidence does not coincide with the optic axes, as shown in Figure 8.13. For the ordinary ray, one has the ordinary Snell's law form,

$$\sin \theta_\iota = n_0 \sin \theta_{r0} \tag{8.11}$$

However, for the extraordinary ray, one can write a similar form, but only after recognizing that the refracting angle depends on the angle of incidence through the dependence of the velocity with angle. One could write this as,

$$\sin \theta_\iota = n\ (\theta_\iota) \sin \theta_e \tag{8.12}$$

where the $n\ (\theta_\iota)$ reflects the dependence of the velocity with angle. The closed expression for the beam-split angle will be given later

8.4.2 Form Birefringence Structures

Alternate thin layers of dielectric materials can develop a high degree of optical anisotropy, proportional to the disparity of the refractive indices of the alternating layers [15]. The definitions of the terms are given in Figure 8.14. Taking the

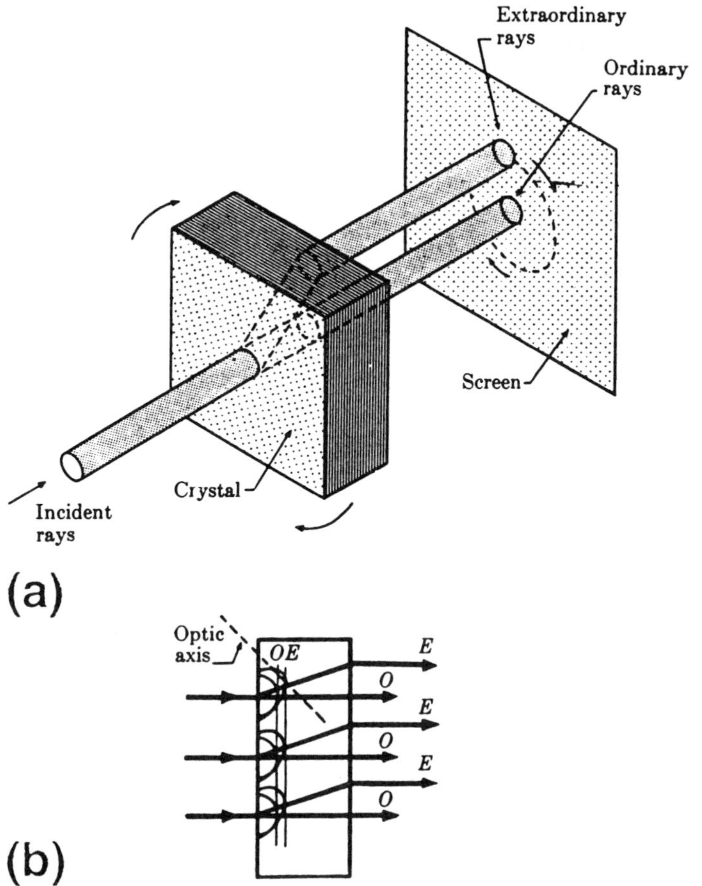

Figure 8.13 Representation showing how double refraction is used to separate the TE and TM components of an incident beam.

ordinary wave as that for which the electric field is parallel to the layers, and the extraordinary wave as that for which electric field is perpendicular to the layers, the refractive indices are given by the following expression [16]:

$$n_0 = [n_1^2 q + n_2^2 (1 - q)]^{1/2}$$

$$n_e = \left[\left(\frac{1}{n_1} \right)^2 q + \left(\frac{1}{n_2} \right)^2 (1 - q) \right]^{-1/2}$$

$$(8.13a)$$

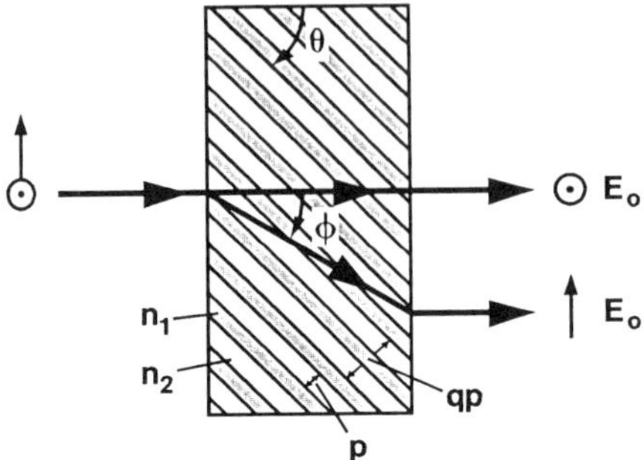

Figure 8.14 Representation of the alternate layer structure of artificial birefringent medium. θ represents the slant angle of the layers and φ is the resulting splitting angle which is dependent on the index difference of the layers and θ.

where q is the ratio of the thickness of the n_1 layer to the period. It should be clear that for this expression to hold the period is much less than the wavelength of light. There is a closed form expression for the beam-split angle, which is given by the following:

$$\tan \phi = \frac{(n_0^2 - n_e^2) \tan \theta}{n_0^2 - n_e^2 \tan^2 \theta} \tag{8.13b}$$

The angle θ is defined relative to the direction of the slant of the layered structure. The split angle reaches the maximum value when the slant angle is given by the expression

$$\theta = \tan^{-1} \left(\frac{n_0}{n_e} \right) \tag{8.14}$$

When this is true then the maximum split angle is just

$$\tan \phi = \frac{n_0^2 - n_e^2}{2 n_0 n_e} \tag{8.15}$$

For small values of Δn, this reduces to the split angle which is equal to Δn in radians. The case of the stretched glass discussed earlier would only split beams by the order of 1.5 min of arc. On the other hand the splitting angle for the alternate dielectric layer structure can be as large as 30 degrees. The comparison between the two common single crystal materials is shown in Figure 8.15.

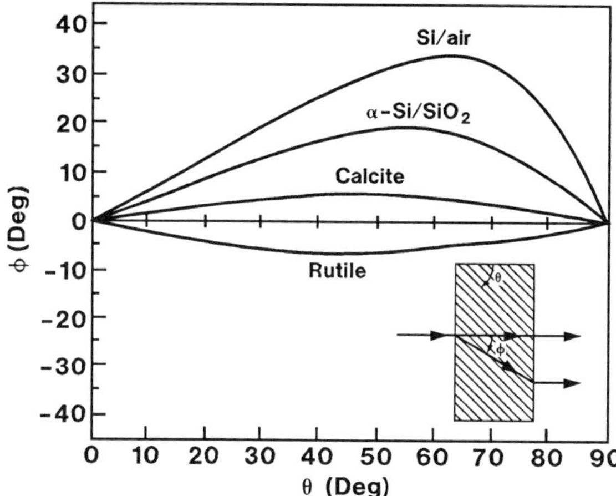

Figure 8.15 Splitting angle as a function of slant angle for two layer structures. A comparison is made to calcite and rutile. (From Ref. 16)

8.5 FARADAY EFFECT

The optical isolator is essentially based on the nonreciprocity of the magnetooptic Faraday effect. The Faraday effect in a material is the rotation of the plane of polarization with a longitudinally applied magnetic field.

$$\theta = V H L \tag{8.16}$$

Here, θ is the rotation of the polarization direction, H is the magnitude of the magnetic field in the direction of the light propagation, L is the sample path length, and V is the constant of proportionality called the *Verdet constant*. All materials exhibit such an effect to some degree. The microscopic origin of the effect is beyond the scope of this book. The reader is referred elsewhere [17] for a discussion of the quantum mechanical nature of the Faraday effect. It will have to suffice to say here that the effect originates from the magnetic field splitting of the atomic energy levels of the constituent atoms (ions), and the corresponding differences that arise in the oscillator strengths of the transitions to the split levels. The transitions in question turn out to have different oscillator strengths for left and right circularly polarized light. One can schematically represent the situation as shown in Figure 8.16. One can see how the circular birefringence arises from the splitting and the corresponding selection rule for the transitions relative to the left or right circular polarization [17]. The example depicted here is for the so-called diamagnetic effect. This simply means that the ground state is nonde-

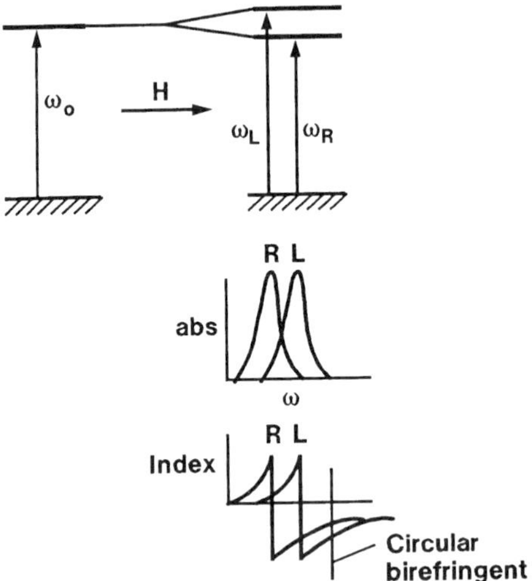

Figure 8.16 Representation of the origin of the Faraday effect. A magnetic field splits the degenerate excited state (diamagnetic effect). The absorption band splits into two states according to left and right circular polarized light. From the dispersion of these two absorption bands, the refractive index difference for left and right circularly polarized light arises.

generate; and thus, only the excited states can be split by the external magnetic field. When the ground state is degenerate, (paramagnetic case), then a similar picture arises, but now, the ground state is also split. The circular dichroim is much larger, however, because the populations of the two split ground states are different as a consequence of the temperature.

The macroscopic phenomenological treatment of the Faraday effect shows that it arises from a nonlinear optical effect from a nonlinear term in the optical frequency polarization as indicated in Eq. (8.17).

$$P_\iota(\omega) = \chi_{\iota j}\, E_j(\omega) + \chi_{\iota jk}\,(\omega;\omega,0)E_jE_k + \cdots + \gamma_{\iota jk}(\omega;\omega,0)E_jH_k + \cdots \quad (8.17)$$

The first term is the ordinary optical dielectric tensor refractive index, the second is the next nonlinear term in the electric field that would lead to such phenomena as the linear electrooptic effect. One can add cubic terms, another such leading to other effects. The next written term is a mixed field term involving both the electric field and the magnetic field. And is responsible for the Faraday effect. One can see this in an approximate way by realizing that the refractive index

$(\varepsilon[\omega])^{1/2}$ is proportional to the derivative of the polarization relative to the electric field.

$$\frac{dP(\omega)}{dE(\omega)} = \chi = \frac{(\varepsilon - 1)}{4\pi} \tag{8.18}$$

To evaluate the nonzero tensor elements one must use the symmetry operations consistent with the overall symmetry of the crystal. It is at this point that the difference between the way the electric field transforms compared with the magnetic field under certain symmetry operations comes into play. The most important and illustrative is the behavior of the respective fields under the inversion operator. For the electric field one uses the simple cartesian unit vectors to demonstrate the inversion, as shown in Figure 8.17. Because the magnetic field is associated with the spin of the electron rather than its vibration, one uses clockwise and counterclockwise spin directions around the unit vector directions to ascertain the behavior of the magnetic field under any given symmetry operation. As one can see from comparing the result for inversion in Figure 8.17a and b, the electric

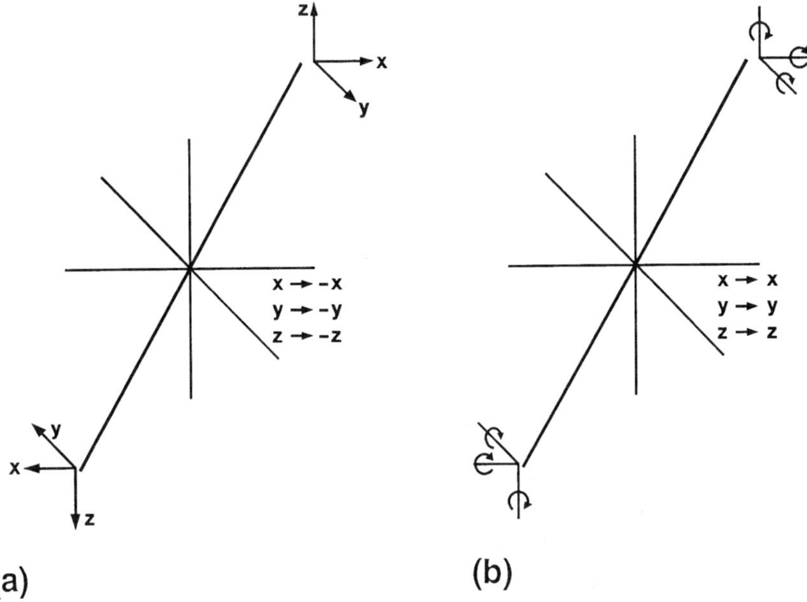

Figure 8.17 The transformation of a polar vector similar to the electric field under the symmetry operation of inversion compared to that of an axial vector similar to the magnetic field.

field is characterized by $E = -E$, whereas $H = H$ under inversion For the case of cubic or isotropic symmetry, one can show that the dielectric tensor for the case of including the first and third terms of Eq. (8.16), has the form

$$\varepsilon = \begin{pmatrix} \varepsilon_0 & -4\pi j\gamma H & \mathbf{0} \\ 4\pi j\gamma H & \varepsilon_0 & \mathbf{0} \\ \mathbf{0} & \mathbf{0} & \varepsilon_0 \end{pmatrix} \tag{8.19}$$

Importantly, the Faraday effect term appears only in the off-diagonal elements. This is directly due to the way the magnetic field transforms under the symmetry operations. The practical significance of these off-diagonal terms is that they influence, not the linearly polarized light, but rather, the circularly polarized light. As we discussed earlier in microscopic explanation, the Faraday effect is really a magnetic–field-induced circular birefringence. In other words, right circularly polarized light travels with a different speed through the material in the presence of a magnetic field than does left circularly polarized light. It is simple to image linearly polarized light as made up of two in-phase circular-polarized components, that is $E = (1 - j, 1 + j)$. If one adds a phase shift δ to one component relative to the other, as the Faraday effect would do, then the result would be a linearly polarized beam rotated at an angle half that of the phase shift.

8.5.1 Reciprocity

Reciprocity is the word often used to describe the symmetry relation between the cause and effect in a medium [19]. If we put it in terms of the present context, it would state if one propagates light in the forward direction through an optical medium, the resultant state of the light will be the same as that if the light had been propagating in the reverse direction. As an example, consider the action of a birefringent sample, the thickness of which makes it a half-waveplate at some wavelength. Consider now that a polarized beam is incident from the left, making an angle ϕ with the optic axis. The effect would be that the polarization direction would rotate to an angle 2ϕ in say the clockwise direction *when viewed in the direction of propagation*. (The actual sense of rotation would depend on whether the optic axis was the fast or slow axis.) Invocation of reciprocity would say that light traveling through the sample from the left would produce the exact same effect. The consequence of this would be that the rotation is just reversed from the forward-traveling effect, when the sense or rotation is reckoned from the original propagating direction. The formal presentation of this problem would consider how the dielectric tensor transforms on the reversal of the propagation direction. The Faraday effect represents the classic nonreciprocal effect because Eq. (8.19) has the dielectric tensor elements that contain the magnetic field. The explanation of the lack of reciprocity can be argued in two ways. The first way is merely to invoke the fact that because the magnetic field is not reversed for the backward-traveling wave, the situations are not equivalent. The second way to understand the lack of reciprocity is outlined in the preceding section where we described the different transformation properties of **H** as compared with *E*. In either case one is left with the situation that for the backward-traveling wave the form of Eq. (8.19) would be such as to reverse the signs in the off-diagonal element.

8.5.2 Verdet Constants

One can classify the nature of the Faraday effect into two categories: diamagnetic and paramagnetic. Moreover, one can distinguish between paramagnetic and ferrimagnetic in the sense that the ferrimagnetic materials are invariably crystals and large magnetization occurs with a relatively low-applied field. In other words, the Faraday effect will be hysteric and saturate at some value of the applied field. In Table 8.2 we will list a representative sampling of the Verdet constant for the various classes of materials. We will break it down into both diamagnetic and paramagnetic glasses or crystals. The final category will be the ferrimagnetic crystals. The units of the Verdet constant will be in degrees of rotation per tesla (10^4Oe) of field per centimeter of length. Similar to all magnetic effects, the units are as diverse as the number of people who write papers. The Verdet constant is certainly no exception. In the early literature, one often sees min/Oe-cm, and lately one sees radians/T-m. They are all relatively easily converted to one another. In the ferrimagnetic crystals, the Verdet constant will be given as the saturated value. That is, it will be at the value achieved at sufficient field to saturate the hystersis loop. The field is not often listed, but is the order of a few kOe for most of the ferrimagnetic crystals. It is desirable, on the one hand, to have them of sufficiently hard magnetic materials, a high coercive field, stable against stray magnetic fields, and on the other hand, not to require a large magnet to provide saturation. As we will shortly see, the optical isolators that employ the Faraday element should be as compact as possible. This favors the high Verdet constant ferrimagnetic crystalline materials, because less than a millimeter of path length is required to produce 45 degrees of rotation.

8.5.3 Other Nonreciprocal Effects

There have been two reported nonreciprocal effects, in addition to the Faraday effect. The first is based on a nonreciprocal phase shift experienced by light propagating in the TM mode of a magnetic waveguide with the external magnetic field *perpendicular* to the propagation direction [20]. The TM mode is the only state that experiences the nonreciprocal phase shift because it has a component of the electric field in the propagation direction. The extent of the phase shift is

Table 8.2 Representative Verdet Constant for Various Materials

Material	V (deg/T-cm)	Wavelength (nm)
Glasses		
FR-123 (Tb-borate)[a]	−67	633
FR-5[b]	−68	633
Pr-phosphate[c]	−38	633
FR-7 (Tb-fluorophos)[c]	−33	633
SF-59[c]	27	633
Tl-gallate[d]	52	633
Paramagnetic crystals		
CdMnTe	2000	633
EuF_3	−250	633
$Tb_3Al_5O_{12}$	−170	633
$Bi_4Ge_3O_{12}$	29	633
ZnSe	112	633
Ferrimagnetic crystals[e] (degrees/cm)		
YIG	870	633
	280	1,064
	210	1,300
$Bi_3Fe_5O_{12}$	−55,000	633
$Gd_{1.6}Bi_{1.4}Fe_5O_{12}$	−28,500	633
	−1,800	1,300
Gd_2Bi_1	−3,300	1,064
	−2,100	1,300
$Y_{1.7}Bi_{1.3}Fe_5O_{12}$	−2100	1300
$Yb_{0.7}Tb_{1.7}Bi_1Fe_5O_{12}$	−1800	1300
	−1200	1550

[a] Kigre

[b] Hoya

[c] Schott

[d] B. G. Aitken and N. F. Borrelli, Thallium-containing glasses with high optical nonlinearity, Proceedings International Symposium on Glass Problems, Vol 1, 527–533, Sept 4–6 Istanbul, Turkey, 1996.

[e] M. J. Weber, ed, *CRC Handbook of Laser Science and Technology*, Suppl 2, *Optical Materials*, CRC Press, Boca Ratan, FL, 1995.

Sources: From SPIE 30th Annual International Technical Symposium on Optical and Optoelectronic Applied Sciences and Engineering. Conference Proceedings 681, Laser and Nonlinear Optical materials; ''Faraday Rotator Materials for Laser Systems,'' M. J. Weber, July 1996.

proportional to the Verdet constant, but much smaller in magnitude. Because E_z changes sign at the center of guide, the waveguide structure must somehow be asymmetrical to avoid cancellation. This is shown in Fig. 8.17. There are various ways that this is suggested both in a planar format as well as a cylindrical fiber structure.

The relevant literature citations that deal with other nonreciprocal structures are Refs. 21a–c. They are all based on planar waveguide structures utilizing some aspect of off-diagonal elements of the dielectric tensor to effect nonreciprocal behavior. Because each is unusual and unique, it is difficult to deal with them here without extensive background. If the reader is interested, consult the original articles.

8.6 ACTUAL OPTICAL ISOLATOR STRUCTURES

The need for ways to eliminate, or capture light traveling back along an optical path has become great in laser diode and fiber-based amplifier systems. Because these are commercial devices that have many other components, to a large extent the isolator function has to fit the device. As a good example of this, the total

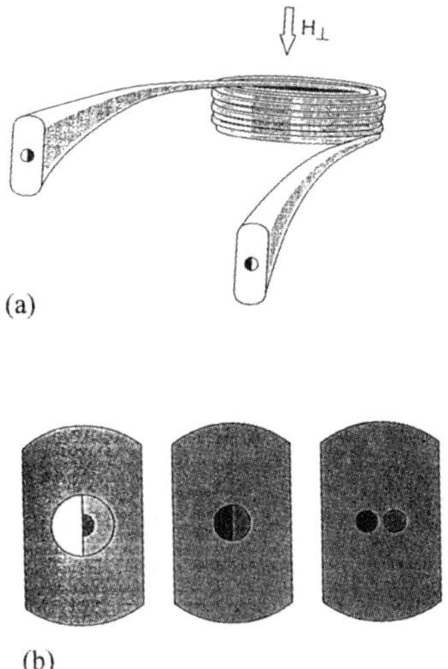

Figure 8.18 (a) Coiled optical fiber with asymmetric core. Magnetic field is applied perpendicular to the axis of the fiber. (b) Three possible core designs to avoid cancellation. In the first, half of the core is air; in the second, the core is made of two different materials with the opposite sign of the Verdet constant; and the third, two coupled cores of different Verdet constants are used. (From Ref. 20 with permission.)

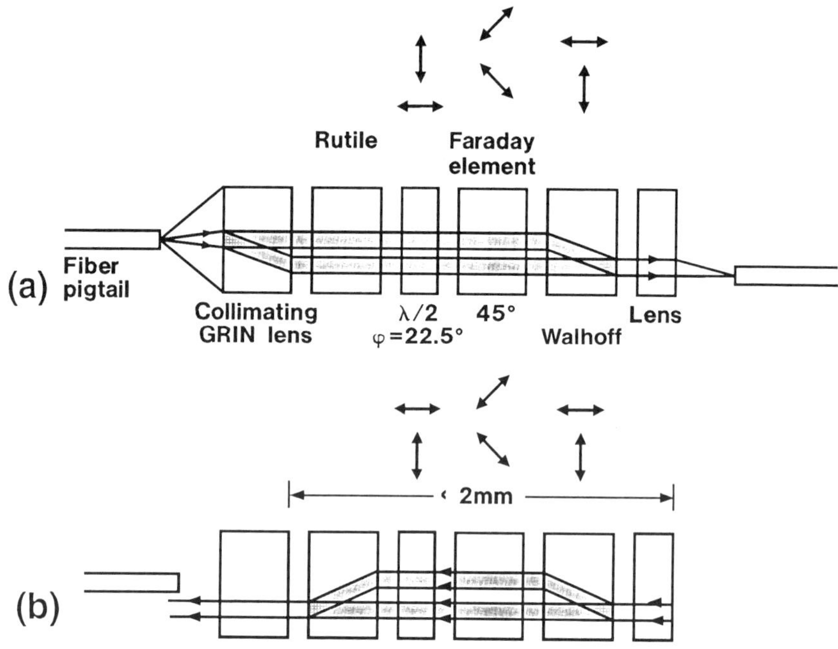

Figure 8.19 Schematic representation of the elements that constitute a polarization independent isolator. The beam paths are shown for the forward and the backward paths. The polarization direction is indicated above the elements.

thickness of the isolator, or circulator package is a major issue, the thinner the better. This means that if the Faraday effect is being used, then the rotation of 45 degrees must be achieved in as thin a sample thickness as possible. This is true for all the functions, beam separation, waveplates, polarizers, or others. This is why the isolators in use predominantly use ferrimagnetic crystal slices as the Faraday rotator because the rotation per unit path length is so much larger.

In this section we will deal with two important structures. The first will be the polarization-insensitive fiber pigtailed device, and the other will be the planar waveguide device.

8.6.1 Polarization-Insensitive Isolator

This is the isolator of choice for optical fiber amplifiers because the state of polarization is not fixed in optical fibers owing to TE–TM mode coupling. We showed this type in Figure 8.1C. The structure is shown in more detail in Figure

8.19. Here we see the lenses that couple the light from the pigtail fibers. The lenses are gradiant index (GRIN) microlenses as described in Chapter 3. An actual $60\times$ SEM photomicrograph of an isolator is shown in Figure 8.20. Each section is identified under the micrograph with the appropriate length scale. The beam separation is effected by a slice of a single crystal of rutile. The half waveplate is quartz, and its optic axis is oriented 22.5 degrees, so that it rotates the polarization by 45 degrees. The Faraday rotator is the ferrimagnetic garnet called BIG ($TbYb_xBi_{3-x}Fe_5O_{12}$) at a thickness of about 0.38 mm. The Verdet constant must be 118 degees/mm because 45 degrees is required. The closest material shown in Table 8.2 is the YBi-garnet. The magnetic field can be either from small external permanent magnets mounted in the package, or with sufficiently

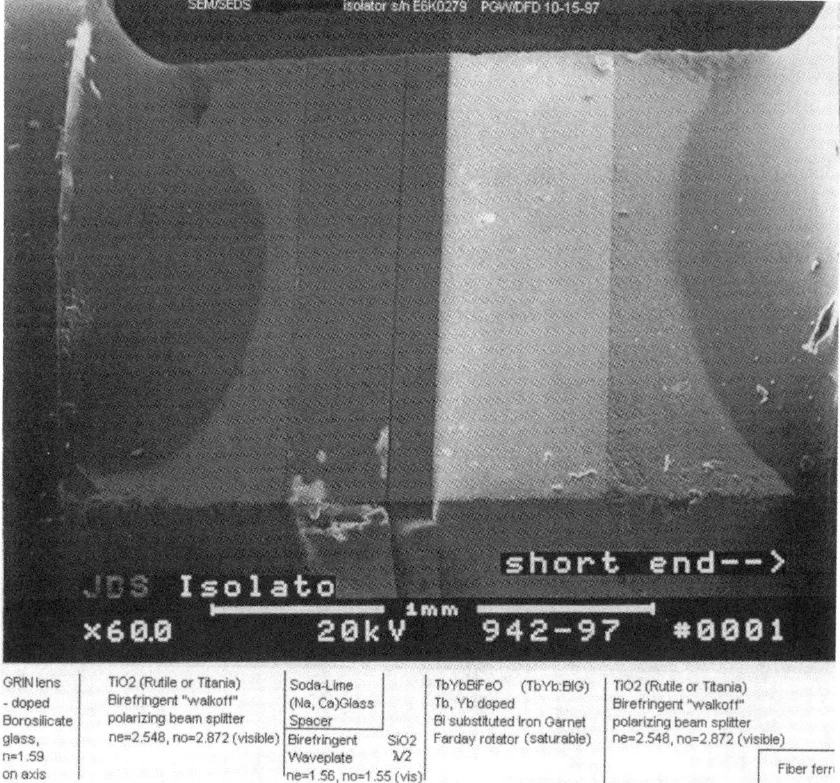

GRIN lens - doped Borosilicate glass, n=1.59 on axis	TiO2 (Rutile or Titania) Birefringent "walkoff" polarizing beam splitter ne=2.548, no=2.872 (visible)	Soda-Lime (Na, Ca)Glass Spacer	Birefringent Waveplate ne=1.56, no=1.55 (vis) SiO2 λ/2	TbYbBiFeO (TbYb:BIG) Tb, Yb doped Bi substituted Iron Garnet Farday rotator (saturable)	TiO2 (Rutile or Titania) Birefringent "walkoff" polarizing beam splitter ne=2.548, no=2.872 (visible) Fiber ferr

Figure 8.20 SEM photograph of an isolator of the design shown in Figure 8.19. The elements are indicated below the photo.

hard magnetic materials, the fully magntized state of the crystal is used. In this case no external magnets are needed. The actual function of each element is shown in the appendix using the Jones matrix method.

The isolation of these isolators (amount of backward light that leaks through) is over 50 dB. Typical insertion losses are 1 dB, or less. Figure 8.21 shows the actual positions of the isolators in the schematic of a forward pumped Er-doped fiber amplifier structure. In the more sophisticated WDM designs circulators are extensively used to separate out the backward-traveling diffracted wave.

What are the disadvantages. At first glance, it would appear then the extreme hybrid construction of the assembly would make it prohibitively expensive. However, the price has decreased as the volume has increased; hence, this may not be a problem, at least for the present applications. The performance is more than adequate, so the move to some other approach is not likely unless the alternative performs at equivalent, or better levels of isolation: It would certainly be desirable to have a more integrated design, such as an all-fiber structure, but this is not likely to happen soon. The more integrated isolators will likely be of a planar nature as described next.

8.6.2 Integrated Planar Isolator

The approach to a fully integrated planar optic circuit, that is an optical equivalent to the silicon chip, will require the ability to fabricate planar optical isolators. The simplest way to approach this is to have the Faraday element act as a guided wave. An example of such a scheme is shown schematically in Figure 8.22. The idea is to fabricate a waveguide in the magnetic film in such a way that it mates

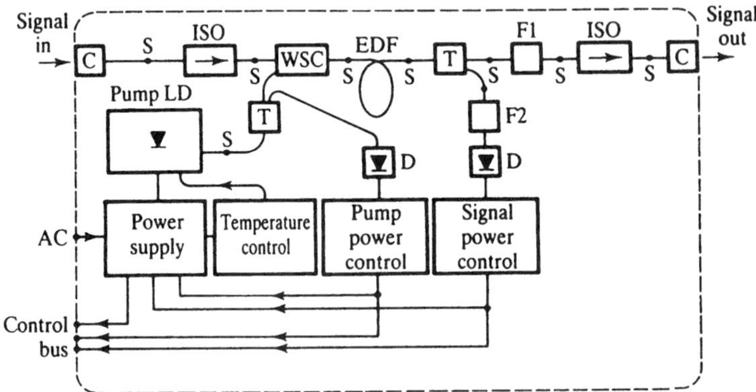

Figure 8.21 Schematic of an Er optical amplifier showing the position of the isolators, marked as ISO.

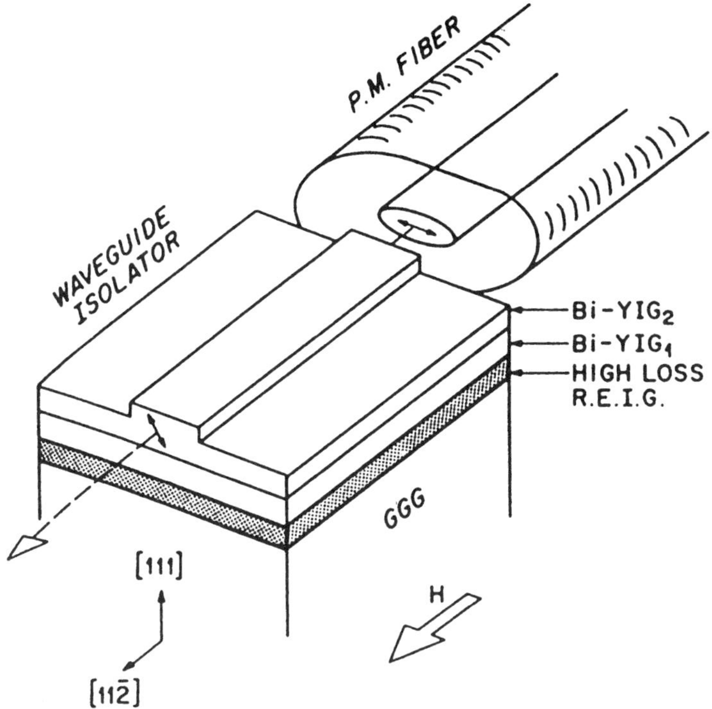

Figure 8.22 Planar waveguide isolator. The Faraday film is fashioned into a channel waveguide as shown. The film is epitaxially grown on GGG and the other layers establish a guiding structure to provide the correct modal field matching the input and output cylindrical guides. (From Ref. 22.)

with low loss to the incoming and outgoing waveguides that are carrying the signal. For a ridge waveguide in the magnetic film, it is to be mated with a cylindrical guide, but it could also be planar. Structures such as this have been made and tested. Wolfe et al. [22] have fabricated the channel guide in a $(BiY)_3(FeGa)_5O_{12}$ material exhibiting a rotation of 133 degrees/cm at 1550 nm. This layer, as well as the other single-crystal layers shown, were epitaxially deposited by LPE on GGG. Isolation of greater than 40 dB was reported. It should be borne in mind that this is not the polarization-insensitive design. To do this, one would have to implement a separate path, as in the device shown in Fig. 1c. This would likely entail some sort of yet unspecified polarization splitter into the beam to create two paths.

A fully implemented design [23] for which even the magnet is deposited as

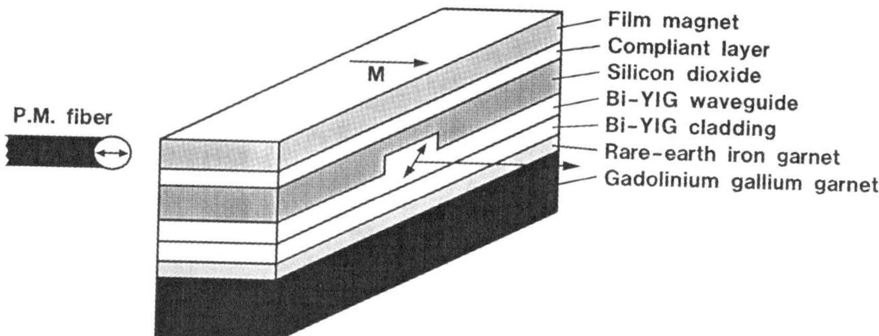

Figure 8.23 Fully implemented planar isolator where even the magnet is deposited as a film. (From Ref. 23)

a thin film, is shown in Figure 8.23. Here too, the garnet layers are epitaxially prepared on a GGG (gadolinium gallium garnet) substrate. To facilitate the waveguiding of the Faraday film, in this case the bismuth-doped YIG, other layers of garnet are deposited. The purpose here is to lower the refractive index between the core and the cladding so that the single-mode thickness of the guiding layer is the order of 4–5 μm. The magnet was a TbCu-type material deposited by sputtering.

APPENDIX: JONES MATRIX

There is a convenient and useful formalism that helps one deal with the sequence of actions that different optical elements perform along a light path. These elements produce things ranging from phase shifts to axis rotations. The basic idea is to define a 2 × 2 matrix for each operation. The sequence of operations becomes nothing more than the product of the matrices in the order that the element is encountered along the path. Since optical isolators contain a number of elements, use of the Jones matrix formalism proves quite useful in describing their overall function.

The light beam is represented by a column vector, the components being the x and y components of the field. The output from the element is then related to the input by the matrix expression,

$$\begin{pmatrix} E_x^{\text{out}} \\ E_y^{\text{out}} \end{pmatrix} = \begin{pmatrix} \alpha_{11} & \alpha_{12} \\ \alpha_{21} & \alpha_{22} \end{pmatrix} \begin{pmatrix} E_x^{\text{in}} \\ E_y^{\text{in}} \end{pmatrix} \tag{A8.1}$$

The matrices for the most common operations will be listed in the following.

Rotation (clockwise angle θ)

$$R = \begin{pmatrix} \cos\theta & -\sin\theta \\ \sin\theta & \cos\theta \end{pmatrix} \tag{A8.2}$$

Phase shift δ with optic axis making an angle φ

$$S = \begin{pmatrix} \cos\phi & -\sin\phi \\ \sin\phi & \cos\phi \end{pmatrix} \begin{pmatrix} \exp(j\delta) & 0 \\ 0 & 1 \end{pmatrix} \begin{pmatrix} \cos\phi & \sin\phi \\ -\sin\phi & \cos\phi \end{pmatrix} \tag{A8.3}$$

Polarizer at angle γ

$$P = \begin{pmatrix} \cos\gamma & -\sin\gamma \\ \sin\gamma & \cos\gamma \end{pmatrix} \begin{pmatrix} p_1 & 0 \\ 0 & 1-p_1 \end{pmatrix} \begin{pmatrix} \cos\gamma & \sin\gamma \\ -\sin\gamma & \cos\gamma \end{pmatrix} \tag{A8.4}$$

The overall effect of a series of optical elements is just the product of the particular matrices representing the function in the order they appear. Reciprocity can be expressed as the equality of the matrix multiplication, applied forward, or in reverse.

For the Faraday rotator, which will have the form of Eq. (A8.2), because of the nonreciprocity attributable to the magnetic field, light traveling in the opposite direction must be represented by the matrix

$$\begin{pmatrix} \cos\theta & \sin\theta \\ -\sin\theta & \cos\theta \end{pmatrix} \tag{A8.5}$$

where $\theta = VHL$.

REFERENCES

1. *Handbook of Laser Science and Technology*, suppl 0002, Optical Materials, M. J. Weber, ed., section 9, Magnetooptic Materials, M. N. Deeter, G. W. Day, and A. H. Rose, CRC Press, Boca Raton, FL 1995, pp. 367–411.
2. D. K. Wilson, Optical isolators cut feedback in visible and near-IR lasers, *Laser Focus, 24*, 103 (1990).
3. R. Clark Jones, A new calculus for the treatment of optical systems, *JOSA, 31*, 488–493 (1941).
4. Y. Fujii, High-isolation polarization-independent optical circulators, *J. Lightwave Technol., 9*, 1238–1243 (1991).
5. N. F. Borrelli, Optical elements derived form stretched glass, *Proc. SPIE*, 1761, 202–212 (1992).
6. POLARCOR Product sheet, Specialty Optical Products, Corning Inc, Corning, N.Y.

7. M. Bass, ed. *Handbook of Optics*, II, *Devices, Measurements and Properties*, 2nd ed., McGraw Hill, New York, 1995.

8. Lamipol product sheet, Sumitomo Osaka Cement Co, Optoelectronics and Electronics Div., Tokyo, Japan.

9. S. Nishino, H. Yamamoto, K. Kasazumi, H. Wada, K. Sano, and T. Saimi, Application of a polarizing holographic optical element to a recordable optical head. *Jpn. J. Appl. Phys. 35*(1B), 357–361 (1996).

10. M. Born and E. Wolfe, *Principles of Optics*, Macmillan, New York, 1964.

11. M. Mansuripur and L. Li, What in the world are surface plasmons, Opt. *Photonic News*, May, 50–57, 1997.

12. J. F. Nye, *Physical Properties of Crystals*, Oxford Press, London, 1957.

13. P. D. Hale and G. W. Day, Stability of birefringence, *Appl. Opt. 27*, 5146–5153 (1988).

14. N. F. Borrelli, Waveplates from stretched glass, *Proc. SPIE*, 1746, 336–342 (1992).

15. K. Shiraishi, T. Sato, and S. Kawakami, Experimental verification of a form–birefringent polarization splitter, *Appl. Phys. Lett., 58*, 211–212 (1990).

16. K. Shiraishi and S. Kawakami, Spatial walk-off polarizer utilizing artificial anisotropic dielectrics, *Opt. Lett., 15*, 516–518 (1990).

17. C. J. Ballhausen, *Introduction to Ligand Field Theory*, Chap. 9, McGraw-Hill, New York, 1962.

18. N. F. Borrelli, Magnetooptic properties of magnetite films, *J. Appl. Phys. 42*, 1120–1123 1971.

19. J. P. Mathieu, *Optics*, parts 1 and 2, Pergamon Press, 1975.

20. R. Wolfe, W.-K. Wang, D. J. DiGiovanni, and A. M. Vengsarkar, An all-fiber magnetooptic isolator based on non-reciprocal phase shift in asymmetric fiber, *Opt. Lett., 20*(16), 1740–1742 (1995).

21. S. Yamamoto, and T. Makimoto, Circuit theory for a class of anisotropic and gyrotropic thin-film optical waveguides and design of nonreciprocal devices for integrated optics. *J. Appl. Phys. 46*(2), 882–889 (1974); F. Auracher, and H. H. Witt, A new design for an integrated optical isolator, *Optics Comm. 13*(4), 435–438 (1975).

22. R. Wolfe, R. A. Lieberman, V. J. Fratello, R. E. Scott, and N. Kopylov, *Appl. Phys. Lett., 56*, 426–430 (1990).

23. M. Levy, R. M. Osgood, H. Hegde, F. J. Cadieu, R. Wolfe, and V. J. Fratello, *IEEE Photonics Technol. Lett., 8*, 903–905 (1996).

9

Photonic Crystals

9.1 BACKGROUND AND INTRODUCTION

Over the past few years more attention has been paid to an approach to light propagation in dense media based on a somewhat different way of looking at the solutions to Maxwell's equations [1,2]. The different look is based on the realization that one can cast Maxwell's equations as a mathematical linear operator and, thereby view the solutions in different terms This is the familiar quantum mechanical formulation for which one defines the hamiltonian operator [3]. The hamiltonian operator acts on a fictitious set of functions called the wavefunctions, in the form

$$H\Psi = E\Psi \tag{9.1}$$

$$H = \frac{-(h/2\pi)^2}{2m}\nabla^2 + V(x) \tag{9.2}$$

Here the periodic potential that the electron sees is contained in the $V(x)$. The particular linear operator H represented by Eq. (9.2) has a number of useful and important properties [2,3]. It is sufficient here to mention that, in general, the set of eigenfunctions of H are orthogonal and complete, allowing one to express any state of the electron as an infinite linear combination of these unique functions. The concept is similar to being able to represent any vector in three spaces as a linear combination of the unit vectors in the x, y, and z directions. When Eqs. (9.1) and (9.2) are applied to electrons, say in a crystal where the electron experiences a periodic potential owing to the charged lattice, one then obtains a map of allowable bands of energy as a function of the wavevector k. For a simple illustrative example in one dimension one writes the wavefunction in the form,

$$\Psi(x) = u(x) \exp(jkx) \tag{9.3}$$

Here the periodicity is contained in the property of $u(x)$ such that $u(x + a) = u(x)$ where a is the period of the potential that the electron is in. The result of this formulation results in an energy (frequency) versus k diagram as shown in Figure 9.1. What one sees is the parabolic shape of the energy bands based on what one would expect for a free electron. [This corresponds to $V(x) = 0$ in Eq. (9.2), and $u(x) = 1$ in Eq. (9.3)]. The symmetry of the structure allows one to draw the extended picture of the bands as shown. For example, the energies must be the same as $k = 0$ and all multiples of $2\pi/a$, where a is the period of the lattice. This will be true for any value of k within the range $-\pi/a < k < \pi/a$ which is termed the reduced Brillouin zone [3]. What is important to see, and forms the particular emphasis in what follows, is that in the presence of a nonzero potential, the bands split as shown. What this means is that there will be forbidden ranges of energies for the electron. These so-called bandgaps play the crucial role in all semiconductor devices. It is the judicious placing of impurities or defects in these materials that provide localized energy levels within the gap. It is the electrons and holes from these localized levels that are manipulated to produce the transistor.

One can write Maxwell's equation in terms of the magnetic field H in this form as well as shown in reference [2].

$$\nabla x \left[\frac{1}{\varepsilon(r)} \nabla x H \right] = \left(\frac{\omega}{c} \right)^2 H(r) \tag{9.4}$$

where ε is the dielectric constant. Following the notation of reference [2] one can define the operator Θ as

$$\Theta = \nabla \mathbf{x} \left[\frac{1}{\varepsilon(r)} \nabla \mathbf{x} \right] \tag{9.5}$$

so that the operator form is

$$\Theta H = \left(\frac{\omega}{c} \right)^2 H \tag{9.6}$$

Aside from the fact that one can now apply the considerable and powerful computation methods developed over the years for large-scale quantum mechanical calculations, based on Eq. (9.1), to an entirely new area, the significance of this

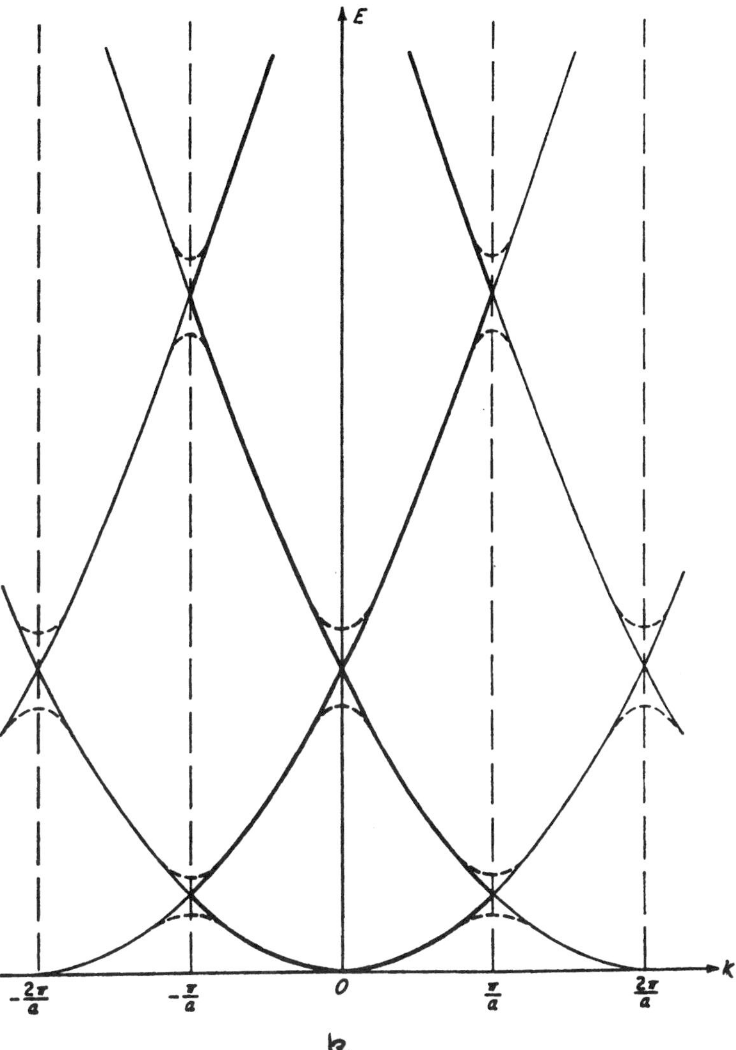

Figure 9.1 Simple one-dimensional reduced Brillouin representation of energy bands for free electrons and the effect of the modifications produced as a consequence of the periodic potential of the nuclei as a solid lattice. The effect is the opening of ''forbidden gaps,'' regions of energy within which the electron cannot reside.

formulation has a more conceptual benefit. The solution of Eq. (9.4) for any given spatial variation in the dielectric constant $\varepsilon(x, y, z)$ (here we mean the optical dielectric constant that is equal to the square of the refractive index) gives rise to an energy versus propagation constant diagram similar to that in Figure 9.1. We will outline an algorithm for the calculation in the next section, but here it is sufficient to realize that the eigenvectors of Eq. (9.4) will be of the form,

$$\boldsymbol{H}_n = \boldsymbol{A}_n \exp[j(\omega t + \boldsymbol{k_n} \cdot \boldsymbol{r})] \tag{9.7}$$

where $\boldsymbol{k_n}$ characterizes the nth propagation mode. We will show a number of these for different geometries in the following sections.

In one sense, it should come as no surprise that a forbidden energy or frequency gap would occur as a result of a periodic pattern in the refractive index. The dielectric multilayer thin film stack, as represented in Figure 9.2a is an example of this phenomenon in a more familiar context. Normally, one would see the transmission, or reflection, plotted against the wavelength for a stack of alternating layers of refractive indicies [4] as shown in Figure 9.3a–c. We have shown the reflection versus wavelength as a function of the number layers to indicate how the reflection maximum begins to rise and flatten as the number of layers increases. It is only in the limit of an infinite number of layers that a flat, totally reflecting spectral region, or gap, would result. It is this condition that corresponds to the photonic bandgap depiction, described in the foregoing, at which frequencies within a certain region are excluded. Plotting the behavior of the multilayer stack in this way leads to the diagram shown in Figure 9.2b. Here one sees that there is a gap in the possible propagating frequencies indicated by the separation on the y-axis. One can compare the calculated gap to that predicted by the conventional approach [for example, see Ref. 4, p. 69] where a transfer matrix approach is used. It can be shown that the reflection will increase exponentially with the number of layers when the argument of the polynomial has an exponential behavior, rather than a sinusoidal one. The condition is thus expressed as Eq. (9.8).

$$\cos^2\beta - (1/2)\left[\frac{1}{n_1} + \frac{1}{n_2}\right] \sin^2\beta < -1 \tag{9.8}$$

Here, $\beta = (2\pi/\lambda)a[(1/n_1) + (1/n_2)]$, a is the period and the ns are the indices of the layers. The gap is then the values of a/λ for which Eq. (9.8) is met. We will see from a practical point of view, the gap is to be viewed as existing for some arbitrarily defined reflectivity, as close to 100% as desired.

It is the extension of this forbidden gap concept into two and three dimensions, in particular, the structures that will be required, that will provide the interesting microoptic applications that we will try to cover in the succeeding sections.

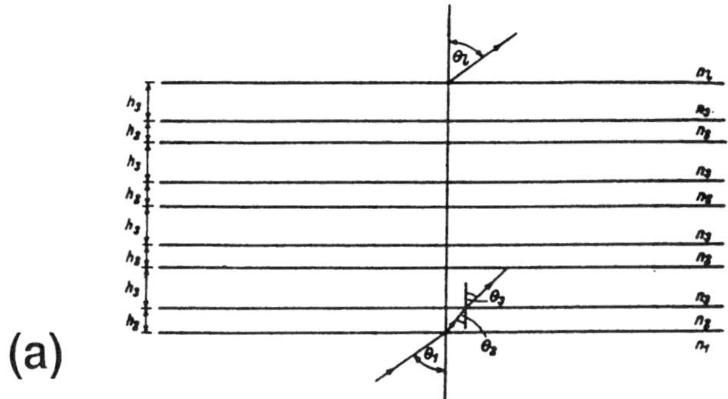

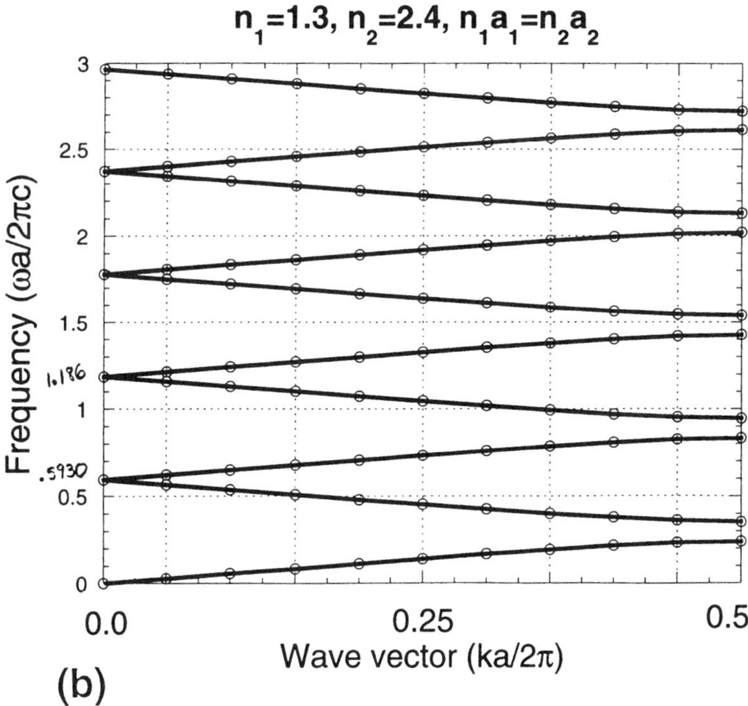

Figure 9.2 (a) Multilayer dielectric stack of thin films. Refractive index alternates from $n1$ to $n2$ in each layer. (b) The calculated infinite layer one-dimensional photonic bandgap; here, the condition of equal optical thickness was used.

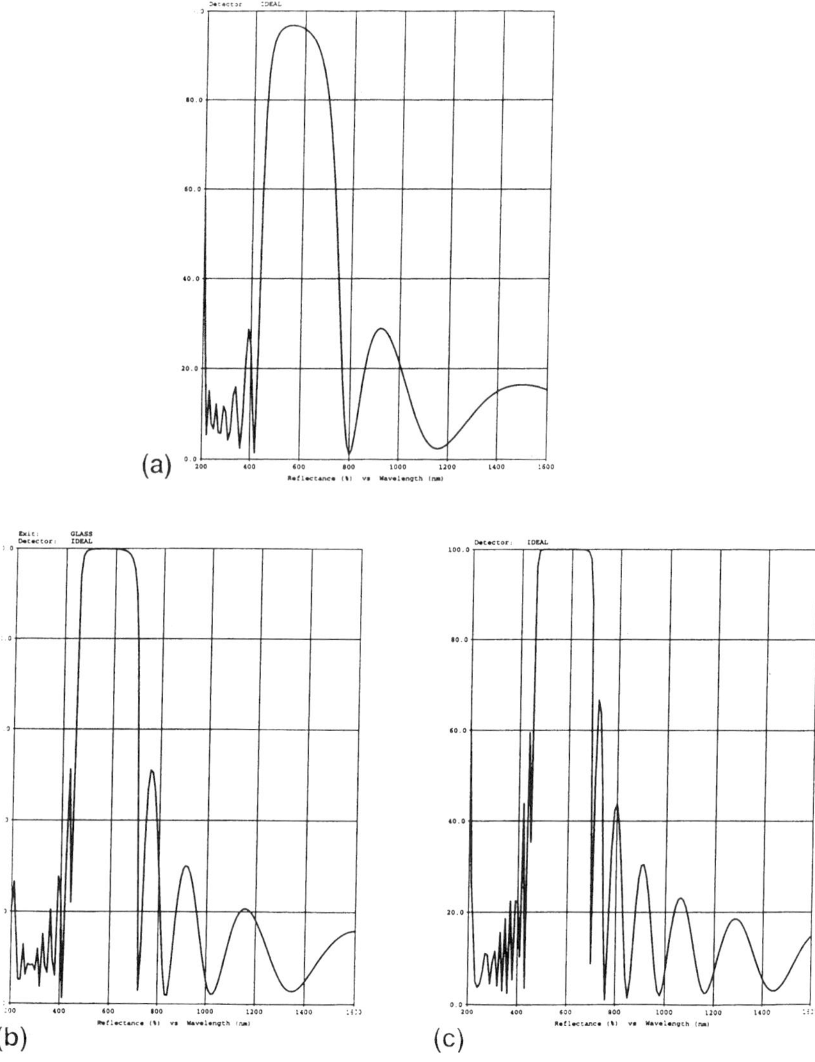

Figure 9.3 Computed reflection spectra for the case given in Figure 9.2 for 9, 15, and 21 layers, respectively.

9.2 MATHEMATICAL FRAMEWORK

The intention is not to present any more mathematics than absolutely necessary, especially in light of several excellent articles that are available [2,5]. Nonetheless, it is valuable to see the formulation, and how the solutions are derived in the photonic crystal framework. This is especially true here because the solution of Maxwell's equations cannot be different, but the approach to obtaining the solution is quite different.

One is looking for a solution to Eq. (9.4) in terms of what are called *plane waves*. What this means is that one wants to write the solution as a infinite series of plane waves of the form $\exp i(k \cdot r)$. Further one uses the periodic repeating nature of the dielectric structure to simplify the writing of the resulting sum. The periodicity of the structure can be characterized in real space and reciprocal space. A simple example of the relation between a structure in real space to that in reciprocal space is shown in Figure 9.4 for a simple two-dimensional (2-D) square structure. The figure on the right is referred to as the Brillouin zone and the shaded portion, the primitive portion thereof. The capital Greek letters refer to different points on, or in, the primitive cell and represent points corresponding to the symmetry they possess [3]. The point Γ, at the center of the zone, contains the complete symmetry of the structure, whereas the other points contain less, depending on the way they transform under the complete set of symmetry operations possessed by the structure. The plane wave expansion of H for any given value of k will be in reciprocal space in the following form [6].

$$H_k(r) = \sum_G h_G \exp i(k + G) \cdot r \tag{9.9}$$

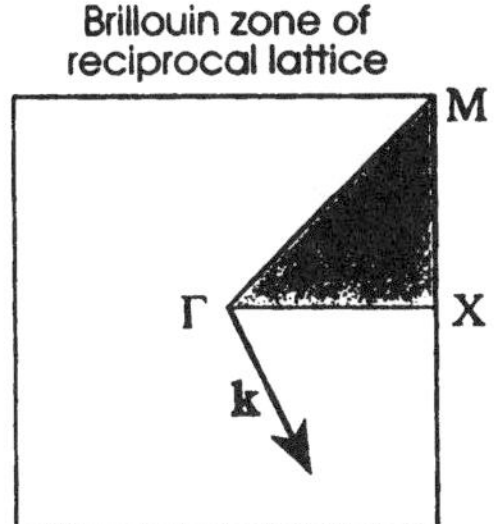

Figure 9.4 Example of the real lattice on the left and the corresponding reciprocal lattice on the right. The symbols refer to positions of the Brillouin zone, center, edge corner, and so on.

The **G** represents an arbitrary set of vectors in reciprocal space. On next expands the $1/\varepsilon$ term in Eq. (9.4) in a similar fashion.

$$1/\varepsilon = \sum_{G} \varepsilon(G)^{-1} \exp i(G \cdot r); \tag{9.10}$$

Substituting Eq. (9.8) and (9.9) into (9.4) leads eventually to the expression

$$\sum_{G'} \{\varepsilon^{-1}(G - G')\}(k + G)\,(k + G')\,h_{G'} = \left(\frac{\omega}{c}\right)^2 h_G \tag{9.11}$$

This constitutes a matrix equation with the matrix of the form

$$\Theta_{G,G'} = (k + G)(k + G')\varepsilon^{-1}(G - G') \tag{9.12}$$

The eigenvectors are the field amplitudes and the eigenvalues are the frequencies. One can summarize the steps in the calculation of the frequency of the allowable modes as follows. Based on the symmetry of the dielectric arrangement construct the Brillouin zone. Choose a basis, that is the number of **G**s that are needed for the respective representations. Compute $\varepsilon^{-1}(G)$ as the Fourier transform of the spatial dielectric structure $\varepsilon^{-1}(r)$. For each value of k, selected symmetric directions in the Brillouin zone, calculate $\Theta_{G,G'}^{k}$. Then solve Eq. (9.12).

9.3   FREQUENCY VERSUS k DIAGRAMS

Following the method described in the previous section, the code written by Joannopoulos et al. [7] was used to calculate the photonic bandgap for a few structures that have two-dimensional periodicity. This is just to familiarize the reader with the types of behavior that can arise from different geometries. The first is that of a square array of dielectric cylinders, as shown in Figure 9.5. Here, there is no overall badgap in the sense that the no mode is everywhere forbidden. This is referred to as a full gap. Such a full gap is exhibited for a structure shown in Figure 9.5b where cylindrical holes are hexangonally arranged. Note that there is a normalized frequency region, roughly between 0.43 and 0.5, at which no modes exist for any direction. The width of this gap is a function of the ratio of the refractive index of the material that constitutes the cylinders and that of the surrounding material. In the example shown, the ratio is 3.6. The dependence of the width of the gap as a function of this ratio is shown in Figure 9.6. This presents an interesting departure from the single-dimension case where there was no threshold for the appearance of a gap. One can understand this difference, at least qualitatively, in that in two dimensions one has both TE and TM modes possible. One can see on inspection of Figure 9.5 (top) that there is a gap for the TM mode, but not for TE. Similarly, if one made the inverse structure holes, rather than posts, one would have seen the opposite situation occur. Suffice to

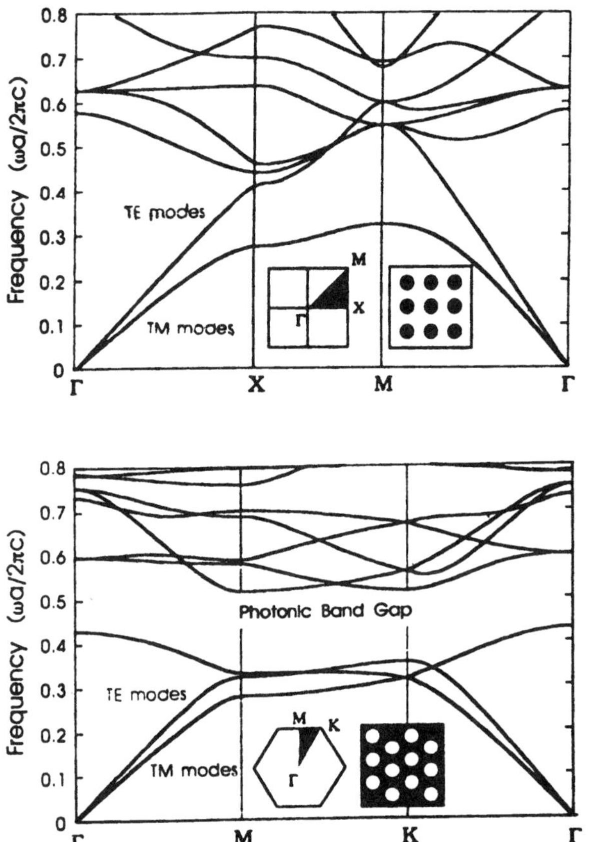

Figure 9.5 Calculated energy diagrams for two simple 2-D lattices as shown. (From Ref. 20.)

say here that with sufficient disparity between the dielectric constant one can force an opening of a full gap.

There is a another very important aspect of the photonic crystal approach. That is the role the defect in the structure plays in the allowable energy states. Identical with the role defects play in the electronic band structure, the result is the formation of localized states that may appear within the forbidden gap. The way that one calculates the consequence of defects is to proceed with the calculation of the infinite periodic array except to remove, one or more of the periodic elements as in the example shown in Figure 9.7a, or it could be a whole line of holes, as shown in Figure 9.7b. The extent that the translational symmetry is

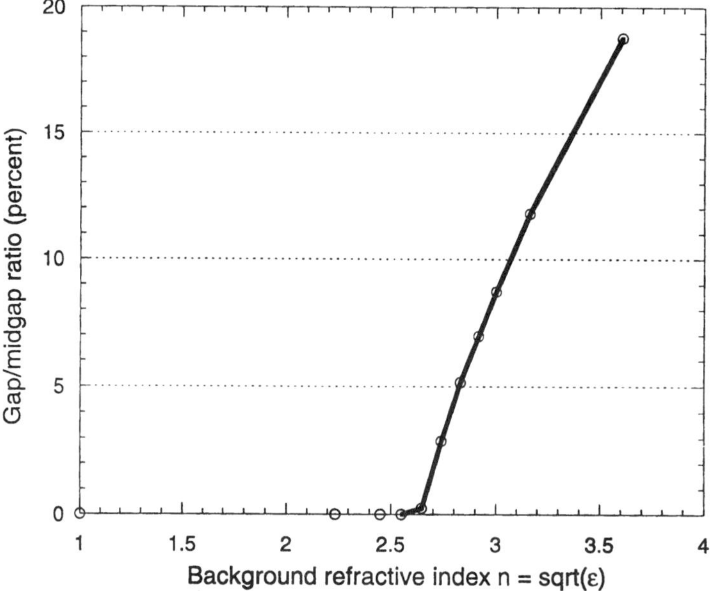

Figure 9.6 Computed value of the refractive index contrast required to produce a full gap for a triangular pattern of holes.

broken by the presence of the defect determines the way one can represent the results. If, for example, the translational symmetry is maintained in the propagation, as would be true if the propagation were along the row of missing elements in Figure 9.7b, then one can still represent the modes relative to the k corresponding to this direction. However, if the propagation were perpendicular to the line, the translational symmetry is broken, and there is no way to define k. There are several ways to display the allowable energy of defect structures. One way to do this is to plot the modes in ascending order of energy. We have done this for the example discussed earlier for which one defect is located in the middle of a 10×10 array for the first four TM modes, as shown in Figure 9.7a. We have done this because a gap exists for the TM modes, as seen from the conventional energy diagram for the infinite (Figure 9.7c). Because there is no way to represent the situation of a defect on this diagram, for there is no **k**, one can circumvent this difficulty; we replot results in the way shown in Figure 9.7d. This is done for the defect-free case where the gap is still a feature of this representation. Figure 9.7e is the case with the defect. For the defect case, one sees not only the gap, but also the mode in the gap, which is the desired result. In general then,

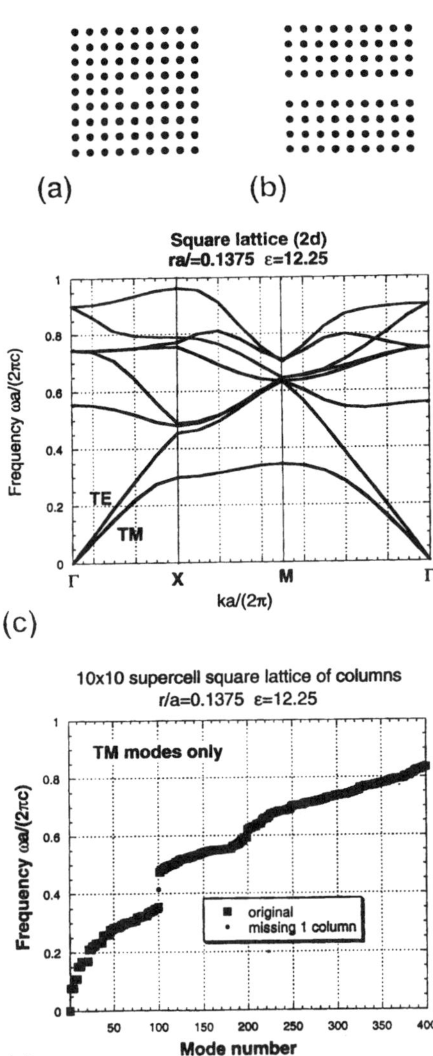

Figure 9.7 Representation of defect states (a) and (b) in a 2-D array. In the former, one element is missing, and in the latter, a whole row is missing. (c) The standard energy vs. k diagram for the perfect array. (d) A representation that shows the consequence of defect structures such as (a) and (b). The energy of the first four TM modes for a 10 × 10 array is plotted in ascending order for both the perfect array and one with a single defect. In this representation, not only is the gap evident, but also the localized state caused by the defect.

the description of cases with defects will have to be represented in this way, or a similar way, or directly in the time domain.

Again, it should be borne in mind that these calculations are based on a 2-D model and thus infinitely long cylinders or holes in the z-direction are assumed, as well as infinite extended in the x–y plane. The diagrams we have shown are for the case in which the light propagation is confined to the x–y plane, that is, perpendicular to the cylinder direction. Joannopoulos et al. [2] have shown that for propagation out of the plane, $\mathbf{k}_z$ not equal to zero, the gaps tend to close. We will see in a later section that gaps can open up in the z-direction for certain structures.

9.4 STRUCTURES

We have shown some simple photonic crystal structures in the previous section. Although they provide some practical properties themselves and can be used to demonstrate certain unique properties, it will be the more complicated 3-D structures that will likely be important. We will briefly discuss these prototype structures so that the reader will have a better appreciation of the fabrication challenges.

9.4.1 Two-Dimensional

What is meant here are structures that have a two-dimensional periodicity, but exist in three dimensions. The structures shown in Figure 9.5 are two of such structures. One can imagine other arrangements of holes or posts, but the energy diagrams are not expected to be significantly different. The truncated version of these idealized structures are what one can really produce. In other words the posts or holes will not extend to infinity, but will be limited by the fabrication process. As we will see, the fabrication of some of these structures are well suited to conventional high-resolution e-beam lithography. Others will require new and unique fabrication techniques.

9.4.2 Three-Dimensional

The extension of periodic dielectric structures into three dimensions is not hard to imagine, and actually the number of possibilities grows. As Joannopoulus and colleagues [2] have pointed out, just take any three-dimensional lattice of points and place a dielectric sphere at the points. A hexagonal close-packed arrangement of spheres would be a simple example of this. One can also use the lattice to construct a photonic crystal by imagining dielectric cylinders connecting the

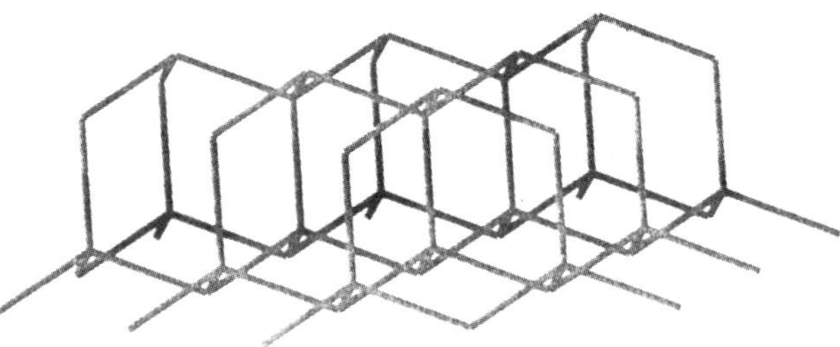

Figure 9.8 Representation of 3-D structures arising from point symmetry arrangements: In the upper figure one puts a sphere where the point would be and in lower one runs tubes from the points. (From Ref. 2.)

points, as shown in Figure 9.8. They do not necessarily have energy gaps of consequence. The Brillouin zone is now in three dimensions and the full gap criterion takes on a more demanding role. Nonetheless, such structures have been proposed. One must also consider the possibility of fabrication. There are some structures that can be made, at least in principle, and the reader is referred elsewhere [8, and the references contained therein] for a more complete discussion. We will primarily be dealing with the 2-D structures and fabrication methods in this book. This is simply because the fabrication of the three-dimensional structure is much more complicated and is not yet to a point at which one can write any synopsis or comparison of them.

9.4.3 Fiber Based

There is a relatively straightforward and proved way to produce a variety of periodic 2-D array geometries at dimensions consistent with the photonic crystal approach. The method is to bundle capillary tubes and redraw them at high temperature micrometer sizes. The periodicity is obviously confined to the *x-y* plane; that is, perpendicular to the fiber axis. We will discuss this process in some detail in a subsequent section [9–12]. Rosenberg et al. [11] have used this method to produce periodic arrays of disparate dielectric media and have measured the transmittance versus wavelength along a number of directions corresponding to

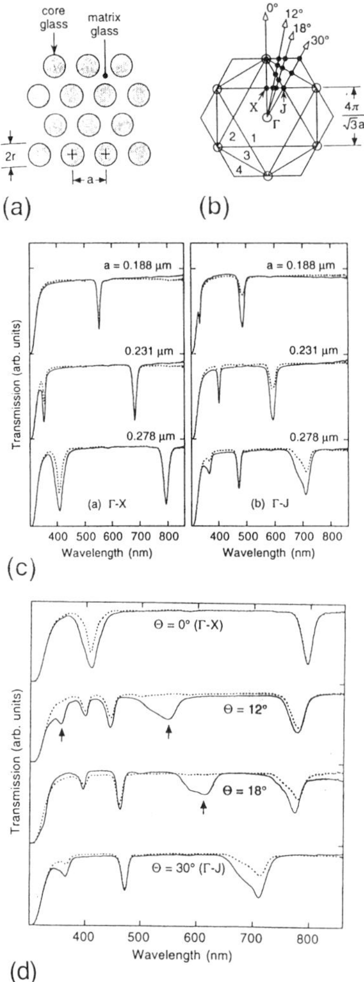

Figure 9.9   (a) The spectral behavior of triangular array of holes pattern. (b) The reference relative to propagation direction. (c) The transmittance as a function of wavelength for the pitch and propagation in the directions indicated. (d) The transmittance for other specific directions. (From Ref. 12.)

positions on the photonic crystal energy diagram in the x–y plane. A representation of this is shown in Figure 9.9a–c. Certain transmittance features correlate with the energy band calculation for the appropriate structure.

Quite surprisingly, and not really understood, is the calculation that energy gaps also exist in the z-direction of such structures. This was first reported by Russell et al. [9]. We [6] show the result of the verification of this calculation in Figure 9.10 using the Joannopoulos code [10]. The gap of interest lies below the 45-degree line which represents the propagation of light in vacuum. (The slope of any line through the origin on this diagram is the reciprocal of the effective refractive index of the mode.) If this turns out to be true, then it might be possible to introduce some defect state into the structure such that a true diffractive waveguide could exist. The defect in this case could be as simple as keeping the center hole solid, or even more interesting, enlarging the center hole.

Yet another unexpected behavior resulted from the periodic fiber structure. This was found by Russell et al. [10,14], for which by considering the situation with a solid hole in the center, one can view this as a normal-type waveguide structure. Normal is in the sense that the region surrounding the solid portion

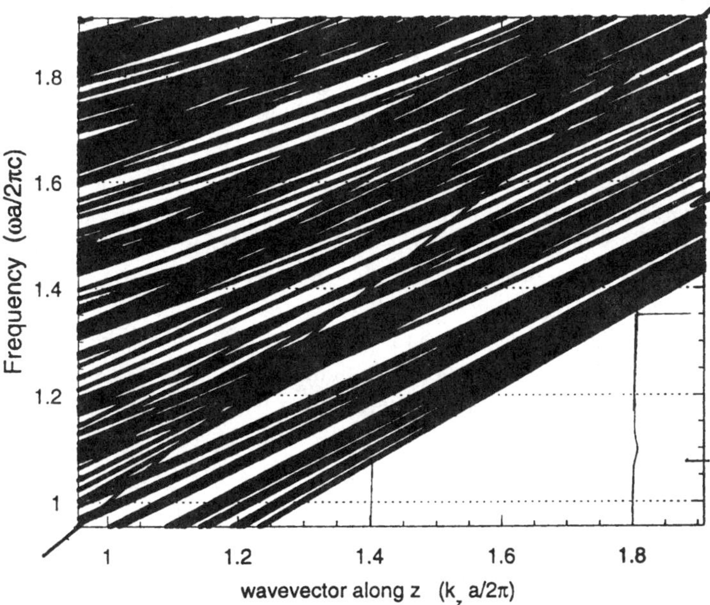

Figure 9.10 Energy versus wavevector in the z-direction (axial direction) of a pattern of a triangular array of holes. Calculation was performed using the code from Ref. 7. The gaps are the open areas.

was made up of composite region of holes and glass with a lower effective refractive index. The interesting feature they found was that one could sustain single mode behavior over a very wide wavelength range, much larger than that of a normal *step waveguide* geometry. (A *step waveguide* is where the refractive index changes at the core radius to some uniform lower radius). The classic condition that relates the core size to the index difference derives from the solution of Maxwell's equation in cylindrical coordinates [15].

$$\frac{2\pi r}{\lambda} (n_c^2 - n_{cl}^2)^{1/2} = V = 2.403 \tag{9.13}$$

The value 2.403 is the point at which the second mode is allowed, so the single-mode condition is for all values of V for which it is less than this value. Because of the wavelength in the denominator, the single mode condition is very wavelength-dependent. What Russell et al. found was that the guiding structure as pictured in Figure 9.11 retained its single-mode behavior over a range of 1000 nm.

The explanation of this effect derives from the way the field redistributes itself as a function of wavelength. Clearly, in the region where the pitch of the holes is much smaller than the wavelength of light, the surrounding medium could simply be treated as a composite medium for which the effective index would depend only on the volume fraction of the holes. It would essentially be the step-index case with $n_{cl} = n_{eff}$. However, as the wavelength becomes longer and comparable with the separation distance of the holes, the effective index assumes a more complicated dependence on structure. They treated the problem in the photonic crystal context to obtain the estimate of n_{eff} as a function of the normalized separation distance. What this amounted to was to consider the unit cell structure shown in Figure 9.11b in which the central portion is a hole surrounded by six holes with the intervening material being glass. They solved the scalar wave equation using the approximate circular geometry shown at the right. The boundary condition that the derivative of the field relative to the radial distance was equal to zero at the boundary b led to the equation that yielded the propagation constant n_{eff} as a function of the separation. The separation distance Λ is related to the radial distance b through the equating of the areas; $b = 1.9\Lambda$ for the geometry shown. The result is plotted as V_{eff}, defined as

$$V_{eff} = \frac{2\pi\Lambda}{\lambda}(n_c^2 - n_{eff}^2)^{1/2} \tag{9.14}$$

in Figure 9.11c. (One could also have used the more exact method of solution using Figure 9.10. The ratio of the x-value to the y-value at any point on the lowest line is n_{eff}). The horizontal line is drawn to represent the possible single-mode cutoff condition. Whether or not this is the correct number, one notes that the extended single-mode behavior would result for values of d/Λ at which the curve becomes nearly horizontal.

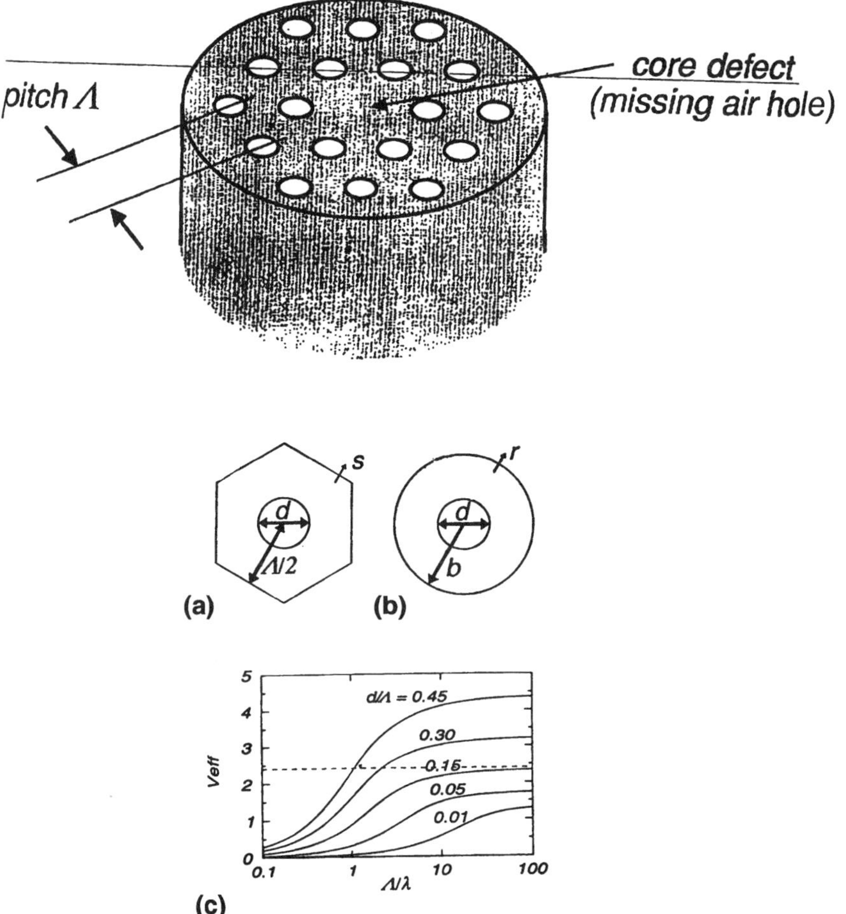

Figure 9.11 Endlessly single-mode structure shown in the upper drawing: (a, b) the approximation used to estimate the space-filling mode, see text; (c) the result of the V_{eff} versus normalized pitch. The flat portion of the curve that occurs beyond certain values of the normalized pitch indicates continued single-mode behavior. (From Ref. 14.)

9.5 FABRICATION METHODS

We will briefly review the fabrication methods that have been used so far to try and make photonic crystal structures. We will cover the simpler 2-D methods as well as the fiber method. We will not discuss any of the 3-D structures except to say that to the extent that they involve multiple lithographic steps, then essentially the same general method applies.

9.5.1 E-Beam Lithography

The major additional problem in the use of the standard state-of-the-art e-beam lithographic method to create a simple two-dimensional photonic crystal pattern, is the depth. The required separation distances are of the order of, $a/\lambda = 0.4$–0.5 as indicated from the energy versus k diagrams of the simple structure. If one is considering telecommunication devices, this translates into separations of 600 nm, which present no real resolution problem. However, the stipulation of how deep, is more difficult to answer. As we pointed out in the earlier sections, the 2-D structure, which is periodic structure in the x–y plane, is assumed to be infinite in the z-direction. The measure of how deep in the z-direction, as well as the related issue of how extended in the x–y direction, must come from some estimate of how fast the field decays. As we will see later on in the application, the devices will all be based on creating a defect state thereby localizing the light to a given region. It will be the field decay in this arrangement that will be the measure of the required dimensions.

In any event, one might anticipate a dimension of something close to ten periods, which would translate into the order of 5 μm if the wavelength regimen were 1500 nm. This is quite deep for conventional resist. In other words, what is important is the etching contrast between the resist and the substrate. Thus, it will also depend on what type of material one is trying to use for the pattern.

An example of a technique that can be used to enhance the contrast of patterning in silicon is to use the resist to pattern a thermal oxide grown on top of the wafer [16]. The oxide material is then used as the resist for the subsequent deep reactive ion-etching step.

There is a novel process involving the use of an electron beam deposition in a scanning electron microscope. Molecules are injected in a gaseous phase and aggregated through the action of the electronic beam. As a result small posts of the organic composite can be precisely positioned [17].

It is too early to tell how the less than ''perfect'' structures will correspond in performance to the idealized mathematical structures. Issues such as roughness of the holes, radius variation, perpendicularity, and other natural consequences of the fabrication process are likely to influence the actual propagation, but to what extent has yet to be determined.

9.5.2 Nanochannel Arrays

An interesting approach to making a two-dimensional array of channels is to use a precision-patterned texturing treatment of the surface of polished Al [18]. This is done just before an anodization step. The molded depressions serve as initiation points and guide the growth of the channels in the oxide film. The initial texturing was accomplished by a special molding process. The process is schematically shown in Figure 9.12a–e. Note that the last step removes the Al, and one obtains

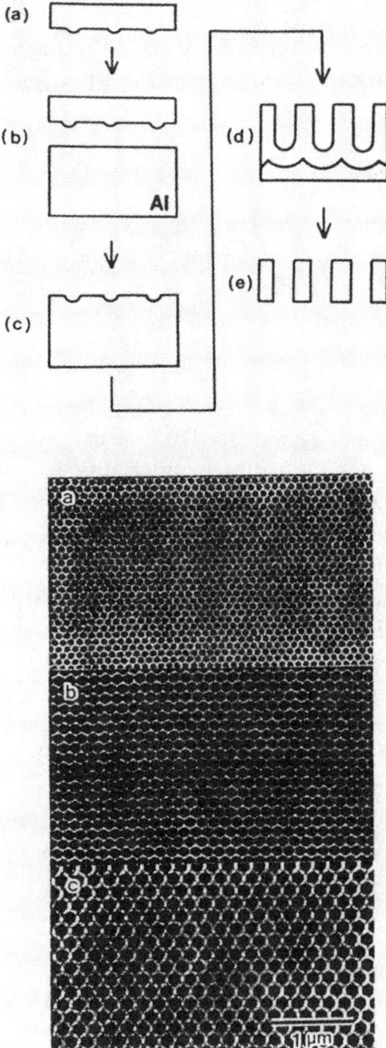

Figure 9.12 Texturizing treatment of Al to produce nanochannel patterns. (a) Mold used to pattern Al; (c) on anodization the textured regions etch differentially; (b,d) the channels; (e) when the Al is removed the holes go all the way through. The lower panels are SEM micrographs of the hole patterns that could be produced. (From Ref. 18.)

holes that go all the way through. Scanning electron microscopic (SEM) pictures of some of the patterns that were produced by the method are shown in Figure 9.12. The authors claim that the method can be applied to other materials, such as semiconductors, by using a two-step process.

9.5.3 Fiber Based

There are various approaches to fabricating fibers that have a periodic structure with the refractive index in the direction normal to the axial direction. The advantage of this structure is that it seems to be compatible with the existing optical fiber technology. We will briefly cover the major approaches to making such fibers based on the photonic crystal.

Capillary Redraw

The process starts with a silica rod with a hole in the middle. The hole may either be drilled to establish the correct radius for the ultimate separation distance, or may be formed by the process from which the blank was made. An example of this is the way optical waveguide blanks are made, often containing a hole in the center. Knight et al. [14] talk about drilling a 16-mm–diameter hole in a 30-mm–diameter silica rod. Six flats were arranged onto the outside of the rod to form a regular hexagon. The rod was then redrawn at 2000°C to produce hexagonal cane. The cane was then bundled together and again redrawn until the center-to-center distance was the order of 50 μm. Finally, a piece of this cane was drawn to yield the final fiber, with a pitch of a few micrometers. The defect was introduced by replacing the center element of a bundle with a fiber that did not contain a hole (Fig. 9.13a).

Recently, a planar version of the endlessly single-mode concept has been reported [19]. Here, a solid region of a silica film is flanked by fins of air, which provide the effective index cladding (see Figure 9.13b).

Multiclad

There is no reason that the index of the included phase of the array has to be 1; in other words, there need not necessarily be holes. In terms of strength and structural integrity, it would be much easier if the included phase were another glass. Clearly, this presents a lower contrast case, but as far as the extended single-mode behavior indicated by Figure 9.11b, it still persists. The effect of the lower refractive index contrast is to move the flattened portion of the curve to larger values of Λ/λ and d/λ. There are methods to make optical fiber couplers by a technique in which multiple cores within a common cladding tube are redrawn to separation distances that are sufficiently close to that of the photonic crystal structures. This process can be extended to make arrays of cores within the same cladding host.

For the reader not familiar with the multicore process, one is referred to Figure

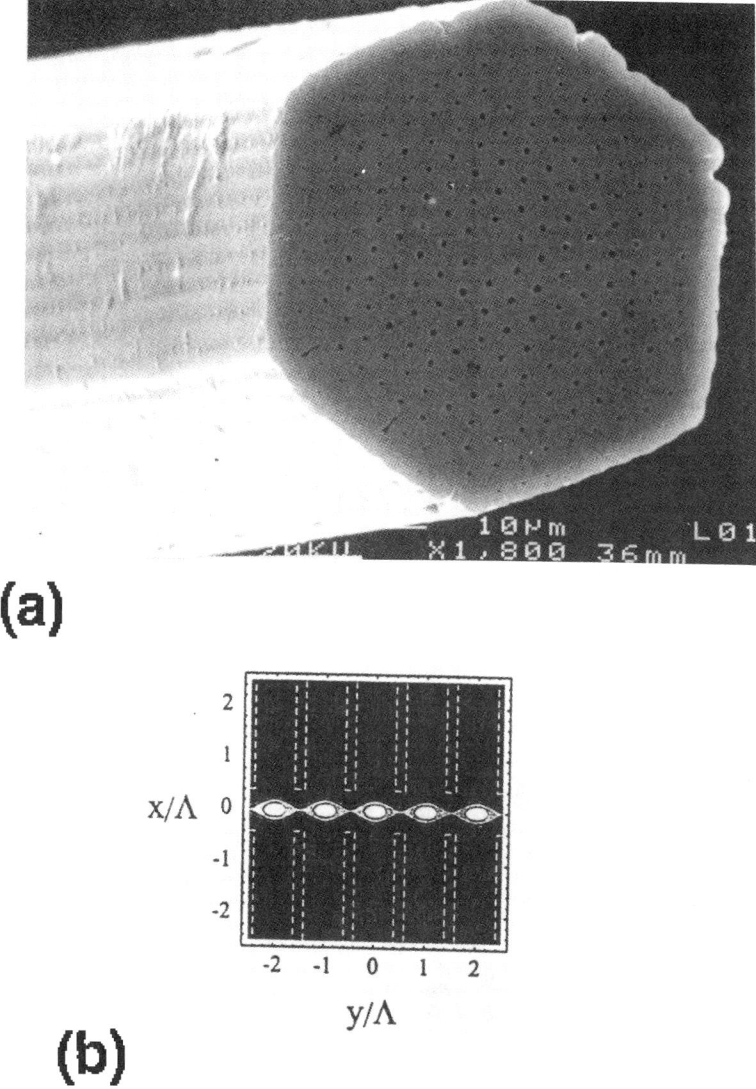

(a)

(b)

Figure 9.13 (a) SEM of fiber made by redrawing capillaries. Note absence of the hole in the center. (b) Planar version of endlessly single-mode structure where fin-like pattern (region enclosed by dotted lines) provides the effective index medium. Mode is propagated as shown. (a, From Ref. 14, b, from Ref. 19.)

9.14 in which the process is shown for a four-core structure [20,21]. There are four preforms shown. These would be made by any of the waveguide processes such as MCVD. For the photonic crystal, they are rods with a core of a different refractive index. The desired spacing would be provided by the starting core to rod diameters. The assembly shown in Figure 9.14a is collapsed under vacuum so that all the gaps disappear, ultimately yielding a structure shown in Figure 9.14b. Bundling these together using a square rod would produce the repeating

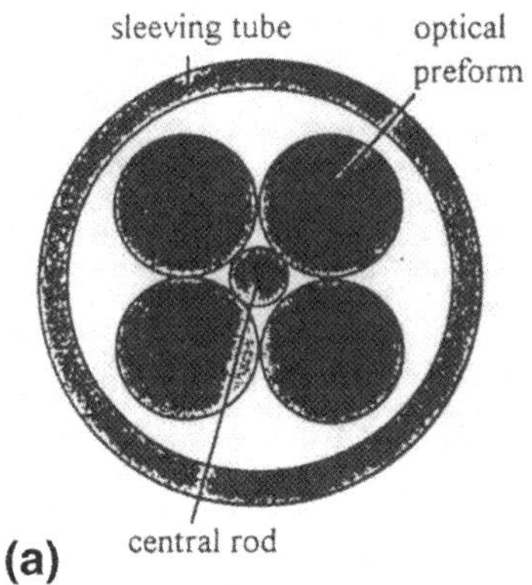

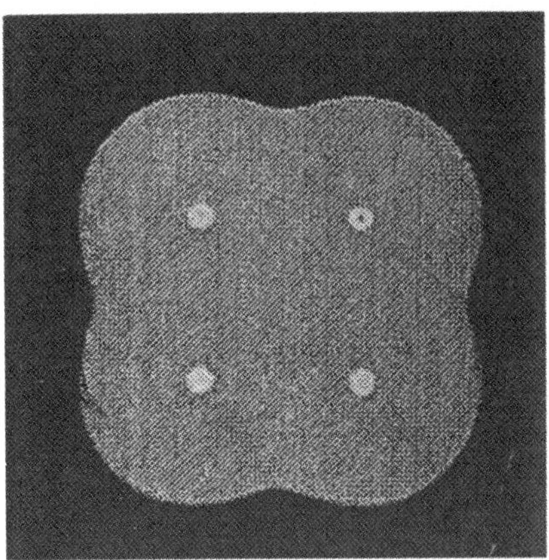

Figure 9.14   Representation of the multicore process: The upper curve shows the initial structure for a four-core device. The lower shows the result after drawing down to fiber size.

extended structure. One should be able to see how this process can be extended to other geometries.

Extrusion

This is a method that involves the extrusion of a powder in a binder through a metal die. The die geometry contains the desired spacing and shape characteristics. One can then heat the extruded body to drive off the binder and further heat it at a higher temperature to consolidate the body to theoretical density. This process was originally developed to produce ceramic honeycomb structures [22], but it can be applied to such materials as glass or silica. Depending on the specific material, the preform can be redrawn to reduce its size. Bundling and further redrawing would be required to reduce it to the 1- to 10-µm range. Examples of extruded ceramic shapes are shown in Figure 9.15. A picture of a silica preform is shown in Figure 9.16

9.6 APPLICATIONS

It is a bit premature to talk about actual applications of photonic crystal structures because few, if any, devices have actually been fabricated for any kind of real evaluation. However, there have been a number of ideas put forth based on the mathematically modeled behavior, and these have been interesting enough to warrant the discussion that follows.

9.6.1 Filters

One can consider the use of the photonic crystal structure as being manifest as narrow-band multilayer optical filters. It employs the defect structure discussed in the foregoing, one period missing, or skipped in the dielectric stack, to create the allowable mode. This is commonly referred to as a Fabry-Perot filter. One can easily understand the name because it is the conventional Faby-Perot etalon structure with the mirrors stemming from the multilayer stacks. The behavior is shown in Figure 9.16 where we have reproduced the reflection spectrum of the 15-and 21-layer stack of Figure 9.3 with the defect (skip one-layer sequence). The spectra are shown in Figure 9.16b and c. The transmission spike is a consequence of the allowed propagation through the localized state in the gap. The position of the transmission maximum, the width of the transmission peak $\Delta\lambda$, and the peak transmission are functions of the mirror reflectivity, hence, of the number of layers, and index contrast and thickness of the alternating layers that make up the stack.

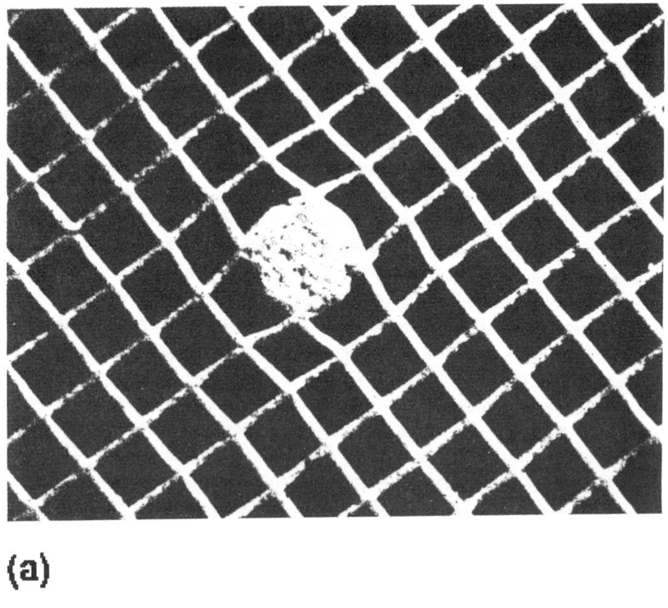

(a)

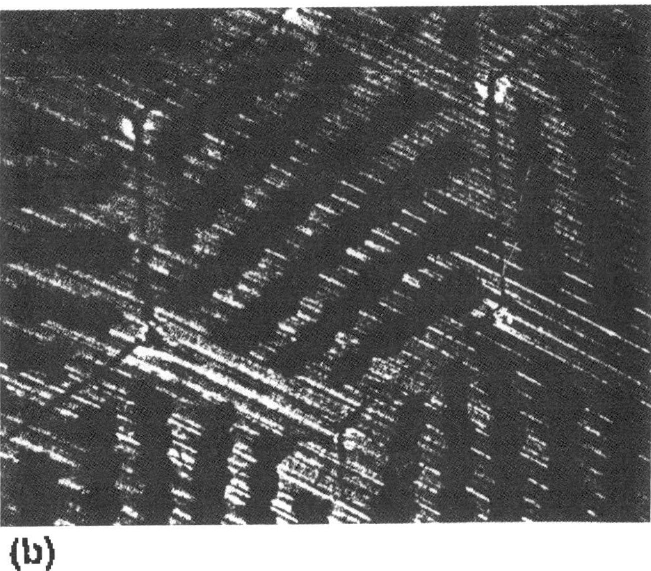

(b)

Figure 9.15　Micrographs (50×) of patterns using extrusion.

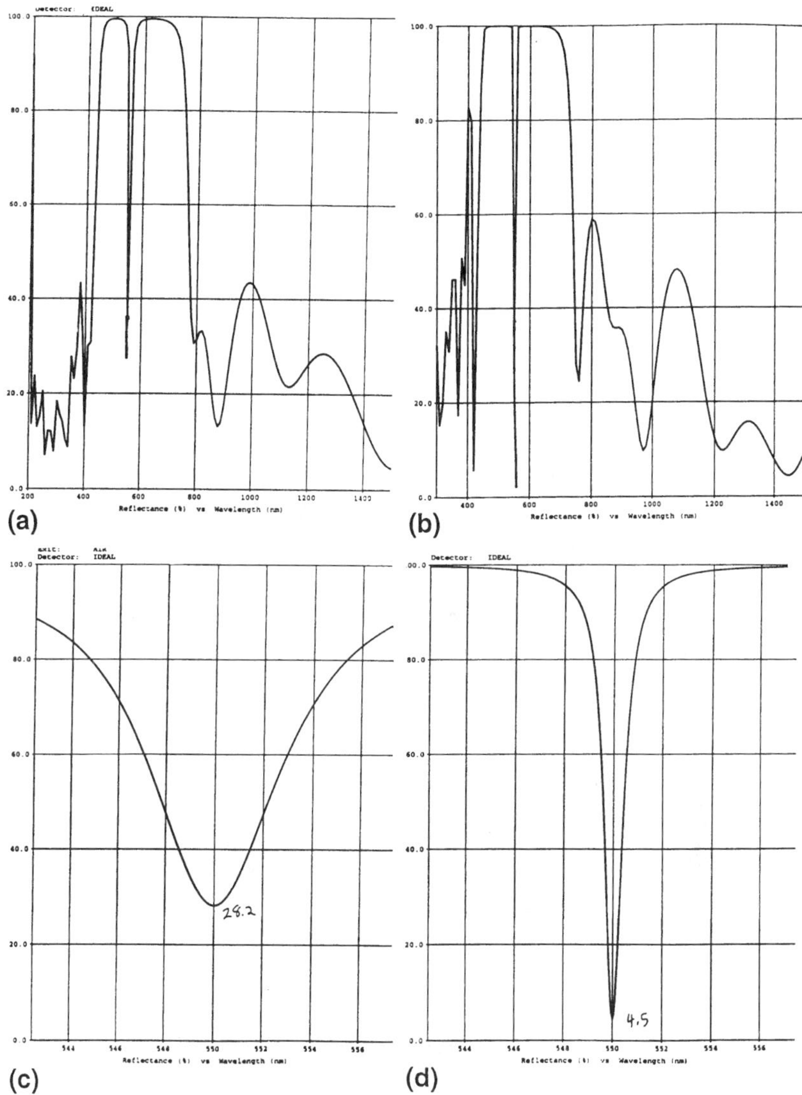

Figure 9.16 Calculated reflectance versus wavelength for a multilayer dielectric stack where one layer is skipped to produce the defect, a Fabry-Perot filter: (a) 15 layers and (b) 21 layers. The lower two traces are a blowup of the the cases showing how the width and depth of the transmission dip depend on the number of layers.

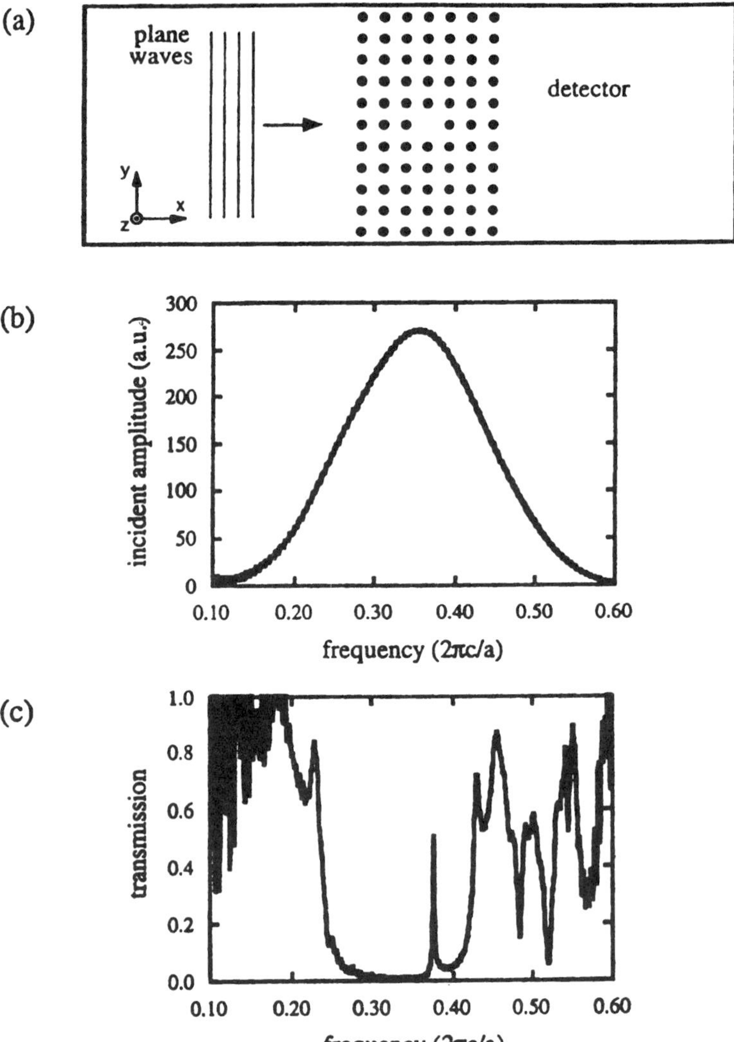

Figure 9.17 Two-dimensional photonic crystal filter arrangement: (a) schematic of the measurement; (b) input pulse; and (c) the transmission. (From Ref. 22.)

9.6.2 Microcavities

In two dimensions, one can have a similar behavior. Here, the defect is referred to as a *microcavity* [23] because the defect is surrounded by the undisturbed periodic structure. Villeneuve et al. [23] show the computed spectral behavior for a simple dielectric structure, as shown in Figure 9.17a. They compute the output transmittance spectrum in Figure 9.17c at a single point, labeled detector in the drawing, for a polarized input pulse, the spectral width of which is shown in Figure 9.17b. An important performance property, namely the quality factor $\omega/\Delta\omega$ was calculated as a function of the size of the crystal, as is shown in Figure 9.18. It is also important from another point of view: the physical size required for the device. The typical pitch is roughly less than a micrometer, so the total dimension of the device in the x–y plane is less than 10 μm. The rods here are assumed infinite for the calculation. If one assumes that the performance as stated will not suffer by truncation to the same level as in the x–y plane, then the entire filter is a 10-μm cube. This issue notwithstanding, one could imagine a spatial array of such elements, each tuned to a different wavelength by the particular choice between the rod diameter and the spacing.

9.6.3 Waveguides

If one extends the defect along a whole line, rather than just a point, one can then have a structure that guides light by diffraction. For example, consider when light is forbidden to propagate in the energy region, as indicated in Figure 9.19a for the TM modes of a square array of dielectric rods. By removing a row of the rods there now appears allowable states within the forbidden gap, as indicated

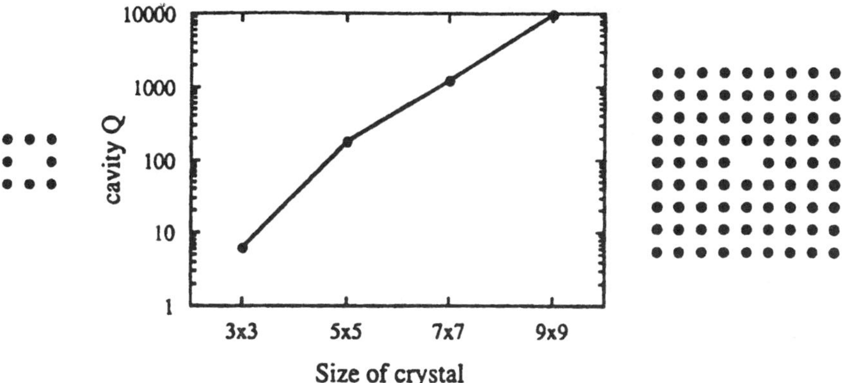

Figure 9.18 Quality factor $\omega/\Delta\omega$ versus number of elements in the array for the structure shown. (From Ref. 22.)

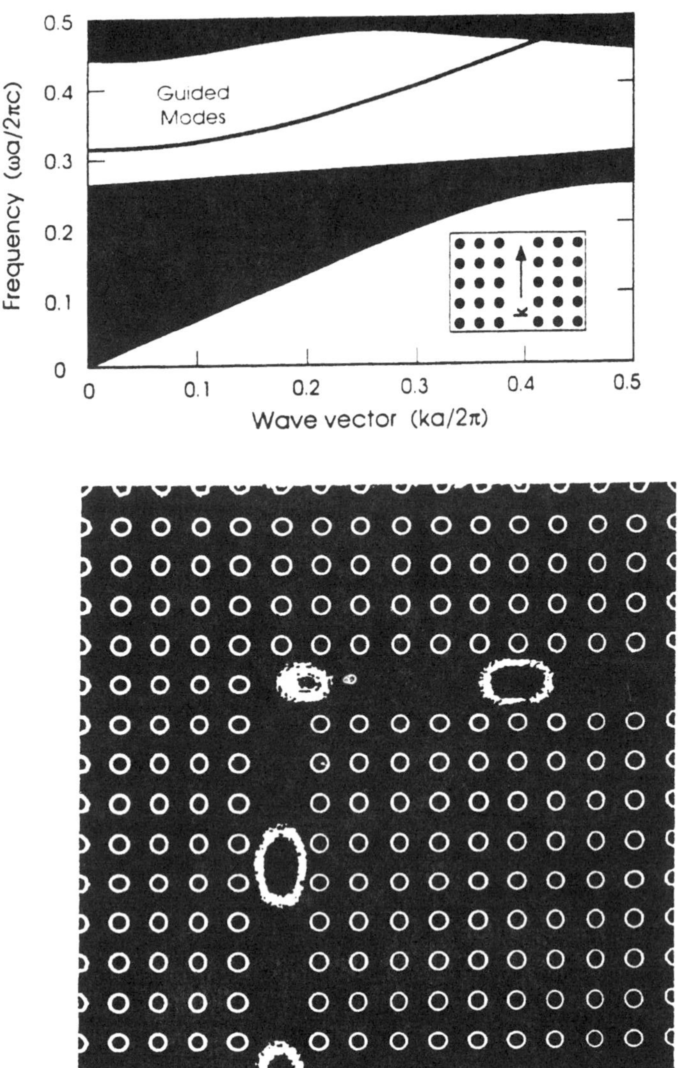

Figure 9.19 Representation of the dispersion of the defect within the forbidden gap of the 2-D structure shown. The lower figure shows the time domain simulation of the propagation around a right-angle bend. (From Ref. 8.)

by the solid line [2]. For this particular structure, the simple removal of a single row is sufficient to produce a single-mode condition. The propagation around a bend is simulated in Figure 9.19b [8]. This is essentially an illustrative example, rather than a representation of an actual structure. It is after all a two-dimensional structure. Nonetheless, it does indicate the possibility of building a true three-dimensional waveguiding structure.

One might question the advantage of guiding light by diffraction, rather than refraction. There certainly is no lack of technology available to construct extensive planar-waveguiding structures, and this has been amply demonstrated [24]. One could argue that the photonic crystal offers possibly smaller structures. However, small is only as good as it allows efficient light coupling in and out. Once one arrives at submicron dimensions this becomes the major problem. However, if the photonic crystal structure were to be totally integrated, source pathways, detectors, all fabricated in the same volume, then the overall size would be an advantage, and coupling once in and once out, even if it were inefficient would not be a severe drawback. If GaAs were the material, this would be within the realm of possibility.

ACKNOWLEDGMENT

The author is grateful to Douglas C. Allan for help in preparing the material contained in this chapter.

REFERENCES

1. E. Yablonovitch, Photonic band-gap structures, *J. Opt. Soc. Am. B*, *10*, 283–295 (1993).
2. J. D. Joannopoulos, R. D. Meade, and J. N. Winn, *Photonic Crystals*, Princeton University Press, Princeton, NJ, 1995.
3. M. Tinkham, *Group Theory and Quantum Mechanics*, McGraw-Hill, New York, 1964.
4. M. Born and E. Wolfe, *Principles of Optics*, Macmillan, New York, 1964.
5. P. R. Villeneuve and M. Piche, Photonic band gaps in two-dimensional square and hexagonal lattices, *Phys. Rev. B*, *46*, 4969–4972 (1992); Photonic band gaps in two-dimensional square lattices, square and circular rods, *Phys. Rev.*, *46*, 4672–4675.
6. D. C. Allan, Corning Inc, private communication.
7. The codes used were provided to us by J. Joannopoulos and his group at MIT.
8. J. D. Joannopoulos, P. R. Villeneuve, and S. Fan, Photonic crytsals: Putting a new list on light, *Nature 386*, 143–149 (1997); S. Fan, P. R. Villeneuve, and J. D. Joannopoulos, *Phys. Rev. B*, *54*, 247–251 (1996).
9. T. A. Birks, P. J. Roberts, P. St. J. Russell, D. M. Atkin, and T. J. Shepherd, Full 2-D photonic bandgaps in silica/air structures, *Elect. Lett. 31*, 1941 (1995).
10. P. St. J. Russell, J. C. Knight, T. A. Birks. R. F. Cregan, and B. J. Mangan, Silica/air

photonic crystal fibers, International Workshop on Structure and Functional Optics; Properties of Silica and Silica-Related Glasses, Shizuoka, Japan, July 10–11, 1997.

11. T. A. Birks, J. C. Knight, and P. St. J. Russell, Endlessly single mode photonic crystal fiber, *Opt. Lett. 22*, 961 (1997).

12. A. Rosenberg, R. J. Tonucci, and E. A. Bolden, Photonic band-structure effects in the visible and near ultraviolet observed in solid-state dielectric arrays, *Appl. Phys. Lett. 69*, 2638 (1996).

13. A. Rosenberg, R. J. Tonnucci, H.-B. Lin, and A. J. Campillo, Near-infrared two-dimensional photonic band-gap materials, *Opt. Lett, 21*, 830 (1996).

14. J. C. Knight, T. A. Birks, P. St. J. Russell, and D. M. Atkin, All-silica single-mode optical fiber with photonic crystal cladding, *Opt. Lett. 23*, 1547 (1996).

15. D. Marcuse, *Light Transmission Optics*, Van Nostrand Reinhold, New York, 1972.

16. Y. Suzuki, Cornell University, private communication.

17. H. W. P. Koops, Photonic crystals built in 3-D additive lithography, *SPIE, 2849*, 248 (1996).

18. H. Masuda, H. Yamada, M. Satoh, and H. Asoh, Highly ordered nanochannel-array architecture in anodic alumina, *Appl. Phys. Lett., 71*, 2770 (1997).

19. E. Silvesrte, P. St. J. Russell, T. A. Birks, and J. C. Knight, Endlessly single-mode heat sink waveguide, CLEO'98 Paper CTh059, Los Angeles, CA.

20. S. Inno, High density multi-core fiber cable, IWCS'79 Proceedings, 1979, pp. 370–384.

21. J.-E. Bourhis, R. Meilleur, P. Nouchi, A. Tardy, and G. Orcel, Manufacturing and characterization of multicore fibers, Proc. 46th Int. Wire and Cable Symp. (IWCS), Philadelphia, PA, Nov., 1997.

22. J. S. Reed, *Principles of Ceramic Processing*, Wiley-Interscience, New York, 1995.

23. P. R. Villeneuve, S. Fan, and J. D. Joannopoulos, Microcavities in photonic crystals: Mode symmetry, tunability, and coupling efficiency, *Phys. Rev., 54*, 7837–7842 (1996).

24. L. D. Hutcheson, ed., *Integtrated Optical Circuits and Components*, Marcel Dekker, New York, 1987.

Index